Advances in Combinatorial Chemistry & High Throughput Screening

Volume 1

Editor

Rathnam Chaguturu

Center for Advanced Drug Research
SRI International
Harrisonburg, VA 22802
USA

CONTENTS

CHAPTERS

Contd…..

FOREWORD

Aiming to accelerate the therapeutic interaction of the expanding chemical space and biological target space to advance drug discovery, this inaugural eBook, *Advances in Combinatorial Chemistry & High Throughput Screening (ACCHTS)*, opens up a new avenue for rapid access to a focused collection of highly regarded contributions in the field. Importantly, ACCHTS not only highlights major conceptual advances and cutting edge technologies, but also serves as a much-needed window to the timely, informative, and enabling resources in its parent journal CCHTS.

Although biological and biomedical advances coupled with genomics studies have uncovered a large number of potential drug targets, the number of novel target drugs that enter clinics remains dismally low at about four per year during the past thirty years. In order to rapidly translate such impressive gains in biomedical research into much-needed therapeutic agents, we must improve each step in the drug discovery process. For example, developing bioassays that faithfully reflect *in vivo* biological processes is essential. Expanding chemical space through innovative combinatorial chemistry and natural products-inspired approaches allows for effective interrogation of high-value targets, such as GPCRs and kinases, as well as traditionally un-druggable targets, such as protein-protein interaction interfaces. Progress in laboratory automation and highly sensitive and versatile detection technologies enables unprecedented high throughput screening operations in both industry and academia to accelerate the process of drug lead discovery and characterization. Chapters selected for this inaugural volume reflect promising developments in these critical areas. These contributions describe innovative assay platforms that include both cell-based phenotypic assays and biochemical assays for efficient HTS, and chemical methodologies for building and assessing chemicals. I hope that these chapters will set the standard for future publications to come.

While technology development is important for accelerating the process of drug discovery, rapid development of the chemical biology field and the emergence of a large number of chemical probes are expected to fuel innovations that will be essential for finding novel target drugs. By integrating new developments in chemical biology and probe discovery into drug discovery, ACCHTS would be in a unique position to catalyze the discovery and development of the next generation of efficacious and safe therapeutic agents.

Haian Fu
Department of Pharmacology
Emory University School of Medicine
Atlanta, GA 30322
USA

PREFACE

There are about 7000 diseases that afflict mankind, treatments are available only for ~200 diseases. With the genetic basis for most of these diseases known, there is a large repertoire of un-drugged therapeutic targets to explore. The process of drug discovery has benefited in the past decade by the advent and rise of high-throughput screening (HTS), an early and vital step of screening small molecule libraries primarily through biochemical and cell-based assays. Central to every HTS endeavor is the compound collection. Industry-style probe discovery has now gained unparalleled momentum with the availability of vendor-supplied chemical libraries and ready access to institutional HTS laboratories. The US NIH Roadmap program and the newly organized EU-Open Screen have greatly influenced the global drug discovery landscape. Likewise, India and China have stepped up their drug discovery research infrastructure. The recently formed International Chemical Biology Society, an independent, nonprofit organization, is dedicating itself to promote research and educational opportunities at the interface of chemistry and biology, by providing an important international forum that brings together cross-disciplinary scientists from academia, nonprofit organizations, government, and industry to communicate new research and help translate the power of chemical biology to advance human health.

Combinatorial Chemistry & High Throughput Screening (*CCHTS*) was the only journal that served the needs of the two disciplines, high throughput screening and combinatorial chemistry, when the very first issue appeared in 1998. The Aims and Scope of CCHTS have never been more relevant than now. A look at the manuscripts published in *Combinatorial Chemistry & High Throughput Screening* is reflective of this commitment and readership's interest. We are making every effort for the journal's success, improving its high standard by publishing quality science articles and guest-edited thematic issues.

This eBook, *Advances in Combinatorial Chemistry & High Throughput Screening* (*ACCHTS*), is another effort to showcase CCHTS authors and a significant milestone in the annals of CCHTS. The wide range of topics represented by these chapters is faithful to the scope and purpose of *CCHTS*, and a testament to *CCHTS* contributions in advancing drug discovery on full throttle. In the words of my good friend, Professor Haian Fu, this eBook *"opens up a new avenue for rapid access to a focused collection of highly regarded contributions in the field."*

My gratitude to all those who have helped in developing *ACCHTS*, the authors, reviewers, section editors and publishing staff, for their time, advice, expertise, and tireless effort.

Rathnam Chaguturu
Center for Advanced Drug Research
SRI International
Harrisonburg, VA 22802
USA
E-mail: rathnam.chaguturu@sri.com

LIST OF CONTRIBUTORS

Lars Carlsen
Awareness Center, Linkøpingvej 35, Trekroner, DK-4000 Roskilde, Denmark

Alexandre Varnek
Department of Physics, Lomonosov Moscow State University, Moscow 119991, Russia

Igor I. Baskin
Laboratory of Chemoinformatics, UMR 7140 CNRS, University of Strasbourg, 4, rue B. Pascal, Strasbourg 67000, France

George T. Hanson
Life Technologies, 501 Charmany Drive, Madison, Wisconsin, 53719, USA

Bonnie J. Hanson
Life Technologies, 501 Charmany Drive, Madison, Wisconsin, 53719, USA

Ye Fang
Biochemical Technologies, Science and Technology Division, Corning Incorporated, Sullivan Park, Corning, NY 14831, USA

Maria Â. Taipa
Department of Bioengineering, Instituto Superior Técnico, Technical University of Lisbon, Av.Rovisco Pais, 1049-001 Lisboa, Portugal

Cassia R. Overk
Department of Neurosciences, School of Medicine, University of California at San Diego, La Jolla, California, CA 92093, USA

Judy L. Bolton
UIC/NIH Center for Botanical Dietary Supplements Research and Department of Medicinal Chemistry and Pharmacognosy, College of Pharmacy, University of Illinois at Chicago, 833 S.Wood St., M/C 781, Chicago, IL 60612, USA

Adam J. Kimple
Department of Pharmacology, Lineberger Comprehensive Cancer Center, and UNC Neuroscience Center, The University of North Carolina at Chapel Hill, Chapel Hill, NC 27599-7365, USA

Adam Yasgar
NIH Chemical Genomics Center, National Center for Advancing Translational Sciences, National Institutes of Health, Bethesda, MD 20892-3370, USA

Mark Hughes
BRITE Institute, North Carolina Central University, Durham, NC 27707, USA

Ajit Jadhav
NIH Chemical Genomics Center, National Center for Advancing Translational Sciences, National Institutes of Health, Bethesda, MD 20892-3370, USA

Francis S. Willard
Department of Pharmacology, Lineberger Comprehensive Cancer Center, and UNC Neuroscience Center, The University of North Carolina at Chapel Hill, Chapel Hill, NC 27599-7365, USA

Robin E. Muller
Department of Pharmacology, Lineberger Comprehensive Cancer Center, and UNC Neuroscience Center, The University of North Carolina at Chapel Hill, Chapel Hill, NC 27599-7365, USA

Christopher P. Austin
NIH Chemical Genomics Center, National Center for Advancing Translational Sciences, National Institutes of Health, Bethesda, MD 20892-3370, USA

James Inglese
NIH Chemical Genomics Center, National Center for Advancing Translational Sciences, National Institutes of Health, Bethesda, MD 20892-3370, USA

Gordon C. Ibeanu
BRITE Institute, North Carolina Central University, Durham, NC 27707, USA

David P. Siderovski
Department of Pharmacology, Lineberger Comprehensive Cancer Center, and UNC Neuroscience Center, The University of North Carolina at Chapel Hill, Chapel Hill, NC 27599-7365, USA

Anton Simeonov
NIH Chemical Genomics Center, National Center for Advancing Translational Sciences, National Institutes of Health, Bethesda, MD 20892-3370, USA

Christopher A. Johnston
Department of Pharmacology, Lineberger Comprehensive Cancer Center, and UNC Neuroscience Center, The University of North Carolina at Chapel Hill, Chapel Hill, NC 27599-7365, USA

Francis S. Willard
Department of Pharmacology, Lineberger Comprehensive Cancer Center, and UNC Neuroscience Center, The University of North Carolina at Chapel Hill, Chapel Hill, NC 27599-7365, USA

Kevin Ramer
Karo Bio USA, Durham, NC, 27703, USA

Rainer Blaesius
Karo Bio USA, Durham, NC, 27703, USA

Natalia Roques
Karo Bio USA, Durham, NC, 27703, USA

David P. Siderovski
Department of Pharmacology, Lineberger Comprehensive Cancer Center, and UNC Neuroscience Center, The University of North Carolina at Chapel Hill, Chapel Hill, NC 27599-7365, USA

2

CHAPTER 1

Assessing Chemicals Using Partial Order Ranking Methodology

Lars Carlsen[*,1,2]

[1]*Awareness Center, Linkøpingvej 35, Trekroner, DK-4000 Roskilde, Denmark;* [2]*Center of Physicochemical Methods of Research and Analysis, al-Farabi Kazakh National University, 96A Tole Bi st, Almaty 050012, Kazakhstan*

Abstract: This review summarizes the use of selected partial order ranking (POR) techniques for the assessment of chemicals. Simple partial order ranking may advantageously be applied to give the single chemicals investigated an identity in relation to other substances. Thus, it constitutes an effective tool for the prioritization of chemicals, *e.g.*, based on their PBT (Persistence, Bioaccumulating, Toxicity) characteristics. In more elaborate cases where a larger number of descriptors are taken into account, *e.g.*, comprising physico-chemical characteristics, atmospheric parameters, geospecific factors and possibly socio-economic factors hierarchical partial order ranking (HPOR) may be applied. Thus, in a first ordering step a series of meta-descriptors are generated that later subsequently being used as descriptors in a subsequent ordering. HPOR allows a sensible ranking model even if a relative high number of descriptors are included. Assessing chemicals can, taken the actual situation into account, be based on a variety of parameters/descriptors. Thus, it might be valuable information to know the mutual importance of the applied descriptor. This information may appropriately be retrieved by a sensitivity study applying a designed module in the PyHasse software package. Finally accumulation partial order ranking (APOR) is illustrated. Accumulating partial APOR is a technique where data from a series of individual tests of various characteristics are aggregated, however, maintaining the basics of the partial order ranking methodology. APOR offers prioritization based on mutual probabilities derived from the aggregated data. Alternatively prioritization may be achieved based on average ranks derived from the APOR. The application APOR is demonstrated by an assessment of a series of potential PBT substances. In contrast to simple ranking techniques, partial order ranking does not automatically lead to an absolute rank of the single substances being studied. However, a weak linear order may in all cases be achieved based on the average ranks of the single substances. Additionally ranking probabilities can be derived based on Monte Carlo simulations.

Keywords: Partial order ranking techniques, hierarchical partial order ranking, accumulation partial order ranking, quantitative structure-activity relationships, assessment of chemicals, prioritization of chemicals, decision support tool, PBT substances.

---

*Address correspondence to Lars Carlsen: Awareness Center, Linkøpingvej 35, Trekroner, DK-4000 Roskilde, Denmark; Tel: +45 2048 0213; E-mail: LC@AwarenessCenter.dk

1. INTRODUCTION

The assessment and regulation of chemicals has over the years developed to a major issue in relation to assuring the human health as well as to protect our environment. However, the vast majority of the chemicals available on the market today has not been properly assessed and regulated. The simple explanation for this is the obvious lack, or at least availability of the necessary data material. For a discussion of data availability see, *e.g.*, Voigt *et al.* [1-3].

A crucial point in the European chemical framework REACH [4] is the handling of substances that meet the criteria for classification as dangerous in accordance to Directive 67/548/EEC [5] or is assessed to be persistent, bioaccumulating and toxic (PBT's) or very persistent and very bioaccumulating (vPvB's). For these substances an assessment including an exposure assessment as well as a risk characterization is mandatory. For the so-called CMR substances (carcinogens, mutagens, and substances toxic to reproduction), endocrine disrupting substances (ED's), persistent organic pollutants (POP's) and substances that are PBT's or vPvB's, authorization must be obtained in order to use these substances ([4], Article 57). However, this applies only in the cases where it is not possible to determine a lower threshold for their action or for substances having PBT or vPvB properties. For the latter type of substances the requirement for substitution apparently is mandatory ([4]; article 57 and 59).

It must be emphasized that the above may well be undermined as authorization of chemicals covered by Articles 55-59 in REACH [4] can be granted if the manufacturer through a socio-economic analysis can show that, *e.g.*, the socio-economic benefits outweigh the risks associated with the use of the substance ([4], Article. 60(4)(b)). It is worthwhile to note that recommendations how to balance the risks and the socio-economic analysis is not included in the REACH framework.

In a proper assessment of the chemical substance, not only the physico-chemical and toxicological characteristics (PBT) should be taken into account. In addition a series of environmental factors may advantageously be considered. Thus, parameters like production tonnage [6], specific release scenarios [6, 7], and geographical and site-specific factors [8] in addition to various substance dependent parameters should be taken into account. Further socio-economic factors may be taken into consideration.

The present review will focus on the use of partial order ranking methods as an attractive supplement to assess chemicals. The study will include 1: a simple ranking of substances based on their physico-chemical characteristics, possibly based on QSAR generated data [9-13], 2: a more elaborate hierarchical partial order ranking (HPOR) [14] method allowing to take a larger variety of parameters, *e.g.*, socio-economic factors into account, and 3: an assessment of specific characteristics of chemical substances, *e.g.*, PBT characteristics applying the accumulating partial order (APOR) approach [15].

Further the use of sensitivity studies to disclose the relative importance of the descriptor brought into play in the assessment of a group of chemicals will be illustrated [16].

2. METHODS

The basic methodology applied for assessing chemical substances is partial order ranking. Thus, in the following the basic concepts of partial order ranking, including deriving linear extensions, ranking probability and average ranks are summarized and the use of sensitivity calculations will be shown to elucidate the relative importance of the single descriptors.. In addition two more elaborate partial order ranking methodologies, *i.e.*, hierarchical partial order ranking (HPOR) and accumulating partial order ranking (APOR) are described.

As experimental data on physico-chemical characteristics and toxicology for chemical substances in many cases are scarce, data for assessing chemicals may often advantageously be derived based on quantitative structure-activity relationships (QSARs). Thus, a short description of QSARs with focus on the so-called 'noise-deficient' QSARs is included.

2.1. Partial Order Ranking

The theory of partial order ranking (POR) is presented elsewhere [17, 18]. In brief, Partial Order Ranking is a simple principle, which a priori includes "≤" as the only mathematical relation. If a system is considered, which can be described by a series of descriptors p_i, a given chemical A, characterized by the descriptors $p_i(A)$ can be compared to another chemical B, characterized by the descriptors $p_i(B)$, through comparison of the single descriptors, respectively. Thus, chemical A will be ranked higher than B, *i.e.*, $B \leq A$, if at least one descriptor for A is higher than the corresponding descriptor for B and no descriptor for A is lower than the corresponding descriptor for B. If, on the other hand, $p_i(A) > p_i(B)$ for descriptor i and $p_j(A) < p_j(B)$ for descriptor j, A and B will be denoted incomparable. Obviously, if all descriptors for A are equal to the corresponding descriptors for B, *i.e.*, $p_i(B) = p_i(A)$ for all i, the two chemicals will have identical rank and will be considered as equivalent, *i.e.*, A = B. In mathematical terms this can be expressed as

$$B \leq A \Leftrightarrow p_i(B) \leq p_i(A) \text{ for all i} \qquad\qquad (1)$$

It further follows that if $A \geq B$ and $B \geq C$ then $A \geq C$. If no rank can be established between A and B these chemicals are denoted as incomparable, *i.e.*, they cannot be assigned a mutual order. Therefore POR is an ideal tool to handle incommensurable attributes.

In partial order ranking - in contrast to standard multidimensional statistical analysis - neither any assumptions about linearity nor any assumptions about distribution properties are made. In this way the partial order ranking can be considered as a non-parametric method. Thus, there is no preference among the descriptors. However, due to the simple mathematics outlined above, it must be emphasized that the method a priori is rather sensitive to noise, since even minor fluctuations in the descriptor values may lead to non-comparability or reversed ordering. A discussion on partial order ranking *vs* multicriteria decision analyses can be found in Bruggemann and Carlsen [19].

A main point is that all descriptors have the same direction, *i.e.*, "high" and "low". As an example bioaccumulation and toxicity can be mentioned. In the case of bioaccumulation, the higher the number the higher a chemical substance tends to bioaccumulate and thus the more problematic the substance, whereas in the case of toxicity, the lower the figure the more toxic the substance. Thus, in order to secure identical directions of the two descriptors, one of them, *e.g.*, the toxicity figures, has to be multiplied by -1. Consequently, both in the case of bioaccumulation and in the case of toxicity higher figures will now correspond to more problematic chemicals.

The graphical representation of the partial ordering is often given in a so-called Hasse diagram [20-23].

The calculations are performed using the PyHasse software that today has substituted the WHasse software [23] as the preferred tool for partial order analyses. PyHasse has been developed (and currently being extended and improved) by dr. R.Bruggemann (Berlin) and consists close to 100 modules. The software package is available on request from R. Bruggemann[1] and should be considered as an experimental, non-professional software. An alternative is the DART (Decision Analysis by Ranking Techniques) software that comprises different kinds of order ranking methods, roughly classified as total - and partial order ranking methods [24]. Beside DART [24], the software package PyHasse is currently the only available one for applying ordinal analyses on data matrices.

2.2. Linear Extensions

The number of incomparable elements in the partial ordering may obviously constitute a limitation in the attempt to rank, *e.g.*, a series of chemical substances based on their potential environmental or human health hazard. To a certain extent this problem can be remedied through the application of the so-called linear extensions of the partial order ranking [25, 26]. A linear extension is a total order, where all comparabilities of the partial order are reproduced [18, 20]. Due to the incomparisons in the partial order ranking, a number of possible linear extensions correspond to one partial order. If all possible linear extensions are found, a ranking probability of the single substances can be calculated. Consequently, if all possible linear extensions are found it is possible to calculate the average ranks of the single elements in a partially ordered set [27, 28].

2.3. Average Ranks

As mentioned above the average rank of the single substances can be established based on the linear extensions as the average of the ranks in all the linear

---

[1]brg_home@web.de

extensions. It has been demonstrated that the average rank of the single element in the Hasse diagram can be obtained through deriving a large number of randomly generated linear extensions [29-32]. The random linear extension approach allows in addition to the determination of the average ranks of the single elements also the determination of the ranking probability distribution of the single elements [12, 13].

Alternatively average rank may be obtained directly. Thus, several approaches, such as the lattice theoretical one [33-35] are developed in algorithmic oriented literature. However, these methods are limited to relatively small datasets due to the necessary computer power. Thus, approximate methods have been developed based on loccal partial orders [36, 37].

2.4. Relative Descriptor Importance

Working with a multi-descriptor system as partial order ranking it is obviously of significant interest to retrieve information about the relative importance of the single indicators.

Hence, the sensitivity expresses how important any single descriptor is for the structure of the partial order as visualized by the Hasse diagram. The PyHasse software comprises a specific module that offers an estimation of the sensitivity (for details on sensitivity analyses, see: [20]) as recently illustrated by a study of 30 chemicals comprising selected organo-phosphorous compounds and industrial products [16] covered by the Stockholm Convention [38].

2.5. Hierarchical POR

Based on the average rank of the single substances, the most probable rank for each element can be obtained and thus the most probably linear rank of the elements studied can be established. This is done for every dataset, *e.g.*, physico-chemical characteristics, atmospheric parameters, geospecific factors and socio-economic factors each leading to an average ranking that subsequently are used as so-called meta-descriptors. Thus, in the second stage the meta-descriptors describe the meta-dimension are subsequently used as descriptors in a consecutive partial order ranking. Obviously, the number of descriptors is significantly reduced and the ranking leading to the possible development of a robust model [39] that in principle will contain all information based on the original set of descriptors [14].

Since the meta-descriptors, as the descriptors, are ordered with the highest rank being denoted "1", the meta-descriptors must all be multiplied by -1 in order to make sure that the elements with the highest rank, *i.e.*, with the lowest attributed number, will be ranked in the top of the Hasse diagram as a result of the ranking based on the meta-descriptors. In Fig. (**1**) a graphical representation of the HPOR approach [14] is depicted.

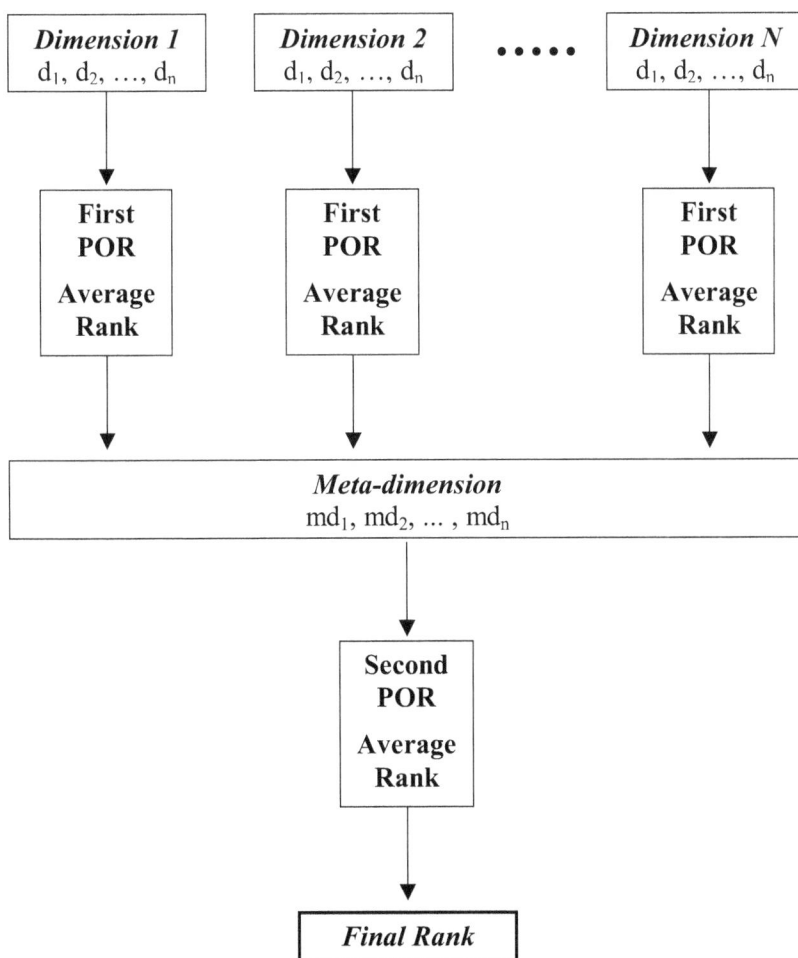

Fig. (1). Graphical representation of the hierarchical partial order ranking.

In the following the methodology will be illustrated by an example in which 16 chemical substances have to be assessed for possible confinement, remediation or clean up.

2.6. Accumulating Partial Order Ranking

In the accumulating partial order ranking a rather simple "true-false", *i.e.,* binary approach, has been adopted. Thus, an object may possess a given characteristic, *e.g.,* a chemical may be toxic, or it may not possess the characteristic, *i.e.,* the chemical being non-toxic or being biodegradable or not as, *e.g.,* derived by the BDP1 and BDP2 approaches in the EPI Suite [40]. Consequently the single descriptors may take only two possible values, *i.e.,* 1 if the characteristic is true and 0 if the characteristic is false, respectively. The resulting posets often are denoted as 2^n. For 2^n posets a linear ranking of the objects can be easily obtained, by applying a function like $h(x) = h(y)+1$ if x covers y (*i.e.,* if $y \leq x$ and there is no other element z with $y < z < x$). Thus, $h(0) = 1$ for 0 as the uniquely defined bottom element of the poset. For example in 2^2 the bottom element 0 would be (0,0) and gets $h(0) = 1$, the two elements x, x', covering 0 would get $h(x)=h(x')=1$ and the maximum element y described by (1,1) would obtain $h(y)=2$. However this ranking includes many ties, therefore it is only useful for a prescreening analysis. To illustrate this further we look at a system where the objects are described by 3 descriptors, resulting in a total of $2^3 = 8$ possible events, *i.e.,* (1,1,1), (1,1,0), (1,0,1), (0,1,1), (1,0,0), (0,1,0), (0,0,1) and (0,0,0), respectively. In general, the event-space built by objects of $\{0,1\}^n$ is isomorphic to the poset, built by an n-object ground set and all its subsets with the inclusion as order relation. Posets of this type are Boolean lattices and it is well known that they are graded and that a function h exists, like that described above.

The resulting Hasse diagram, the so-called event space is depicted in Fig. (**2**).

We assume a test battery with *M* tests, any single test leading to n descriptors. Thus, a given object *Q* is for any single test characterized by the set of *n* descriptors ($p_1(Q)$, $p_2(Q)$, ..., $p_i(Q)$, ..., $p_n(Q)$), all values being 1 or 0 denoting true and false, respectively. Hence, if *M* individual tests are carried out for each of the *n* descriptors a (*M* x *n*) matrix is developed.

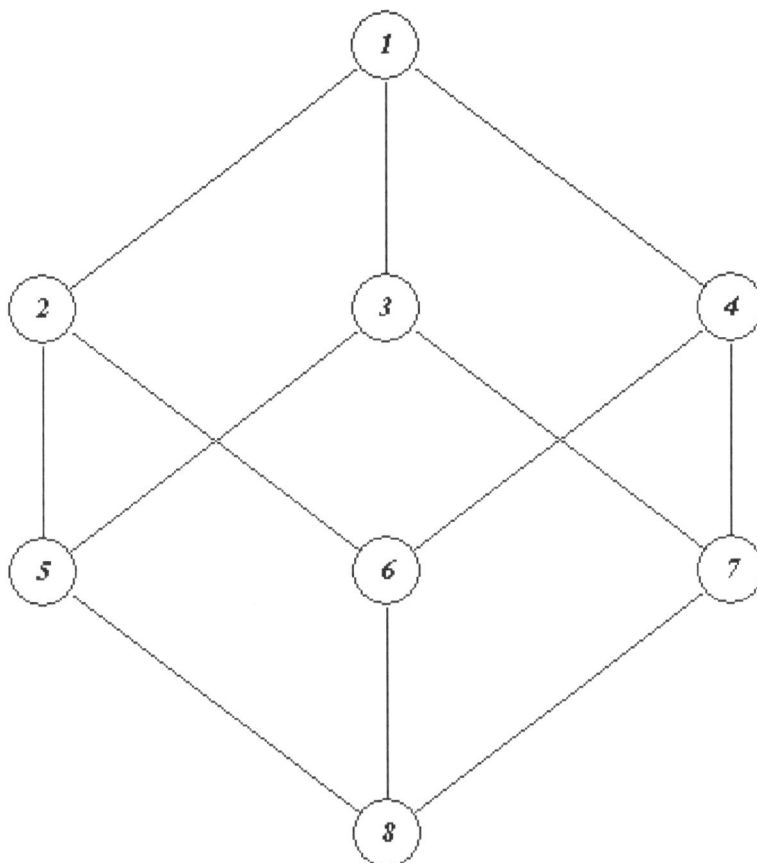

Fig. (2). Event space corresponding to 3 descriptors.

The entry $p_{i,j}$ describes the ith descriptor of a single test j. Summing $\sum_{j=1}^{M} p_{i,j}$ leads to the result for the ith aspect (*e.g.*, bioaccumulation, evaluated by M test variants). In order words, the "accumulated descriptors" p_i^{Acc} is obtained simply by adding the M single descriptor values resulting from the individual tests. The eventual set of accumulated descriptors $(p_1^{Acc}, p_2^{Acc}, ...p_i^{Acc}, ...p_n^{Acc})$ for the object Q is consequently given by

$$(p_1^{Acc}, p_2^{Acc}, ..., p_i^{Acc}, ..., p_n^{Acc}) = (\sum_{j=1}^{M} p_{1,j}(Q), \sum_{J=1}^{M} p_{2,j}(Q), ..., \sum_{j=1}^{M} p_{i,j}(Q), ..., \sum_{j=1}^{M} p_{n,j}(Q)) \quad (2)$$

where M is the number of individual tests performed. Thus, the accumulated descriptors contain the combined information of the single characteristics of object Q.

It is immediately noted that the single accumulated descriptors, $\sum_{j=1}^{M} p_{i,j}(Q)$, may take any value from 0 to M. In order to give the single objects an identity in the resulting ranking, we include, if not already present, an artificial reference maximum and minimum objects characterized by the sets of descriptors of $(M, M, M,....)$ and $(0, 0, 0, ...)$, respectively, are introduced. Unambiguously, the accumulation process will enrich the resulting Hasse diagrams (*vide infra*), because the procedure of summing up single descriptors is order preserving.

It is obvious that the successful use of accumulating partial order ranking (APOR) obviously depends on the quality of the single techniques applied in the M test of the single compounds.

2.6.1. Probabilities Based on Accumulated Ranks

The probability, S, that a given object, characterized by n accumulated descriptors originating from M individual tests possess all the characteristics included in the study can be calculated according to eqn. 4

$$S = \frac{(\sum_{j=1}^{M} p_{1,j}(Q) \cdot \sum_{J=1}^{M} p_{2,j}(Q) \cdot ..... \cdot \sum_{j=1}^{M} p_{i,j}(Q) \cdot ..... \cdot \sum_{j=1}^{M} p_{n,j}(Q))}{M^{n}} \tag{3}$$

Obviously, following a single test only the value of S maybe 0 or 1 only, whereas the inclusion of further tests leads to an increased discriminatory power.

2.7. Quantitative Structure-Activity Relationships

The basic concept of QSARs can in its simplest form be expressed as the development of correlations between a given physico-chemical property or biological activity (endpoint), P, and a set of parameters (descriptors), D_i, that are inherent characteristics for the compounds under investigation

$$P = f(D_i) \tag{4}$$

The properties (endpoints), P that has been subjected to QSAR modeling comprises physico-chemical properties as well as biological activities.

In general models that describe/calculate key properties of chemical compounds are composed of three types of inherent characteristics of the molecule, *i.e.,* structural, electronic and hydrophobic characteristics. Depending on the actual model few or many of these descriptors may be taken into account. Thus, eqn. 5 can be rewritten as

$$P = f(D_{structural}, D_{electronic}, D_{hydrophobic}, D_x) + e \qquad\qquad (5)$$

The descriptors reflecting structural characteristics may, *e.g.,* be element of the actual composition and 3-dimensional configuration of the molecule, whereas descriptors reflecting the electronic characteristics may, *e.g.,* be charge densities, dipole moment *etc.* The descriptors reflecting the hydrophobic characteristics are related to the distribution of the compound between a biological, hydrophobic phase, and an aqueous phase. The fourth type of characteristics, D_x, accounts for possible underlying characteristics that may be known or unknown, such as environmental or experimental parameters as, *e.g.,* temperature, salt content *etc.* The data may often be associated with a certain amount of systematic and non-quantifiable variability in combination with uncertainties. These unknown variations are expressed as "noise". Thus, the parameter, e, account for possible noise in the system, *i.e.,* the variation in the property that cannot be explained by the model.

In principle all types of QSAR models can be used to generate descriptors for subsequent use in partial order ranking, *i.e.,* commercially available generally applicable QSARs as well as more specialized custom made QSARs. However, as partial order ranking due to its inherent nature only focusing on the relation ">" (*vide infra*) may be hampered by random fluctuations in the descriptors, the so-called 'noise-deficient' QSARs [9-13] advantageously can be applied. 'Noise-deficient'

QSARs are QSAR models where the natural variation in both the experimental data and the primary model data has been suppressed in a subsequent modeling step. Thus, recent studies on organophosphates appear as an illustrative example on the application of 'noise-deficient' QSAR-derived endpoints as input for a subsequent partial order ranking. The descriptors are generated through QSAR modeling, the EPI Suite [40] being the primary tool [9-13].

Based on the EPI generated values new linear QSAR models are build by estimating the relationships between the EPI generated data and available experimental data, the general formula for the descriptors, D_i, to be used being

$$D_i = a_i \times D_{EPI} + b_i \qquad (6)$$

D_{EPI} being the EPI generated descriptor value and a_i and b_i being constants. The log K_{OW} values generated in this way may subsequently be used to generate log *BCF* values according to the Connell formula [41]

$$log\ BCF = 6.9 \times 10^{-3} \times (log\ K_{ow})^4 - 1.85 \times 10^{-1} \times (log\ K_{ow})^3$$
$$+ 1.55 \times (log\ K_{ow})^2 - 4.18 \times log\ K_{ow} + 4.72 \qquad (7)$$

The model was somewhat modified [9] to handle the calculation of BCF values corresponding to low log K_{OW} values. Thus, a linear decrease of log *BCF* with log K_{OW} was assumed in the range $1 < log\ K_{OW} < 2.33$, the log $BCF = 0.5$ for log $K_{OW} \leq 1$, the latter value being in accordance with BCFWin [40].

Subsequently, these QSAR generated endpoints may be applied for a partial order ranking of the substances using two or more of the endpoints as descriptors for the ranking exercise.

3. RESULTS

In the following the simple partial order ranking, the hierarchical partial order ranking and the accumulating partial order ranking methodologies will be illustrated in relation to the assessment of chemicals. Further the concepts of average ranks and relative descriptor importance will be illustrated.

3.1. The Simple Approach

In the simple approach [9-13, 42, 43] a selection of compounds is partially ordered based on a set of descriptors, like, *e.g.*, persistence, bioaccumulation and toxicity. This approach appears appropriate either as an ordering and thus a possible basis for prioritizing a series of substances as illustrated by the analyses of a series of substituted anilines [13]. This group of substances was characterized by their octanol-water partitioning coefficient (log K_{OW}), their vapor pressure (log

VP), their possible ready/non-ready biodegradability as well as their aquatic toxicity as disclosed by the population growth impairment of *Tetrahymena pyriformis* [44]. Missing data for log K_{OW} and log *VP*, respectively, was derived using the EPI Suite [40] and subsequently the combination of experimentally and EPI Win derived log K_{OW} and log *VP*, respectively, was used to generate so-called 'noise-deficient' QSARs for log K_{OW} and log *VP*, respectively (vide supra). The ready/non-ready biodegradability is estimated using the BioWin software that is an integrated part of the EPI Suite [40]. In Table **1** the derived descriptors for the assessment of 45 anilines are given and the resulting ranking, depicted as a Hasse diagram are shown in Fig. (**3**). It should be noted that in order to have the same direction of all descriptors, the log *VP* and the *PNEC* were both multiplied by -1 (*vide supra*) prior to calculation as low *VP* corresponds to high persistence and low *PNEC* values to high toxicity.

The diagram consists of 11 levels, the top level, comprising the anilines 21, 29, 30, 44 and 45 is associated with the environmentally more hazardous species whereas the bottom level, comprising the anilines 1, 3 and 36 the less hazardous species. Thus, in terms of risk assessment the 11 levels can *a priori* be regarded as a classification of the 45 anilines in 11 classes according to their potential environmental impact simultaneously taking into account water-octanol partitioning, vapor pressure, biodegradability and toxicity [13].

Another application of the simple partial order ranking is the possibility of given certain chemicals and identity in relative to other possibly well-investigated chemicals. This has been illustrated in a series of studies on chemical warfare agents of the organo phosphorous type (nerve agents) [9-11]. These substances share structural elements with a wide range of organo phosphorous-based insecticides the latter being well characterized. Thus, it is possible to order the nerve agents with the set of insecticides and thus giving them an identity. This idea is illustrated in Fig. (**4**), where a Hasse diagram (Fig. **4A**) displays the ranking of 10 compounds using three descriptors, *e.g.*, the PBT characteristics of the compounds and a second diagram (Fig. **4B**) display the ranking of the same 10 compounds plus one new compound X.

Table 1.   Descriptors for Ordering 45 Anilines Based on 'Noise- Deficient' Values for Octanol Water Partitioning and Vapour Pressure Together with Ready/Non-Ready Biodegradability and Toxicity

CAS No	Name Anilin	ID	log Kow (nd)	log VP (nd) Pa	$BDP2^a$	PNEC (µg/L)
62-53-3	H	1	1.03	1.90	1	158.16
95-53-4	2-Methyl	2	1.59	1.55	1	154.89
108-44-1	3-Methyl	3	1.59	1.48	1	204.19
106-49-0	4-Methyl	4	1.59	1.33	1	120.24
578-54-4	2-Ethyl	5	2.09	1.34	2	201.11
587-02-0	3-Ethyl	6	2.09	1.24	2	129.85
589-16-2	4-Ethyl	7	2.09	1.16	2	113.09
643-28-7	2-iso-Propyl	8	2.53	1.09	2	102.57
99-88-7	4-iso-Propyl	9	2.53	0.99	2	81.47
104-13-2	4-Butyl	10	3.12	-0.02	1	12.70
30273-11-1	4-sec-Butyl	11	3.03	0.60	2	36.63
769-92-6	4-tert-Butyl	12	3.00	0.63	2	65.15
87-59-2	2,3-Dimethyl	13	2.15	1.07	1	326.16
95-68-1	2,4-Dimethyl	14	2.15	1.24	1	236.28
95-78-3	2,5-Dimethyl	15	2.15	1.24	1	259.08
87-62-7	2,6-Dimethyl	16	2.15	1.22	1	326.16
95-64-7	3,4-Dimethyl	17	2.15	0.66	1	175.16
108-69-0	3,5-Dimethyl	18	2.15	1.12	1	277.61
88-05-1	2,4,6-Trimethyl	19	2.72	0.82	1	151.71
579-66-8	2,6-Diethyl	20	3.17	0.76	1	73.09
24544-04-5	2,6-di-iso-Propyl	21	4.04	0.25	2	30.81
95-51-2	2-Chloro	22	1.69	1.36	2	188.69
108-42-9	3-Chloro	23	1.69	0.87	2	76.87
106-47-8	4-Chloro	24	1.69	0.36	2	113.70
554-00-7	2,4-Dichloro	25	2.36	0.16	2	44.62
95-82-9	2,5-Dichloro	26	2.36	0.15	2	42.62
626-43-7	3,5-Dichloro	27	2.36	-0.09	2	31.59
636-30-6	2,4,5-Trichloro	28	3.02	-0.74	2	9.85
3481-20-7	2,3,5,6-Tetrachloro	29	3.69	-1.49	2	4.01
634-83-3	2,3,4,5-Tetrachloro	30	3.69	-1.29	2	2.53

(Table 1) contd.....

CAS No	Name Anilin	ID	log *Kow* (nd)	log *VP* (nd) Pa	BDP2[a]	PNEC (µg/L)
615-65-6	2-Chlor-4-Methyl	31	2.26	1.11	2	93.55
95-69-2	4-Chlor-2-Methyl	32	2.26	0.50	2	63.25
95-79-4	5-Chloro-2-Methyl	33	2.26	0.65	2	44.78
95-74-9	3-Chloro-4-Methyl	34	2.26	0.56	2	57.69
87-60-5	3-Chloro-2-Methyl	35	2.26	0.54	2	59.03
348-54-9	2-Fluor	36	1.23	2.09	2	260.49
372-19-0	3-Fluor	37	1.23	1.81	2	139.89
771-60-8	penta-Fluor	38	2.06	2.42	2	100.61
615-43-0	2-Iodo	39	2.23	-0.18	2	97.84
626-01-7	3-Iodo	40	2.23	0.04	2	49.03
2237-30-1	3-Cyano	41	1.12	-0.76	1	348.66
88-74-4	2-Nitro	42	2.00	-0.81	2	114.89
99-09-2	3-Nitro	43	1.43	-1.76	2	128.91
97-02-9	2,4-Dinitro	44	1.81	-3.16	2	34.89
606-22-4	2,6-Dinitro	45	1.24	-2.88	2	26.47

[a] *BDP2* = 1 denotes 'ready biodegradable', *BDP2* = 2 denotes 'non-ready biodegradable'.

The revised Hasse diagram (Fig. **4B**), now including 11 compounds immediately discloses that compound X has now obtained an identity in comparison to the original 10 compounds. Assuming that the original 10 compounds were all well-characterized, we can conclude that compound X on a cumulative basis taking both P, B and T characteristics into account is less environmentally harmful than compounds 4 and 7, but more harmful than compound 10. Thus, through the partial order ranking the compound, X, has obtained an identity in the scenario with regard to its potential environmental impact [9, 12]. Hence, in a decision support scenario this analysis can be used to evaluate the potential impact on the environment of X based on previously verified impacts of compounds 4, 7 and 10, *i.e.*, the impact of X is expected to be less than those of compounds 4 and 7, respectively, but higher that that of compound 10.

It is immediately seen (Fig. **4B**) that the compounds 4 and 7 are incomparable, *i.e.*, looking just for these two compounds it cannot from the Hasse diagram be

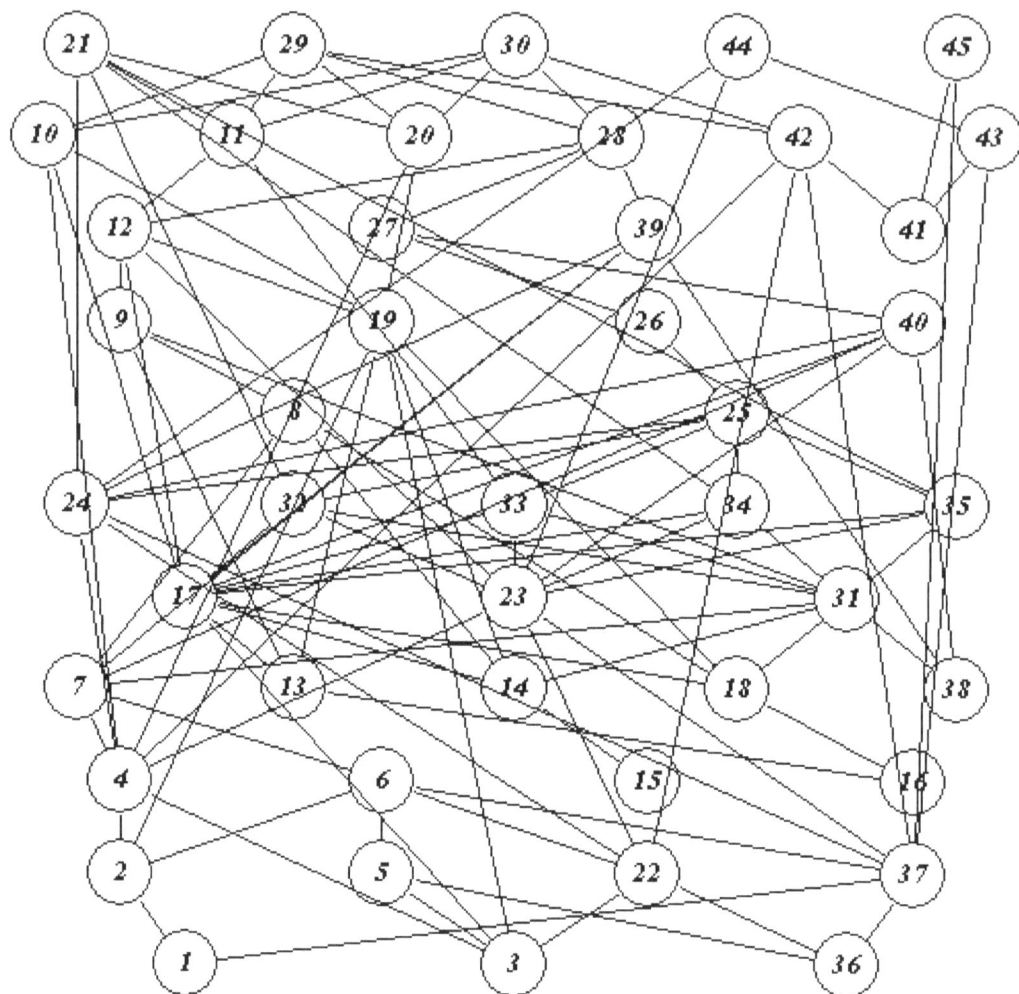

Fig. (3). Hasse diagram of the 45 anilines based on the 4 descriptors given in Table **1**. The single compounds are identified through their ID (cf. Table **1**).

concluded which of them will be the environmentally more problematic. However, based on the linear extensions the probability for having a certain absolute rank can be derived. In Fig. (**5A**) the probability distribution for the compounds 4 and 7 for the possible absolute ranks is visualized whereas the probability distribution for compounds 10 is shown in Fig. (**5B**). From the former it can be concluded that comparing compounds 4 and 7, the most probable absolute ranking will place compound 4 above compound 7, whereas compound 10, as seen from Fig. (**4B**) will be place at the lowest ranks [12].

A

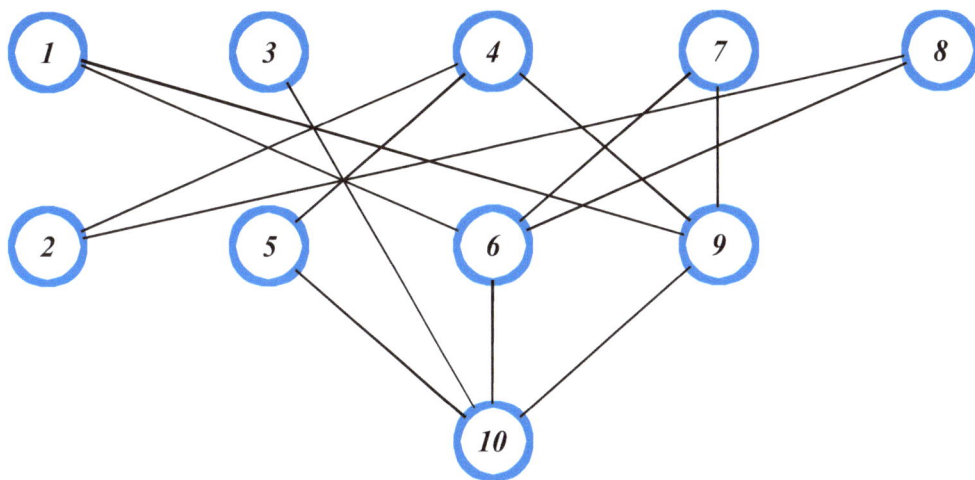

B

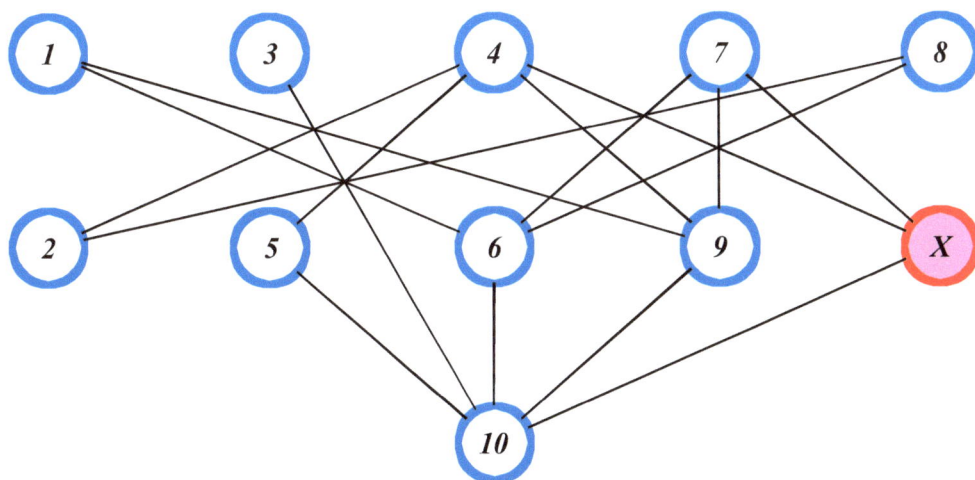

Fig. (4). Illustrative Hasse diagram of A: 10 compounds using three descriptors and B: the same 10 compounds plus one new compound X.

A

B

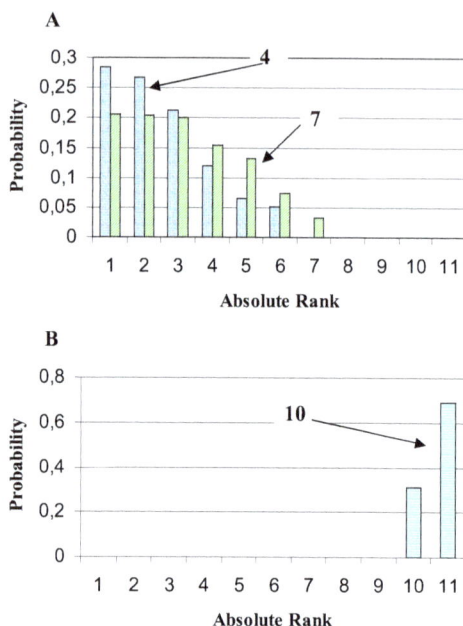

Fig. (5). Probability distribution of A: compounds 4 and 7 and B: compound 10 to occupy specific absolute ranks (rank 1 and 11 is top and bottom rank respectively).

The compound, X (Fig. **4B**) is comparable with compound 4, 7 and 10 and incomparable with the remaining 7 compounds in the scenario, the high number of incomparisons being reflected in the relative broad probability distribution for compound X (Fig. **6**).

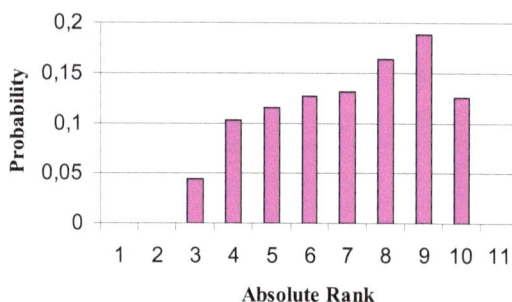

Fig. (6). Probability distribution of compound X to occupy specific absolute ranks.

The probability distribution of compounds X, 4, 7, 10 is depicted in Fig. (**7**) strongly indicating that compound X must be located between compounds 4 and 7 and compound 10 (cf. Fig. **4B**) [12].

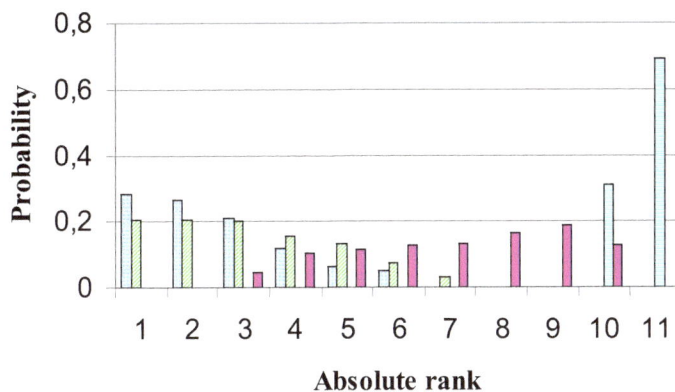

Fig. (7). Probability distribution of compound X in relation to compounds 4, 7 and 10 to occupy specific absolute ranks.

The behavior of the single compounds under investigation on a combined basis, *e.g.*, taking all descriptors simultaneously into account was further elucidated applying the concept of average rank, in the present case as calculated by the approximate local linear order model (LPOMext) [37]. In Table **2** the average rank calculated according to eqn. 4 is given together with the 'uncertainty', *i.e.*, the range wherein the true/absolute rank are to be found.

Table 2. Average Ranks of the 11 Compounds Included in the Hasse Diagram Displayed in Fig. (4B) Calculated by the Local Partial Order Model (LPOMext) [37]

Compound No.	Average Rank (Rk_{av})	Uncertainty
1	3,1	1...8
2	8,4	3...11
3	5,3	1...10
4	2,0	1...6
5	6,7	2...10
6	8,4	4...10
7	2,5	1...7
8	3,0	1...8
9	8,3	4...10
10	10,9	10...11
X	7,6	3...10

The average ranks (Table **2**) is in complete agreement with the above conclusion that the new compound X would be located between the compounds 4 and 7 and compound 10 (cf. Fig. **4B**) [12]. Further the relative broad range ('uncertainty') of the Rkav for X is in perfect agreement with the fact that X is comparable only to 4, 7 and 10 and thus a significant number of incomparisons prevails.

If the average ranks, Rk_{av}, of two compounds are close, the two compounds will on an average basis display similar characteristics based on the set of descriptors applied. Thus, the study of Carlsen [12] disclosed (Table **2**) that compound X most closely resembles compounds 2, 6 and 9. Consequently, compound X has in this way obtained an identity compared to the basis set of compounds, *i.e.*, compounds 1-10.

3.2. Relative Descriptor Importance

To illustrate the relative importance of the descriptors used for a partial order ranking, *e.g.*, the sensitivity analysis, we return to the above described study on the 45 anilines. The result of the sensitivity analysis is depicted in Fig. (**8**).

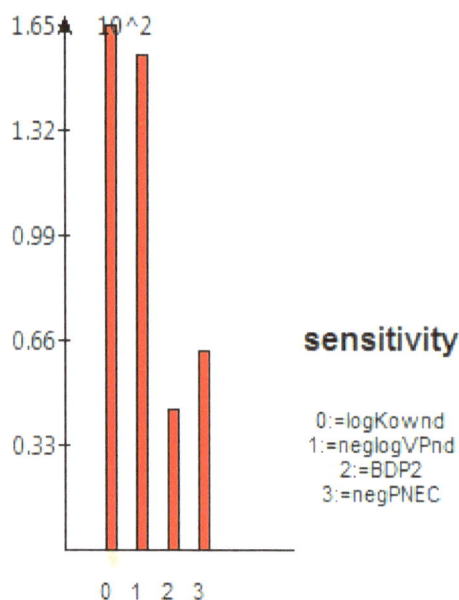

Fig. (8). Relative importance of the 4 descriptors used to characterize the environmental behaviour of the 45 anilines.

From Fig. (**8**) it appear clear that in the present example the descriptors log *Kow*, reflecting bioaccumulation and log *VP*, reflecting persistence apparently are the most important for the overall ranking the individual relative weights being 0.39 and 0.36, respectively. The relative weights for the biodegradation potential (*BDP2*) and the toxicity component, expressed as the *PNEC* value was found to be 0.10 and 0.15, respectively. The results are in good agreement with recent studies on the ranking of a series of PBT compounds [45].

3.3. Hierarchical Partial Order Ranking

Hasse diagrams are characterized by the presence of a number of comparisons and incomparisons. Sørensen *et al.* [39] demonstrated that the actual number of incomparisons is a result of interplay between the number of compounds and the number of descriptors. Thus, increasing the number of descriptors will, for the same number of compounds, increase the number of incomparisons. In a simple example looking at 16 compounds applying 9 descriptors unambiguously shows the lack of robustness, the Hasse diagram displays a complete anti-chain (Fig. **9**).

Fig. (9). Hasse diagram visualizing the partial order ranking of 16 compounds applying 9 descriptors.

Carlsen [14, 46] showed that the adoption of the more elaborate approach, hierarchical partial order ranking (HPOR) [14] might remedy this problem leading to a robust model [39].

HPOR involves primary analyses of specific sets of connected descriptors, leading to latent variables, meta-descriptors derived as the average ranks. Subsequently, the set of meta-parameters are used for a second partial order ranking analysis leading to a final assessment. Thus, HPOR may advantageously be applied to problems involving a high number of parameters and possibly a relatively low number of element (here chemicals).

To illustrate the HPOR concept, let the above mentioned 9 descriptors characterizing the 16 compounds could correspond to the following: Persistence,

Bioaccumulation potential, Toxicity, octanol water partitioning, vapor pressure, Henry's Law Constant, production volume, production cost and costs of substitution, respectively. These 9 descriptors may now be collected in 3 sets, *i.e.*, PBT characteristics, factors related to solubility and factors related to production and possible substitution. Subsequent ranking of the 16 compounds based on the 3 sets of 3 descriptors, respectively are illustrated by the Hasse diagrams depicted in Fig. (**10**), the corresponding meta-descriptors derived as average ranks according to eqn. 4 are found in Table **3**

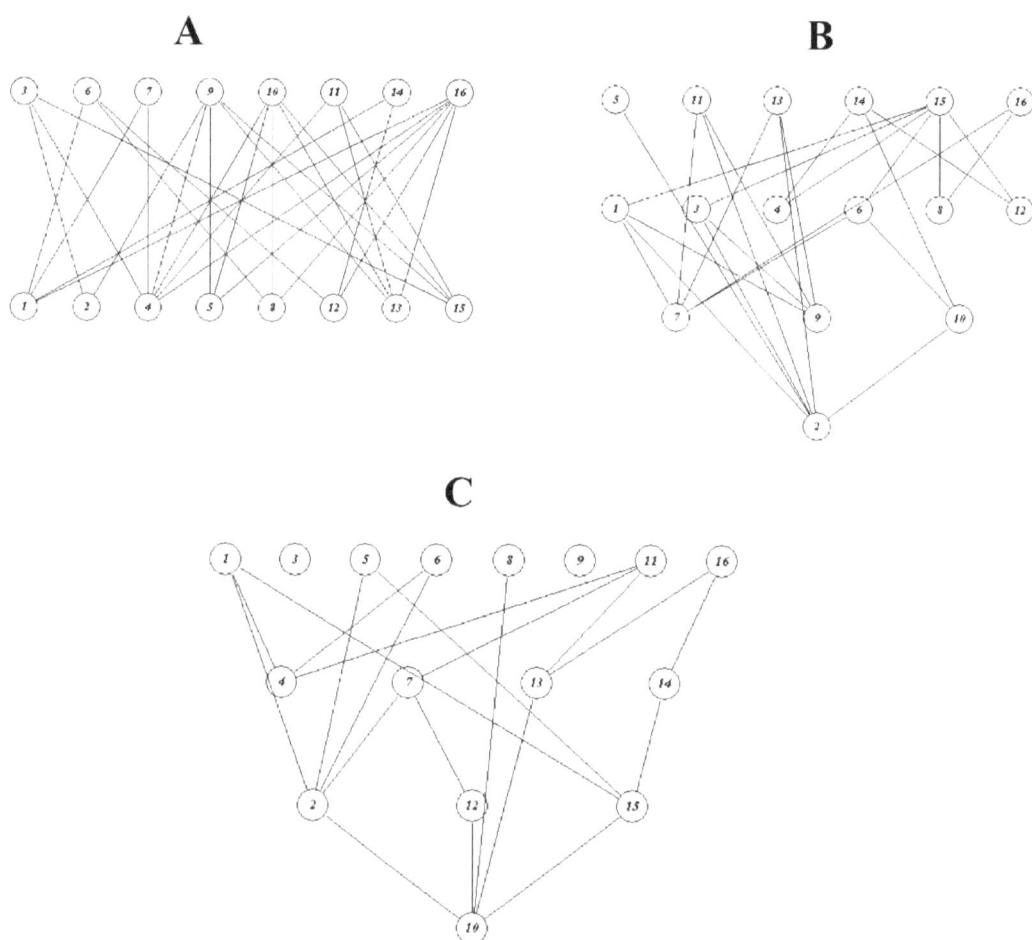

Fig. (10). Partial order ranking of the 16 compounds applying based on the 3 sets of 3 descriptors, *e.g.*, representing A: Persistence, Bioaccumulation potential and Toxicity, B: octanol water partitioning, vapor pressure and Henry's Law Constant, and C: Production volume, Production cost and costs of substitution.

The subsequent ranking based on the 3 meta-descriptors (Table **3**) is visualized in the Hasse diagram in Fig. (**10**) and the final average ranking being found in Table **4**.

Table 3.   **Meta-Descriptors Derived as Averaged Ranks (cf. Fig. 9)**

Compound No.	Meta A (Rk$_{av}$)	Meta B (Rk$_{av}$)	Meta C (Rk$_{av}$)
1	14,2	5,7	2,8
2	12,8	15,5	12,8
3	3,4	6,8	8,5
4	14,9	12,8	13,6
5	13,6	5,7	3,4
6	3,4	5,7	3,4
7	4,3	14,9	5,7
8	13,6	12,8	5,7
9	2,4	14,6	8,5
10	2,4	11,3	15,8
11	3,4	3,4	2,1
12	13,6	12,8	10,2
13	14,2	3,4	10,2
14	4,3	2,8	6,8
15	14,2	1,4	12,1
16	2,1	4,3	2,8

3.4. Accumulating Partial Order Ranking

The assessment of specific characteristics for chemical substances, like PBT characteristics may advantageously be conducted applying the accumulating partial order methodology [15]. In order to increase the basis for the eventual assessment of the chemical under investigation several "tests" will often be conducted. Thus, the combined - or accumulated - knowledge based on the results from the test battery will constitute the eventual support for the eventual assessment. One possibility to do so would be to adopt the concepts of Bayesian statistics [47, 48] increasing the predictability of QSARs as a risk assessment tool. However, applying this methodology to multicriteria analysis is not directly intelligible. In contrast to this it appears that partial order ranking can be applied in a straightforward way.

Table 4.   Final Averaged Ranks Following HPOR (cf. Fig. 10)

Compound No.	Final Averaged Ranks (Rk$_{av}$)
1	10,2
2	15,1
3	8,5
4	15,7
5	8,5
6	4,6
7	11,3
8	10,6
9	8,5
10	11,3
11	1,4
12	13,6
13	10,2
14	2,8
15	5,7
16	1,3

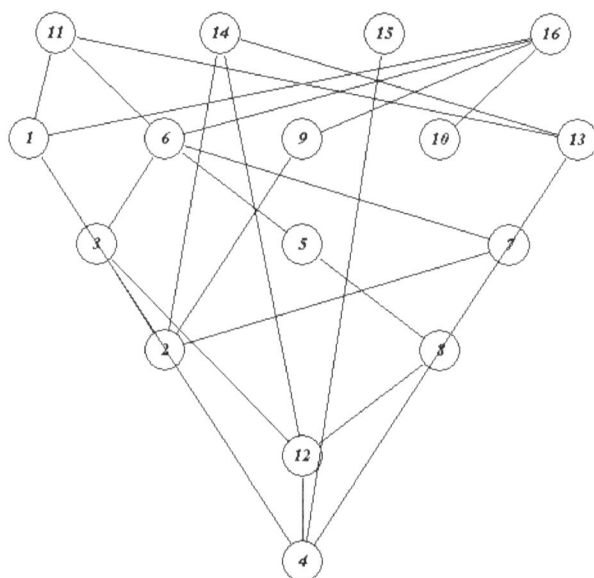

Fig. (11). Partial order ranking of the 16 compounds based on the 3 meta-descriptors given in Table **3**.

The applicability of APOR was demonstrated in a study investigating a series of compounds to establish their possible classification as PBT compounds [15] based on 5 different tests. To be classified as a PBT compound, the compound must be environmentally persistent (P), bioaccumulating (B) and toxic (T) [49]. Consequently, based a single test for a given compound can be classify the compound as PBT only if P, B and T, respectively, are found "true", *i.e.*, all three descriptors, p_i, b_i and t_i, (vide supra) are assigned the value 1. The Hasse diagrams of the single tests will obviously be equal to that of the event space (Fig. **2**), the single elements in the diagram reflecting equivalent classes of compounds. Obviously the analysis, based on only one single test is rather rigid as the single compounds are denoted PBT ($S = 1$), *i.e.,* all three demands are fulfilled or not ($S = 0$), where one or more of the demands are not fulfilled, respectively (cf. eqn. 4).

The APOR approach follows partially the recently published evaluation method METEOR [50, 51], where the descriptors are step-wise aggregated; hence the effect of weights can be controlled and traced back. The advantage of APOR and METEOR is that there is an enrichment of order relations. The disadvantage of both methods is that now some descriptors can be considered as substitutes for others.

In this case only compounds placed in the top level of the Hasse diagram, *i.e.*, the (1, 1, 1) equivalent class will in this case be depicted as PBT's. A much more subtle picture develops following accumulation of the results obtained through the further tests, which is obvious from the Hasse diagrams displayed in Fig (**12**) corresponding to accumulating data from 2, 3, 4 and 5 tests, respectively, the results being adopted from a recent study including 30 compounds subjected to 5 tests (for illustrative purposes the corresponding data were randomly generated) [15].

The enrichments of the Hasse diagrams (Fig. **12**) compared to the event space (Fig. **2**) are obvious. The accumulated descriptors may take any value between 0 and *M*.

The analyses of the 4 partial ordered rankings visualized by the Hasse diagrams (Fig. **12**) are summarized in Table **5**, reflecting the much more subtle approach to the evaluation of PBT substances. The average ranks of the single compounds as well as the probabilities, ranging between 0 and 1 (cf. eqn. 4), for a given compound to be classified as PBT.

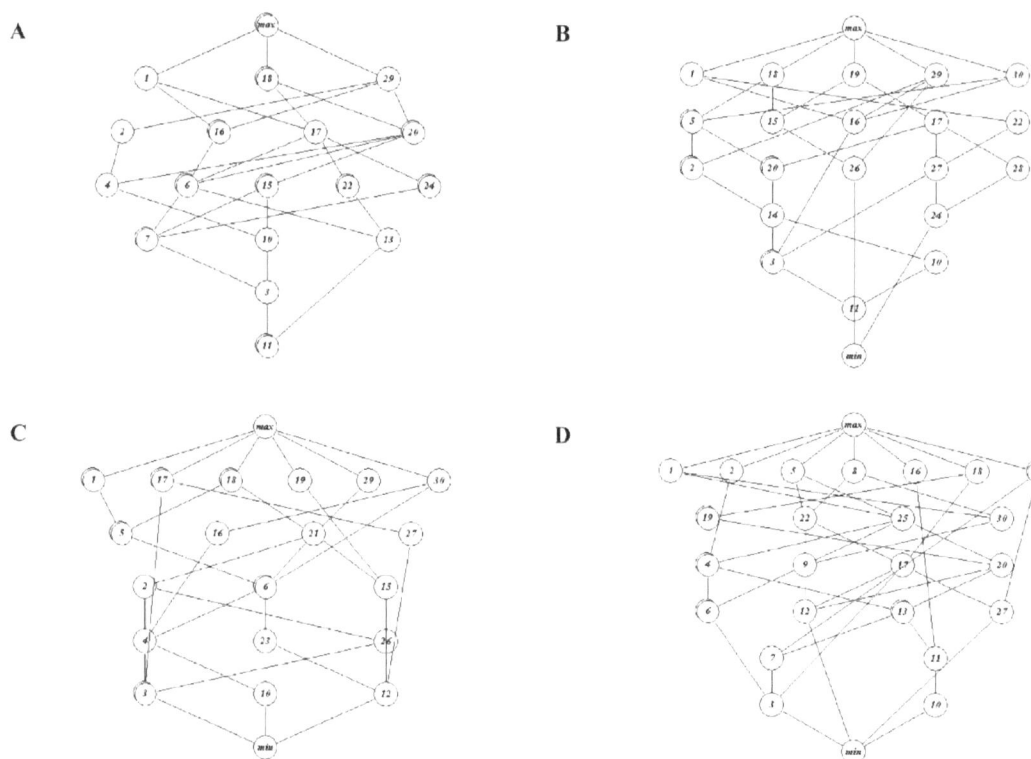

Fig. (12). Hasse diagrams following consecutive inclusion of Tests-2 to Test-5: A: After Tests 1 and 2, B: After Tests 1, 2 and 3, C: After Tests 1, 2, 3 and 4, and D: After Tests 1, 2, 3, 4 and 5.

4. DISCUSSION

The above discussion has illustrated how partial order ranking methodologies advantageously may be applied as decision support tools in relation to the assessment of chemicals. However, it is important to stress that partial order ranking only constitutes as a decision support help, as the techniques unambiguously suffer from potential drawbacks.

Obviously the number of incomparisons constitutes a major problem applying partial order ranking. This is typically a result of the interplay between the number

Table 5. **Ranking Analysis of 30 Arbitrary Compounds Plus Artificially Introduced Maximum and Minimum References After Accumulation**

ID	After Tests 1 and 2			After Tests 1, 2 and 3			After Tests 1, 2, 3 and 4			After Tests 1, 2, 3, 4 and 5		
	HD Level[a]	Rk_{av}[b]	S[c]	HD Level[a]	Rk_{av}[b]	S[c]	HD Level[a]	Rk_{av}[b]	S[c]	HD Level[a]	Rk_{av}[b]	S[c]
max	1	2.0	1.000	1	1.0	1.000	1	1.0	1.000	1	1.0	1
1	2	4.3	0.500	2	4.4	0.333	2	5.2	0.281	2	3.1	0.288
2	3	13.2	0.000	4	15.7	0.148	4	13.2	0.094	2	4.1	0.192
3	6	29.7	0.000	6	29.5	0.037	6	30.7	0.031	7	30.6	0.032
4	4	21.2	0.000	4	15.7	0.148	5	23.4	0.063	4	12.2	0.144
5	1	2.0	1.000	3	9.0	0.296	3	11.0	0.188	2	2.9	0.288
6	4	23.4	0.125	6	29.5	0.037	4	19.3	0.125	5	27.2	0.096
7	5	28.1	0.000	6	29.5	0.037	6	30.7	0.031	6	28.1	0.048
8	4	23.4	0.125	3	9.0	0.296	2	5.2	0.281	2	4.7	0.256
9	4	23.4	0.125	4	15.7	0.148	4	19.3	0.125	4	17.6	0.144
10	5	24.8	0.000	6	27.8	0.000	6	29.7	0.000	7	29.5	0.000
11	7	32.0	0.000	7	30.7	0.000	6	30.7	0.031	6	27.8	0.048
12	4	23.4	0.125	6	29.5	0.037	6	30.3	0.031	5	28.9	0.048
13	5	28.7	0.000	6	29.5	0.037	6	30.7	0.031	5	23.1	0.096
14	4	23.4	0.125	5	19.8	0.074	4	19.3	0.125	5	27.2	0.096
15	4	18.6	0.000	3	18.9	0.000	4	15.4	0.094	5	23.1	0.096
16	3	10.4	0.250	3	11.8	0.111	3	9.0	0.125	2	11.0	0.120
17	3	8.6	0.250	3	5.2	0.222	2	11.0	0.125	5	18.0	0.128
18	2	4.7	0.500	2	2.9	0.444	2	4.3	0.281	2	3.9	0.256
19	2	4.7	0.500	2	3.1	0.333	2	6.0	0.188	3	8.8	0.192
20	3	9.6	0.250	4	15.7	0.148	4	19.3	0.125	4	14.7	0.144
21	3	9.6	0.250	4	15.7	0.148	3	7.5	0.188	3	8.8	0.192
22	4	22.0	0.000	3	7.1	0.222	3	11.0	0.188	3	9.4	0.192
23	5	28.1	0.000	6	29.5	0.037	5	27.5	0.063	5	27.2	0.096
24	4	18.9	0.000	5	26.4	0.000	2	11.0	0.125	2	12.4	0.120
25	4	23.4	0.125	3	9.0	0.296	2	4.3	0.281	3	6.3	0.216
26	4	18.6	0.000	4	24.8	0.000	5	19.8	0.047	5	23.1	0.096
27	4	22.0	0.000	4	12.4	0.111	3	20.6	0.063	5	27.0	0.032
28	4	18.9	0.000	4	18.9	0.000	2	11.0	0.125	2	12.4	0.12
29	2	4.0	0.500	2	3.7	0.333	2	3.3	0.281	4	12.2	0.144
30	3	10.4	0.250	2	3.0	0.444	2	3.9	0.250	3	11.0	0.192
min	7	32.0	0.000	8	32.0	0.000	7	32.0	0.000	8	32.0	0

[a] cf. Hasse diagram Fig. (**1**), [b] calculated by eqn. 3, [c] calculated by eqn. 4

of elements studied and the number of parameters taken into account [39]. Thus, a too high number of parameters unequivocally result in a non-robust model with a high number of incomparisons eventually leading to a complete anti-chain (cf. Fig. **9**). As has been demonstrated by Carlsen [14] this problem may be remedied by application of the HPOR methodology grouping the parameters and in a first step generating so-called meta-descriptors that subsequently are used for the final ranking.

Further, the application of ranking probabilities and average ranks appears as advantageous tools in relation to circumvent the problems associated with incomparisons. In this connection it is worthwhile to mention that deriving ranking probabilities will give the regulators a tool where they can base their decisions on probabilities for certain events.

A second major problem is the sensitivity of the methods to uncertainties or fluctuations in the data material as generally seen in experimentally generated data. This applies both to physico-chemical data but to an even higher extent to biologically based data. This is a simple consequence of the fact that POR basically is based only on the "$\leq$" relation. Thus, whether A = 1.001 and B = 0.999 or A = 1001 and B = 0.999 the ranking will be A > B. However, in the first case small uncertainties, or even "experimental noise" in the data may cause a reversal of the ranking, which obviously is not the case in the second case. To circumvent this problem the use of so-called noise-deficient QSAR generated data has been suggested [9-13]. A somewhat similar approach for the evaluation of biomonitoring data has recently been suggested by Helm [52], the scheme involving splitting of the data in subset that after generation of a total order for each subset eventually are merged.

Immediately one could imagine that applying a binary scheme like APOR automatically would circumvent the problems associated with uncertainties and fluctuations in the data material. Unfortunately this is not true, as the underlying tests leading to "yes/no" decisions obviously may be defective. However, the APOR scheme, where the eventual decision will be based on a series of test will obviously to some extent remedy this problem. Nevertheless, the selection, and subsequently judgment of the results of the single test should advantageously be based on expert judgments to increase the validity of the final assessment retrieved by the APOR analysis.

Finally it should be noted that the partial order methodologies illustrated above obviously are not limited to be used in the assessment of chemicals. Hence, the simple partial order ranking techniques including the application of linear extensions and average ranks as well as the more elaborate methodology HPOR may be used as decision support tools in a wide variety of situations where an assessment and subsequent prioritization are to be carried out based on a multitude of parameters, like, *e.g.*, the prioritization of polluted sites for potential remediation or clean up [46] or of compounds, as described here. The APOR scheme likewise has a broad range of applications to strengthen the eventual assessment based on data from a multitude of approaches.

5. CONCLUSIONS AND OUTLOOK

The application of partial order ranking methodologies, comprising a simple one-step scheme as well as more elaborate assessment schemes like hierarchical partial order ranking and accumulating partial order ranking for the assessment of chemicals has been illustrated by selective examples. The partial order ranking methodologies constitute, together with the application of linear extensions and average ranks an effective decision support tool that will enable decision makers and regulators to draw significant conclusions concerning the characteristics hazards of groups of chemicals being investigated as well as about their potential impact on the environment, ecosystems and the humans health.

ACKNOWLEDGEMENTS

The author is grateful to valuable discussion with Dr. Rainer Brüggemann and Dr. Peter B. Sørensen over the years in connection with the work being the basis for this chapter.

CONFLICT OF INTEREST

The authors state that there is no conflict of interest.

DISCLOSURE

This chapter is an update of our previous publication in CCHTS "Assessment of Chemicals Applying Partial Order Ranking Techniques" published in *'Combinatorial Chemistry & High Throughput Screening'*, Volume 11, Issue 10, 2008, pp. 794 to 805.

REFERENCES

[1]   Voigt, K., Brüggemann, R., Pudenz, S. Chemical Databases Evaluated by Order Theoretical Tools. *Anal Bioanal Chem*, 2004, **380**, 467-474.

[2]   Voigt, K., Brüggemann R., Pudenz S. Information Quality of Environmental and Chemical Databases Exemplified by High Production Volume Chemicals and Pharmaceuticals. *Online Information Review*, 2006a, **30**, 1, pp. 8-23.

[3]   Voigt, K., Brüggemann, R., Pudenz, S. A multi-criteria evaluation of environmental databases using the Hasse Diagram Technique (ProRank) software. *Environmental Modeling & Software*. 2006b, **21**, 1587-1597.

[4]   European Commission Regulation (EC) No 1907/2006 of the European Parliament and of the Council of 18 December 2006 concerning the Registration, Evaluation, Authorisation and Restriction of Chemicals (REACH), establishing a European Chemicals Agency, amending Directive 1999/45/EC and repealing Council Regulation (EEC) No 793/93 and Commission Regulation (EC) No 1488/94 as well as Council Directive 76/769/EEC and Commission Directives 91/155/EEC, 93/67/EEC, 93/105/EC and 2000/21/EC; Article 57d,e; http://eur-lex.europa.eu/LexUriServ/site/en/oj/2006/l_396/l_39620061230en000108 49.pdf; 2006 (Accessed July 2012).

[5]   European Economic Community. Council Directive 67/548/EEC of 27 June 1967 on the approximation of laws, regulations and administrative provisions relating to the classification, packaging and labelling of dangerous substances, as well as later adopted annexes; http://eur-lex.europa.eu/LexUriServ/LexUriServ.do?uri=CELEX:31967L0548:en: NOT 1967 (Accessed July 2012)

[6]   Verdonck, F.A.M., Boeije, G., Vandenberghe, V., Comber, M., de Wolf, W., Feijtel, T., Holt, M., Koch, K., Lecloux, A., Siebel-Sauer, A., Vanrolleghem, P.A. A rule-based screening environmental risk assessment tool derived from EUSES. *Chemosphere*, 2005, **58** 1169-1176.

[7]   Sørensen P.B., Brüggemann R., Carlsen, L., Mogensen, B.B., Kreuger, J., Pudenz, S. Analysis of monitoring data of pesticide residues in surface waters using partial order ranking theory. *Environ.Toxicol.Chem*. 2003, **22**, 661-670.

[8]   Carlsen, L. Evaluating the state of polluted sites. Application of partial order ranking techniques, Proc. 3rd Int. Scientific and Practical Conf.: Innovative Development Problems in Oil & Gas Industry. *Almaty, Febr*. 2010, Vol **I**, pp. 14-21.

[9]   Carlsen, L. Giving molecules an identity. On the interplay between QSARs and Partial Order Ranking. *Molecules*, 2004, **9**, 1010-1018 (http://www.mdpi.org/molecules/papers/91 201010.pdf) (Accessed July, 2012)

[10] Carlsen, L. A QSAR Approach to Physico-Chemical Data for Organophosphates with Special Focus on Known and Potential Nerve Agents, Internet Electron. *J. Mol. Des.*, 2005a, **4**, 355-366 (http://www.biochempress.com/av04_0355.html) (Accessed July, 2012)

[11] Carlsen, L. Partial Order Ranking of Organophosphates with Special Emphasis on Nerve Agents. *MATCH Communications in Mathematical and Computer Chemistry*, 2005b, **54**, 519-34.

[12] Carlsen, L. Interpolation Schemes in QSAR. In: *'Partial Order in Environmental Sciences and Chemistry'*, Brüggemann, R. and Carlsen, L., Eds., Springer, Berlin, 2006, pp 163-180

[13] Carlsen, L. A Combined QSAR and Partial Order Ranking Approach to Risk Assessment. *SAR QSAR Environ.Res.* 2006b, **17**, 133-146.

[14] Carlsen, L. Hierarchical partial order ranking, *Environ. Pollut.* 2008, **155**, 247-253.

[15] Carlsen, L., Brüggemann, R. Accumulating partial order ranking. *Environ. Model. Software* 2008, **23**, 986-993.

[16] Carlsen, L., Bruggemann, R., Sailaukhanuly, Y. Application of Selected Partial Order Tools to Analyse. Fate and Toxicity Indicators of Environmentally Hazardous Chemicals. *Ecol.Ind.* 2013, **29**, 191-202.

[17] Brüggemann, R., Carlsen, L., Eds. *Partial Order in Environmental Sciences and Chemistry.* Springer, Berlin, 2006.

[18] Davey, B.A., Priestley, H.A. *Introduction to lattices and Order.* Cambridge University Press, Cambridge (UK) 1990.

[19] Brüggemann, R., Carlsen, L., Multicriteria decision analyses. Viewing MCDA in terms of both process and aggregation methods: some thoughts, motivated by the paper of Huang, Keisler and Linkov. *Sci. Total Environ.*, 2012, **425**, 293-295.

[20] Brüggemann, R., Halfon, E., Welzl, G., Voigt, K., Steinberg, C.E.W. Applying the concept of partially ordered sets on the ranking of near-shore sediments by a battery of tests. *J. Chem. Inf. Comput. Sci.*, 2001, **41**, 918-925.

[21] Halfon, E. and Reggiani, M.G., 1986, On the ranking of chemicals for environmental hazard, Environ. Sci. Technol, **20**, 1173-1179.

[22] Hasse, H. Über die Klassenzahl abelscher Zahlkörper. Akademie Verlag, Berlin 1952.

[23] Brüggemann, R., Halfon , E., Bücherl, C., Theoretical base of the program "Hasse", GSF-Bericht 20/95, Neuherberg (GER); The software may be obtained by contacting Dr. R. Brüggemann, Institute of Freshwater Ecology and Inland Fisheries, Berlin 1995, (brg_home@web.de)

[24] Talente srl. DART - Decision Analysis by Ranking Techniques (ver. 2.05), http://www.talete.mi.it/products/dart_description.htm 2007 (Accessed July 2012)

[25] Fishburn, P.C. On the family of linear extensions of a partial order. *J.Combinat.Theory*, 1974, **17**, 240-243.

[26] Graham, R.L. Linear Extensions of Partial Orders and the FKG Inequality. In *Ordered Sets*, (Rival, I., ed.), Reidel Publishing Company, Dordrecht, (NL), 1982, pp. 213-236.

[27] Winkler, P.M. Average height in a partially ordered set. *Discrete Mathematics*, 1982, **39**, 337-341.

[28] Winkler, P.M., Correlation among partial orders. *Siam.J.Alg.Disc.Meth.*, 1983, **4**, 1-7.

[29] Sørensen, P.B., Lerche, D.B., Carlsen, L., Brüggemann, R. Statistically approach for estimating the total set of linear orders. A possible way for analysing larger partial order sets. In: *Order Theoretical Tools in Environmental Science and Decision Systems*

(Brüggemann, R., Pudenz, S. and Lühr, H.-P. Eds), Berichte des IGB, Leibniz-Institut of Freshwater Ecology and Inland Fisheries, Berlin, Heft 14, Sonderheft IV, 2001, pp. 87-97.

[30] Lerche, D., Brüggemann, R., Sørensen, P., Carlsen, L., Nielsen O.J. A comparison of partial order technique with three methods of multi-criteria analysis for ranking of chemical substances. *J.Chem.Inf.Comput.Sci.,* 2002, **42**, 1086-1098.

[31] Lerche, D., Sørensen, P.B., Brüggemann, R. Improved estimation of ranking probabilities in partial orders using random linear extensions by approximation of the mutual ranking probability. *J.Chem.Inf.Comput.Sci.*, 2003, **43**, 1471-1480.

[32] Sørensen, P.B., Thomsen, M., Fauser, P., Münier, B.,. The Usefulness of a Stochastic Approach for Multi-Criteria Selection. In: Hryniewicz,O., J.Studzinski, and A.Szediw (eds), *Environmental Informatics and Systems Research*, Volume 2: Workshop and Application papers. Shaker Verlag, Aachen, 2007, 187-194.

[33] Wienand, O. lcell, http://bio.math.berkeley.edu/ranktests/lcell/ 2001 (Accessed, July 2012)

[34] De Loof, K., De Meyer, H., De Baets, B.,. Exploiting the Lattice of Ideals Representation of a Poset. *Fundamenta Informaticae* 2006, **71**, 309-321.

[35] Morton, J., Pachter, L., Shiu, A., Sturmfels, B., Wienand, O. Convex Rank Tests and Semigraphoids. *SIAM J.Discrete Math*. 2009. **23**:1117-1134.

[36] Brüggemann, R., Lerche, D., Sørensen, P.B. Carlsen, L. Estimation of average ranks by a local partial order model. *J.Chem.Inf.Comput.Sci.*, 2004, **44**, 618-625.

[37] Bruggemann, R., Carlsen, L. An Improved Estimation of Averaged Ranks of Partially Orders. Match - *Commun.Math.Comput.Chem.,* 2011, **65**, 383-41.

[38] Stockholm Convention. http://chm.pops.int/Home/tabid/2121/language/en-GB/Default.aspx 2011 (Accessed July 2012)

[39] Sørensen, P.B., Mogensen, B.B., Carlsen, L. Thomsen M., The influence of partial order ranking from input parameter uncertainty. Definition of a robustness parameter. *Chemosphere*, 2000, **41**, 595-601.

[40] EPA. Estimation Program Interface (EPI) Suite, ver. 4.10 http://www.epa.gov/oppt/expo sure/pubs/episuite.htm 2011 (Accessed July. 2012).

[41] Connell D.W., Hawker D.W., Use of polynomial expressions to describe the bioconcentration of hydrophobic chemicals in fish, Ecotox. *Environ. Safety*, 1988, **16**:242-257.

[42] Carlsen, L. and Walker, J.D. QSARs for Prioritizing PBT Substances to Promote Pollution Prevention. *QSAR Comb.Sci..,* 2003, **22**, 49-57.

[43] Carlsen, L., Walker, J.D. Prioritizing PBT substances, in '*Partial Order in Environmental Sciences and Chemistry*', Brüggemann, R. and Carlsen, L., Eds., Springer, Berlin, 2006, pp 153-160.

[44] Schultz, T.W. TETRATOX: *Tetrahymena pyriformis* population growth impairment endpoint - A surrogate for fish lethality. *Toxicol.Methods*, 1997, **7** 289-309.

[45] Sailaukhanuly, Y. Zhakupbekova, A., Amutova, F., Carlsen, L. On the Ranking of Chemicals based on their PBT Characteristics. Comparison of Different Ranking Methodologies Using Selected POPs as an Illustrative Example. *Chemosphere*, 2013, **90**, 112-117.

[46] Carlsen; L. Assessing polluted sites using partial order ranking techniques. Journal of KazNU, Almaty, *Chemistry*, 2007, Ser. No. 5(**49**), 18-21, 2007.

[47]   Walker, J.D., Carlsen, L., Jaworska, J. Improving Opportunities for Regulatory Acceptance of QSARs: The Importance of Model Domain, Uncertainty, Validity and Predictability. *QSAR Comb.Sci.*, 2003, **22,** 346-350.

[48]   Eriksson, L. Jaworsksa, J., Worth, A.P., Cronin, M.T.D., McDowell, R.M., Gramatica, P. Methods for reliability and uncertainty assessment and for applicability evaluations of classification- and regression-based QSARs. *Environ Health Perspect.*, 2003, **111**, 1361-1375.

[49]   Walker, J.D. Carlsen, L. QSARs for Identifying and Prioritizing Substances with Persistence and Bioconcentration Potential. *SAR QSAR Environ.Res.*, 2002, **13**, 713-726.

[50]   Simon , U., Brüggemann, R. Mey, S. Pudenz, S., 2005. METEOR - application of a decision support tool based on discrete mathematics. MATCH - communications in mathematical and in computer chemistry **54**, 623-642.

[51]   Voigt, K., Brüggemann, R. Ranking of Pharmaceuticals Detected in the Environment: Aggregation and Weighting Procedures, submitted to Comb.Chem.High Throughp Screen. 2008.

[52]   Helm, D. Evaluation of biomonitoring data, in: Partial order in environmental sciences and chemistry, R. Brüggemann and L. Carlsen, Eds., Springer Berlin, 2006, 286-307.

Send Orders for Reprints to reprints@benthamscience.net

CHAPTER 2

Building a Chemical Space Based on Fragment Descriptors

Igor I. Baskin[1] and Alexandre Varnek[*,2]

[1]Department of Physics, Lomonosov Moscow State University, Moscow 119991, Russia; [2]Laboratory of Chemoinformatics, UMR 7140 CNRS, University of Strasbourg, 4, rue B. Pascal, Strasbourg 67000, France

Abstract. This article reviews the application of fragment descriptors at different stages of virtual screening: filtering, similarity search, and direct activity assessment using QSAR/QSPR models. Several case studies are considered. It is demonstrated that the power of fragment descriptors stems from their universality, very high computational efficiency, simplicity of interpretation and versatility.

Keywords: Fragmental approach, fragment descriptors, QSAR, QSPR, filtering, similarity, virtual screening, *in silico* design.

1. INTRODUCTION

Chemogenomics aims to discover active and/or selective ligands for biologically related targets by conducting screening, ideally, of all possible compounds against all possible targets, or at least, in practice, available libraries of compounds against main target families [1]. One can hardly imagine to screen experimentally the chemical universe containing from 10^{12} to 10^{180} druglike compounds [2] against biological target universe. While the number of compounds that can be screened in the course of high-throughput screening experiments can hardly be imagined to exceed several millions per biological target even in the largest pharmaceutical corporations, the number of molecules that can be screened *in silico* in a single non-expensive computational study currently reaches 10^{12}, and this number can grow sharply in the nearest future. Therefore, the virtual, or *in silico*, screening approaches play a key role in chemogenomics.

---

*Address correspondence to Alexandre Varnek:** Laboratory of Chemoinformatics, UMR 7140 CNRS, University of Strasbourg, 4, rue B. Pascal, Strasbourg 67000, France; Tel: +33-3-68851560; Fax: +33-3- 68 85 16 52, E-mail: varnek@unistra.fr

Rathnam Chaguturu (Ed.)

Virtual screening is usually defined as a process in which large libraries of compounds are automatically evaluated using computational techniques [3]. Its goal is to discover putative hits in large databases of chemical compounds (usually ligands for biological targets) and remove molecules predicted to be toxic or those possessing unfavorable pharmacodynamic or pharmacokinetic properties. Generally, two types of virtual screening are known: structure-based and ligand-based. The former explicitly uses 3D structure of a biological target at the stage of hit detection, whereas the latter uses only information about structure of small molecules and their properties (activities), see [4-6] and references therein. Structure-based virtual screening (docking, 3D pharmacophores) has been described in series of review articles, see [7-9] and references therein.

In this paper mostly ligand-based virtual screening involving fragment descriptors is considered. Fragment descriptors, represent selected substructures (fragments) of 2D molecular graphs and their occurrences in molecules; they constitute one of the most important types of molecular descriptors [10]). Their main advantage is related to simplicity of their calculation, storage and interpretation (see review articles [11-15] and ebook chapter [16]). From the mathematical point of view, they constitute a basis of invariants of labeled molecular graphs, and therefore any molecular property and any similarity measure for molecular graphs can be expressed through them [17-19]. Fragment descriptors are information-based ones [20] which tend to code the information stored in molecular structures. This contrasts with knowledge-based (or semiempirical) descriptors issued from the consideration of the mechanism of action. Selected descriptors form a "chemical space" in which each molecule is represented as a vector [21]. Due to their versatility, fragment descriptors could be efficiently used to create a chemical space which separates active and non-active compounds.

Historically, molecular fragments were used in first additive schemes developed in 1950-ies to estimate physicochemical properties of organic compounds by Tatevskii [22, 23], Bernstein [24], Laidler [25], Benson and Buss [26] and others. The Free-Wilson method [27], one of the first QSAR approaches invented in 1960-ies, is based on the assumption of the additivity of contributions of structural fragments to the biological activity of the whole molecule. Later on, fragment descriptors were successfully used in expert systems able to classify chemical compounds as active or inactive with respect to certain type of biological activity. Hiller [28, 29], Golender and Rosenblit [30, 31], Piruzyan, Avidon *et al.* [32, 33], Cramer [34], Brugger, Stuper and Jurs [35, 36], and Hodes *et al.* [37] pioneered in this field.

An important class of fragmental descriptors, so-called *screens* (structural *keys*, *fingerprints*), has been developed in seventies [38-42]. As a rule, they represent bit strings which can effectively be stored and processed by computers. Although their primary role is to provide efficient substructure searching capabilities in large chemical databases, they are efficiently used for similarity searching [43, 44], to cluster large data sets [45, 46], to assess chemical diversity [47], as well as to conduct SAR [48] and QSAR [49] studies. Nowadays, application of modern machine-learning techniques significantly improves predictive performance of structure-property models based on fragment descriptors [50].

This paper briefly reviews the application of fragment descriptors in virtual screening of large libraries of organic compounds focusing mostly on its three approaches: (i) filtering, (ii) similarity search, and (iii) direct activity/property assessment using QSAR/QSPR models.

2. TYPES OF FRAGMENT DESCRIPTORS

Due to their enormous diversity, one could hardly review all types of 2D fragment descriptors used for structural search in chemical database or in SAR/QSAR/ QSPR studies. Here, we focus on some of them which are the most efficiently used in virtual screening and *in silico* design of organic compounds.

Generally, molecular fragments can be classified with respect to their topology (atom-based, chains, cycles, polycycles, *etc.*), information content of vertices in molecular graphs (atoms, groups of atoms, pharmacophores, descriptor centers) and the level of abstraction when some information concerning atom and bond types is omitted.

Purely structural fragments are used as descriptors in ACD/Labs [51], NASAWIN [52], ISIDA [15] and some other programs. These are 2D subgraphs in which all atoms and/or bonds are represented explicitly and no information about their properties is used. Their typical examples are the sequences of atoms and/or bonds of variable length, branch fragments, saturated and aromatic cycles (polycycles) and atom-centered fragments (ACF). The latter consist of a single central atom surrounded by one or several shells of atoms with the same topological distance from the central one. The ACF were invented by Tatevskii [22] and Benson and Buss [26] in 1950-ies as elements of additive schemes for predicting physicochemical properties of organic compounds. In earlier seventies, Adamson [53] investigated the distribution of one shell ACF in some chemical databases with respect to their possible application as screens. Hodes reinvented one shell ACF as descriptors in SAR studies under the name *augmented atoms* [37], and also suggested *ganglia augmented atoms* [54] representing two shells ACF with generalized second-shell atoms. Later on, one shell ACF were implemented by Baskin *et al.* in the NASAWIN [52] software and by Varnek *et al.* in ISIDA [15, 55] package. Atom-centered fragments with arbitrary number of shells were implemented by Poroikov *et al.* in the PASS [56] program under the name *multilevel neighborhoods of atoms* [57], by Xing and Glen under the name *tree structured fingerprints* [58] (sometimes referred to by Bender, Glen *et al.* as *atom environments* [59, 60] and *circular fingerprints* [61-63]), and by Faulon under the name *molecular signatures* [64-66].

It has been found that characterizing atoms only by element types is too specific for similarity searching and therefore doesn't provide sufficient flexibility needed for large-scaled virtual screening. For that reason, numerous studies were devoted to increase an informational content of fragment descriptors by adding some useful empirical information and/or by representing a part of molecular graph implicitly. The simplest representatives of those descriptors were *atom pairs and topological multiplets* based on the notion of *descriptor center* representing an atom or a group of atoms which could serve as centers of intermolecular interactions. Usually, descriptor centers include heteroatoms, unsaturated bonds and aromatic cycles. An *atom pair* is defined as a pair of atoms (**AT**) or descriptor centers separated by a fixed topological distance: $AT_i\text{-}AT_j\text{-}Dist$, where $Dist_{ij}$ is the shortest path (the number of bonds) between AT_i and AT_j. Analogously, a *topological multiplet* is defined as a multiplet (usually triplet) of descriptor centers and topological distances between each pair of them. In most of cases, these descriptors are used in binary form in order to indicate the presence or absence of the corresponding features in studied chemical structures.

The atom pairs were first suggested for SAR studies by Avidon under the name *SSFN* (*Substructure Superposition Fragment Notation*) [33, 67]. Then they were independently reinvented by Carhart and co-authors [68] for similarity and trend vector analysis. In contrast to SSFN, Carhart's atom pairs are not necessarily composed only of descriptor centers, but account for the information about element type, the number of bonded non-hydrogen neighbors and the number of π electrons. Nowadays, Carhart's atom pairs are rather popular for conducting virtual screening. Modified Carharts's atom pairs, in which descriptors centers are represented by atoms coded with the help of a hierarchical classification scheme [52], have recently been used in virtual screening based on the one-class classification technique [69]. *Topological Fuzzy Bipolar Pharmacophore Autocorrelograms* (*TFBPA*) [70] by Horvath are based on atom pairs, in which real atoms are replaced by pharmacophore sites (hydrophobic, aromatic, hydrogen bond acceptor, hydrogen bond donor, cation, anion), while $Dist_{ij}$ corresponds to different ranges of topological distances between pharmacophores. These descriptors were successfully applied in virtual screening against a panel of 42 biological targets using similarity search based on several fuzzy and non-fuzzy metrics [71], performing only slightly less well than their 3D counterparts [70].

Fuzzy Pharmacophore Triplets (*FPT*) by Horvath [72] is an extention of (*Fuzzy Bipolar Pharmacophore Fingerprints*) *FBPF* [71] for three sites pharmacophores. An important innovation in the *FPT* concerns accounting for proteolytic equilibrium as a function of pH [72]. Due to this feature, even small structural modifications leading to a pK_a shift, may have a profound effect on the fuzzy pharmocophore triples. As a result, these descriptors efficiently discriminate structurally similar compounds exhibiting significantly different activities [72].

Some other topological triplets should be mentioned. Thus, *Similog pharmacophoric keys* by Jacoby [73] represent triplets of binary coded types of atoms (pharmacophoric centers) and topological distances between them. Atomic types are generalized by four features (represented as four bits per atom): potential hydrogen bond donor or acceptor; bulkiness and electropositivity. The *topological pharmacophore-point triangles* implemented in the MOE software [74] represent triplets of MOE atom types separated by binned topological distances. Structure-property models obtained by support vector machine method with these descriptors have been successfully used for virtual screening of COX-2 inhibitors [75] and D_3 dopamine receptor ligands [76].

Topological torsions by Nilakantan *et al.* [77] is a sequence of four consecutively bonded atoms AT_i-AT_j-AT_k-AT_l, where each atom is characterized by a number of parameters similarly to atoms in Carhart's pairs. In order to enhance efficiency of virtual screening, Kearsley *et al.* [78] suggested to assign atoms in the Carhart's atom pairs and Nilakantan's topological torsions to one of seven classes: cations, anions, neutral hydrogen bond donors, neutral hydrogen bond acceptors, polar atoms, hydrophobic atoms and other.

In the frameworks of the Simplex representation of molecular structure (SiRMS) by Kuz'min *et al.* [79] any molecule can be represented as an ensemble of different 4-atomic fragments - simplexes. The occurrence number of identical simplexes in a molecule is a descriptor value. Subgraphs corresponding to simplexes can be either connected or disconnected. Not only atom type, but also some other physical-chemical characteristics of atoms (partial charge, lipophilicity, refraction and atom's ability for being a donor/acceptor in hydrogen-bond formation, *etc.*) can be used for atom labeling. Fragment descriptors based on SiRMS were used for building QSAR models and performing virtual screening and molecular design of compounds with strong antiviral activity [80].

ISIDA Property-Labeled Fragment Descriptors (IPLF) were introduced as counts of specific subgraphs in which atom vertices are colored with respect to some local property [81]. Various coloring strategies (notably pH-dependent pharmacophore and electrostatic potential-based flagging) can be combined in the framework of this approach with various fragmentation schemes (chains, atom pairs, augmented atoms, trees, *etc.*). IPLF showed excellent results in similarity-based virtual screening for analogue protease inhibitors and in building highly predictive octanol-water partition coefficient and hERG channel inhibition models [81]. IPLF were also advantageously used in combination with the Generative

Topographic Mapping (GTM) method for chemical space visualization, structure-activity modeling and database comparison [82].

In contrast to QSPR studies based mostly on the use of complete (containing all atoms) or hydrogen-suppressed molecular graphs, handling biological activity at the qualitative level, often demands more abstractions. Namely, it is rather convenient to approximate chemical structures by *reduced graphs*, in which each vertex is an atom or a group of atoms (descriptor or pharmacophoric center), whereas each edge is a topological distance $Dist_{ij}$. Such biology-oriented representation of chemical structures was suggested by Avidon *et al.* as descriptor center connection graphs [33]. Gillet, Willett and Bradshaw have proposed the GWB-reduced graphs which use the hierarchical organization of vertex labels. This allows one to control the level of their generalization which may explain their high efficiency in similarity searching.

3. APPLICATION OF FRAGMENT DESCRIPTORS IN VIRTUAL SCREENING AND *IN SILICO* DESIGN

In this chapter application of fragment descriptors at different stages of virtual screening is considered.

3.1. Filtering

Filtering is a rule-based approach aimed to perform fast assessment of usefulness molecules in the given context. In drug design area, the filtering is used to eliminate compounds with unfavorable pharmacodynamic or pharmacokinetic properties as well as toxic compounds. Pharmacodynamics considers binding drug-like organic molecules (ligands) to chosen biological target. Since the efficiency of ligand-target interactions depends on spatial complementarity of their binding sites, the filtering is usually performed with 3D-pharmacophores, representing "optimal" spatial arrangements of steric and electronic features of ligands [83, 84]. Pharmacokinetics is mostly related to absorption, distribution, metabolism and excretion (ADME) related properties: octanol-water partition coefficients (*log P*), solubility in water (*log S*), blood-brain coefficient (*log BB*), partition coefficient between different tissues, skin penetration coefficient, *etc.*

Fragment descriptors are widely used for early ADME/Tox prediction both explicitly and implicitly. The easiest way to filter large databases concerns detecting undesirable molecular fragments (*structural alerts*). Appropriate lists of structural alerts are published for toxicity [85], mutagenicity [86], and carcinogenicity [87]. Klopman *et al.* were the first to recognize the potency of using fragmental descriptors for this purpose [88-90]. Their programs CASE [88], MultiCASE [91, 92], as well as more recent MCASE QSAR expert systems [93], proved to be effective tools to assess mutagenicity [89, 92, 93] and carcinogenicity [90, 92] of organic compounds. In these programs, sets of biophores (analogs of structural alerts) were identified and used for activity predictions. A number of more sophisticated fragment-based expert systems of toxicity assessment - DEREK [94], TopKat [95] and Rex [96] – have been developed. DEREK is a knowledge-based system operating with human-coded or automatically generated [97] rules about toxicophores. Fragments in the DEREK knowledge base are defined by means of linear notation language PATRAN which codes the information about atom, bonds and stereochemistry. TopKat uses a large predefined set of fragment descriptors, whereas Rex implements a special kind of atom-pairs descriptors (*links*). To read more information about fragment-based computational assessment of toxicity, including mutagenicity and carcinogenicity, see review [98] and references therein.

The most popular filter used in drug design area is based on the Lipinsky "rule of five" [99], which takes into account the molecular weight, the number of hydrogen bond donors and acceptors, along with the octanol-water partition coefficient *logP*, to assess the bioavailability of oral drugs. Similar rules of "drug-likeness" or "lead-likeness" were later proposed by by Oprea [100], Veber [101] and Hann [102]. Formally, fragment descriptors are not explicitly involved there. However, many computational approaches to assess *logP* are fragment-based [51, 103, 104]; wheras H-donors and acceptor sites are simplest molecular fragments.

3.2. Similarity Search

The similarity-based virtual screening is based on an assumption that all compounds in a chemical database, which are similar to a query compound, could also have similar biological activity. Although this hypothesis is not always valid (see discussion in [105]), quite often the set of retrieved compounds is enriched by actives [106].

To achieve high efficacy of similarity-based screening of databases containing millions compounds, molecular structures are usually represented by *screens* (structural keys) or fixed-size or variable-size *fingerprints*. Screens and fingerprints can contain both 2D- and 3D-information. However, the 2D-fingerprints, which are a kind of binary fragment descriptors, dominate in this area. Fragment-based structural keys, like MDL keys [48], are sufficiently good for handling small and medium-sized chemical databases, whereas processing of large databases is performed with fingerprints having much higher information density. Fragment-based Daylight [107], BCI [108] and UNITY 2D [109] fingerprints are the most known examples.

The most popular similarity measure for comparing chemical structures represented by means of fingerprints is the Tanimoto (or Jaccard) coefficient T [110]. Two structures are usually considered similar if $T>0.85$ [106] (for Daylight fingerprints [107]). Using this threshold, Taylor estimated a probability to retrieve actives as 0.012-0.50 [111], whereas according to Delaney this number raises to 0.40-0.60 [112] (using Daylight fingerprints [107]). These computer experiments confirm usefulness of the similarity approach as an instrument of virtual screening.

Schneider *et al.* have developed a special technique for performing virtual screening referred to as CATS (Chemically Advanced Template Search) [113]. In its framework chemical structures are described by means of so-called correlation vectors, each component of which equals to the number of times a certain atom pair is contained in a chemical structure divided by the total number of non-hydrogen atoms in it. Each atom in an atom pair is specified as belonging to one of five classes (hydrogen-bond donor, hydrogen-bond acceptor, positively charged, negatively charged, and lipophilic), while topological distances of up to 10 bonds are also considered in the atom-pair specification. The similarity between molecules is assessed in this approach using Euclidean distance between the corresponding correlation vectors. CATS was shown to outperform MERLIN with Daylight fingerprints [107] for retrieving thrombin inhibitors in a virtual screening experiment [113].

Hull *et al.* have developed the *Latent Semantic Structure Indexing* (LaSSI) approach to perform similarity search in low-dimensional chemical space [114] [115]. To reduce the dimension of initial chemical space, the singular value decomposition method is applied for the descriptor-molecule matrix. Ranking molecules by similarity to a query molecule was performed in the reduced space using the cosine similarity measure [116], whereas the Carhart's atom pairs [68] and the Nilakantan's topological torsions [77] were used as descriptors. The authors claim that this approach "has several advantages over analogous ranking in the original descriptor space: matching latent structures is more robust than matching discrete descriptors, choosing the number of singular values provides a rational way to vary the 'fuzziness' of the search" [114].

The issue of "fuzzification" of similarity search was addressed by Horvath *et al.* [70-72]. The first fuzzy similarity metric suggested in work [70] relies on partial similarity scores calculated with respect to the inter-atomic distances distributions for each pharmacophore pair. In this case the "fuzziness" enables to compare pairs of pharmacophores with different topological or 3D distances. Similar results [71] were achieved using fuzzy and weighted modified Dice similarity metric [116]. Fuzzy pharmacophore triplets FPT (see section 2) can be gradually mapped onto related basis triplets, thus minimizing binary classification artifacts [72]. In new similarity scoring index introduced in reference [72], the simultaneous absence of a pharmacophore triplet in two molecules is taken into account. However, this is a less-constraining indicator of similarity than simultaneous presence of triplets.

Most of similarity search approaches require only a single reference structure. However, in practice several compounds with the same type of biological activity are often available. This motivated Hert *et al.* [117] to develop the *data fusion method* which allows one to screen a database using all available reference structures. Then, the similarity scores are combined for all retrieved structures using selected fusion rules. Searches conducted on the MDL Drug Data Report database using fragment-based UNITY 2D [109], BCI [108], and Daylight [107] fingerprints have proved the effectiveness of this approach.

The main drawback of the conventional similarity search concerns an inability to use experimental information on biological activity to adjust similarity measures. This results in inability to discriminate between relevant and non-relevant fragment descriptors being used for computing similarity measures. To tackle this problem, Cramer *et al.* [34] developed *substructural analysis* in which each fragment (represented as a bit in a fingerprint) is weighted by taking into account its occurrence in active and in inactive compounds. Later on, many similar approaches have been described in the literature [118].

One-class classification [119] (or novelty detection [120, 121]) is a novel promising approach to conducting similarity-based virtual screening. It considers two types of compounds: active ("object class") and inactive ("outliers"). A new compound is predicted to be active if it lies in the dense area of the point cloud formed by active compounds contained in the training set and inactive if outside. Thus, this new compound is viewed as being similar to the set of active compounds. The fundamental difference between the one-class classification and the conventional similarity search is the ability of the former to use the whole training set instead of a single query compound in order to learn the optimal similarity measure for virtual screening. In accordance with this methodology, Karpov *et al.* [69, 122] used the auto-associative neural networks and the one-class Support Vector Machines (1-SVM) in virtual screening (using fragment descriptors) against various biological targets.

One more way to conduct a similarity-based virtual screening is to retrieve the structures containing a user-defined set of "pharmacophoric" features. In *Dynamic Mapping of Consensus positions* (DMC) algorithm [123] those features are selected by finding common positions in bit strings for all active compounds. The *potency-scaled DMC* algorithm (POT-DMC) [124] is a modification of DMC in which compounds activities are taken into account. The latter two methods may be considered as intermediate between conventional similarity search and probabilistic SAR approaches.

Batista, Godden and Bajorath developed the MolBlaster method [125], in which molecular similarity is assessed by *Differential Shannon Entropy* [126] computed from populations of randomly generated fragments. For the range $0.64 < T < 0.99$, this similarity measure provides with the same ranking as the Tanimoto index T. However for the smaller values of T the entropy-based index is a more sensitive, since it distinguishes between pairs of molecules having almost identical T. To adapt this methodology for large-scale virtual screening, the *Proportional Shannon Entropy* (PSE) metrics was introduced [127]. A key feature of this approach is that class-specific PSE of random fragment distributions enables the identification of the molecules sharing with known active compounds a significant number of signature substructures.

Similarity search methods developed for individual compounds are difficult to apply directly for chemical reactions involving many species subdividing by two types: reactants and products. To overcome this problem, Varnek *et al.* [15] suggested to condense all participating in reaction species in one only molecular graph (*Condensed Graphs of Reactions* (*CGR*) [15]) followed by its fragmentation and application of developed fingerprints in "classical" similarity search. Besides conventional chemical bonds (simple, double, aromatic, *etc.*), a CGR contains dynamical bonds corresponding to created, broken or transformed bonds. This approach could be efficiently used for screening of large reaction databases.

It should be noted that the similarity concepts are widely used in selecting of diverse sets of compounds (see reviews [128-132] and references therein).

3.3. SAR/QSAR/QSPR Models

Simplistic and heuristic similarity-based approaches can hardly produce as good predictive models as modern statistical and machine learning methods able to assess quantitatively biological or physicochemical properties [50]. QSAR-based virtual screening consists in direct assessment of activity values (numerical or binary) of all compounds in the database followed by selection of hits possessing desirable activity. Mathematical methods used for models preparation could be subdivided into *probabilistic* and *regression* approaches. The former assesses a probability that a given compound is active or not active whereas the latter numerically evaluate the activity values. A limited size of this paper doesn't allow us to cite all successful stories related to application of probabilistic and regression models in virtual screening; only some examples will be presented.

Harper *et al.* [133] have demonstrated a much better performance of probabilistic *binary kernel discrimination* method to screen large databases compared to backpropagation neural networks or conventional similarity search. The Carhart's atom-pairs [68] and Nilakantan's topological torsions [77] were used as descriptors in that study.

Aiming to discover new cognition enhancers, Geronikaki *et al.* [134] applied the PASS program [56], which implements a probabilistic Bayesian-based approach, and the DEREK rule-based system [94] to screen a database of highly diverse chemical compounds. Eight compounds with the highest probability of cognition-enhancing effect were selected. Experimental tests have shown that all of them possessed a pronounced antiamnesic effect.

Bender, Glen *et al.* have applied [59-63] several probabilistic machine learning methods (naïve Bayesian classifier, inductive logic programming, and support vector inductive learning programming) in conjunction with circular fingerprints for making classification of bioactive chemical compounds and performing virtual screening on several biological targets. The latter of these three methods (*i.e.* support vector inductive learning programming) was shown to perform significantly better than the other two methods [63]. Advantages of using circular fingerprints were pointed out [61].

Regression QSAR/QSPR models are used to assess ADME/Tox properties or to detect "hit" molecules capable to bind a certain biological target. Available in the literature fragments based QSAR models for blood-brain barrier [135], skin permeation rate [136], blood-air [137] and tissue-air partition coefficients [137] could be mentioned as examples. Many theoretical approaches of calculation of octanol/water partition coefficient log P involve fragment descriptors. The methods by Rekker [138, 139], Leo and Hansch (CLOGP) [103, 140], Ghose-Crippen (ALOGP) [141-143], Wildman and Crippen [144], Suzuki and Kudo (CHEMICALC-2) [145], Convard (SMILOGP) [146], Wang (XLOGP) [147, 148] represent just a few modern examples. Fragment-based predictive models for estimation solubility in water [149] and DMSO [149] are available.

Benchmarking studies performed in references [135-137, 150] show that QSAR/QSPR models for various biological and physicochemical properties involving fragment descriptors are, at least, as robust as those involving topological, quantum, electrostatic and other types of descriptors.

3.4. *In Silico* **Design**

In this section we consider examples of virtual screening performed on a database containing only virtual (still non-synthesized or unavailable) compounds. Generation of virtual libraries is usually performed using combinatorial chemistry approaches [151-153]. One of simplest ways is to attach systematically user-defined substituents $R_1, R_2, ..., R_N$ to a given scaffold. If the list for the substituent R_i contains n_i candidates, the total number of generated structures is $N = \prod n_i$, although taking symmetry into account could reduce the library's size.[i] The number of substituents R_i (n_i) should be carefully selected in order to avoid a generation of too large set of structures (combinatorial explosion). The "optimal" substituents could be prepared using fragments selected at the QSAR stage, since their contributions into activity (for linear models) allow one to estimate an impact of combining the fragment into larger species (R_i). In such a way, a focused combinatorial library could be generated.

The technology based on combining QSAR, generation of virtual libraries and screening stages has been implemented into ISIDA program and applied to computer-aided design of new uranyl binders belonging to two different families of organic molecules: phosphoryl containing podands [154] and monoamides [155]. QSAR models have been developed using different machine-learning methods (multi-linear regression analysis, associative neural networks [156] and support vector machines [157]) and fragment descriptors (atom/bond sequences and augmented atoms). Then, these models were used to screen virtual combinatorial libraries containing up to 11000 compounds. Selected hits were synthesized and tested experimentally. Experimental data well correspond to predicted the uranyl binding affinity. Thus, initial data sets were significantly enriched with new efficient uranyl binders, and one of new molecules was found more efficient than previously studied compounds. A similar study was conducted for development of new 1-[2-hydroxyethoxy)methyl]-6-(phenylthio)thymine (HEPT) derivatives potentially possessing high anti-HIV activity [158]. This demonstrates universality of fragment descriptors and broad perspectives of their use in virtual screening and *in silico* design.

CONCLUSION

The power of fragment descriptors originates from their universality, very high computational efficacy, simplicity of interpretation, as well as their high diversity and versatility. The latest challenges in chemogenomics and high throughput virtual screening have raised their role in effective processing of huge amounts of relevant data and computer-aided design of new compounds.

ACKNOWLEDGEMENTS

The authors thank Dr. I. Tetko and Dr. G. Marcou for fruitful discussion.

CONFLICT OF INTEREST

The authors state that there is no conflict of interest.

DISCLOSURE

This chapter is an updated version of the publication: Baskin, I.; Varnek, A. Building a chemical space based on fragment descriptors. *Comb. Chem. High Throughput Scr.* **2008**, *11*, pp. 661-668.

REFERENCES

[1]   Kubinyi, H.; Muler, G. *Chemogenomics in Drug Discovery.* Wiley-VCH Publishers: Weinheim, 2004.
[2]   Gorse, A. D. Diversity in medicinal chemistry space. *Curr. Top. Med. Chem.* **2006**, *6*, 3-18.
[3]   Walters, W. P.; Stahl, M. T.; Murcko, M. A. Virtual screening - an overview. *Drug Discov. Today* **1998**, *3*, 160-178.
[4]   Geppert, H.; Vogt, M.; Bajorath, J. Current Trends in Ligand-Based Virtual Screening: Molecular Representations, Data Mining Methods, New Application Areas, and Performance Evaluation. *J. Chem. Inf. Mod.* **2010**, *50*, 205-216.

[5]   Ripphausen, P.; Nisius, B.; Bajorath, J. State-of-the-art in ligand-based virtual screening. *Drug Discovery Today* **2011**, *16*, 372-376.

[6]   Ripphausen, P.; Nisius, B.; Peltason, L.; Bajorath, J. Quo vadis, virtual screening? A comprehensive survey of prospective applications. *J. Med. Chem.* **2010**, *53*, 8461-8467.

[7]   Seifert, M. H.; Kraus, J.; Kramer, B. Virtual high-throughput screening of molecular databases. *Curr. Opin. Drug. Discov. Devel.* **2007**, *10*, 298-307.

[8]   Cavasotto, C. N.; Orry, A. J. Ligand docking and structure-based virtual screening in drug discovery. *Curr. Top. Med. Chem.* **2007**, *7*, 1006-1014.

[9]   Ghosh, S.; Nie, A.; An, J.; Huang, Z. Structure-based virtual screening of chemical libraries for drug discovery. *Curr. Opin. Chem. Biol.* **2006**, *10*, 194-202.

[10]   Todeschini, R.; Consonni, V. *Molecular Descriptors for Chemoinformatics*. Wiley-VCH: Weinheim, 2009.

[11]   Zefirov, N. S.; Palyulin, V. A. Fragmental Approach in QSPR. *J. Chem. Inf. Comput. Sci.* **2002**, *42*, 1112-1122.

[12]   Japertas, P.; Didziapetris, R.; Petrauskas, A. Fragmental methods in the design of new compounds. Applications of The Advanced Algorithm Builder. *Quant. Struct.-Act. Relat.* **2002**, *21*, 23-37.

[13]   Artemenko, N. V.; Baskin, I. I.; Palyulin, V. A.; Zefirov, N. S. Artificial neural network and fragmental approach in prediction of physicochemical properties of organic compounds. *Russ. Chem. Bull.* **2003**, *52*, 20-29.

[14]   Merlot, C.; Domine, D.; Church, D. J. Fragment analysis in small molecule discovery. *Curr. Opin. Drug Discov. Devel.* **2002**, *5*, 391-399.

[15]   Varnek, A.; Fourches, D.; Hoonakker, F.; Solov'ev, V. P. Substructural fragments: an universal language to encode reactions, molecular and supramolecular structures. *J. Comput. Aided Mol. Des.* **2005**, *19*, 693-703.

[16]   Baskin, I.; Varnek, A., Fragment Descriptors in SAR/QSAR/QSPR Studies, Molecular Similarity Analysis and in Virtual Screening. In *Chemoinformatics Approaches to Virtual Screening* Varnek, A.; Tropsha, A., Eds. RSC Publisher: Cambridge, 2008; pp 1-43.

[17]   Baskin, I. I.; Skvortsova, M. I.; Stankevich, I. V.; Zefirov, N. S. On the Basis of Invariants of Labeled Molecular Graphs. *J. Chem. Inf. Comput. Sci.* **1995**, *35*, 527-531.

[18]   Skvortsova, M. I.; Baskin, I. I.; Stankevich, I. V.; Palyulin, V. A.; Zefirov, N. S. Molecular similarity. 1. Analytical description of the set of graph similarity measures. *J. Chem. Inf. Comput. Sci.* **1998**, *38*, 785-790.

[19]   Skvortsova, M. I.; Baskin, I. I.; Skvortsov, L. A.; Palyulin, V. A.; Zefirov, N. S.; Stankevich, I. V. Chemical graphs and their basis invariants. *Theochem* **1999**, *466*, 211-217.

[20]   Jelfs, S.; Ertl, P.; Selzer, P. Estimation of pKa for Druglike Compounds Using Semiempirical and Information-Based Descriptors. *J. Chem. Inf. Model.* **2007**, *47*, 450-459.

[21]   Varnek, A.; Baskin, I. I. Chemoinformatics as a Theoretical Chemistry Discipline. *Mol. Inf.* **2011**, *30*, 20-32.

[22]   Tatevskii, V. M. Chemical structure of hydrocarbons and their heats of formation. *Dokl. Akad. Nauk SSSR* **1950**, *75*, 819-822.

[23]   Tatevskii, V. M.; Mendzheritskii, E. A.; Korobov, V. The additive scheme of the heat of formation of hydrocarbons and the problem of the heat of sublimation of graphite. *Vestnik Moskovskogo Universiteta* **1951**, *6*, 83-86.

[24] Bernstein, H. J. The Physical Properties of Molecules in Relation to Their Structure. I. Relations between Additive Molecular Properties in Several Homologous Series. *J. Chem. Phys.* **1952**, *20*, 263-269.

[25] Laidler, K. J. System of Molecular Thermochemistry for Organic Gases and Liquids. *Canadian J. Chem.* **1956**, *34*, 626-648.

[26] Benson, S. W.; Buss, J. H. Additivity Rules for the Estimation of Molecular Properties. Thermodynamic Properties. *J. Chem. Phys.* **1958**, *29*, 546-572.

[27] Free, S. M., Jr.; Wilson, J. W. A Mathematical Contribution to Structure-Activity Studies. *J. Med. Chem.* **1964**, *7*, 395-399.

[28] Hiller, S. A.; Golender, V. E.; Rosenblit, A. B.; Rastrigin, L. A.; Glaz, A. B. Cybernetic methods of drug design. I. Statement of the problem--the perceptron approach. *Comput. Biomed. Res.* **1973**, *6*, 411-421.

[29] Hiller, S. A.; Glaz, A. B.; Rastrigin, L. A.; Rosenblit, A. B. Recognition of phisiological activity of chemical compounds on perceptron with random adaptation of structure. *Dokl. Akad. Nauk SSSR* **1971**, *199*, 851-853.

[30] Golender, V. E.; Rozenblit, A. B. Interactive system for recognition of biological activity features in complex chemical compounds. *Avtomatika i Telemekhanika* **1974**, 99-105.

[31] Golender, V. E.; Rozenblit, A. B. Logico-structural approach to computer-assisted drug design. *Med. Chem. (Academic Press)* **1980**, *11*, 299-337.

[32] Piruzyan, L. A.; Avidon, V. V.; Rozenblit, A. B.; Arolovich, V. S.; Golender, V. E.; Kozlova, S. P.; Mikhailovskii, E. M.; Gavrishchuk, E. G. Statistical study of an information file on biologically active compounds. Data bank of the structure and activity of chemical compounds. *Khimiko-Farmatsevticheskii Zhurnal* **1977**, *11*, 35-40.

[33] Avidon, V. V.; Pomerantsev, I. A.; Golender, V. E.; Rozenblit, A. B. Structure-Activity Relationship Oriented Languages for Chemical Structure Representation. *J. Chem. Inf. Comput. Sci.* **1982**, *22*, 207-214.

[34] Cramer, R. D., 3rd; Redl, G.; Berkoff, C. E. Substructural analysis. A novel approach to the problem of drug design. *J. Med. Chem.* **1974**, *17*, 533-535.

[35] Stuper, A. J.; Jurs, P. C. ADAPT: A Computer System for Automated Data Analysis Using Pattern Recognition Techniques. *J. Chem. Inf. Model.* **1976**, *16*, 99-105.

[36] Brugger, W. E.; Stuper, A. J.; Jurs, P. C. Generation of Descriptors from Molecular Structures. *J. Chem. Inf. Model.* **1976**, *16*, 105-110.

[37] Hodes, L.; Hazard, G. F.; Geran, R. I.; Richman, S. A statistical-heuristic methods for automated selection of drugs for screening. *J. Med. Chem.* **1977**, *20*, 469-475.

[38] Milne, M.; Lefkovitz, D.; Hill, H.; Powers, R. Search of CA Registry (1.25 Million Compounds) with the Topological Screens System. *J. Chem. Doc.* **1972**, *12*, 183-189.

[39] Adamson, G. W.; Cowell, J.; Lynch, M. F.; McLure, A. H. W.; Town, W. G.; Yapp, A. M. Strategic Considerations in the Design of a Screening System for Substructure Searches of Chemical Structure Files. *J. Chem. Doc.* **1973**, *13*, 153-157.

[40] Feldman, A.; Hodes, L. An Efficient Design for Chemical Structure Searching. I. The Screens. *J. Chem. Inf. Model.* **1975**, *15*, 147-152.

[41] Willett, P. A Screen Set Generation Algorithm. *J. Chem. Inf. Model.* **1979**, *19*, 159-162.

[42] Willett, P. The Effect of Screen Set Size on Retrieval from Chemical Substructure Search Systems. *J. Chem. Inf. Model.* **1979**, *19*, 253-255.

[43] Willett, P.; Winterman, V.; Bawden, D. Implementation of nearest-neighbor searching in an online chemical structure search system. *J. Chem. Inf. Model.* **1986**, *26*, 36-41.

[44]　Fisanick, W.; Lipkus, A. H.; Rusinko, A. Similarity searching on CAS Registry substances. 2. 2D structural similarity. *J. Chem. Inf. Model.* **1994**, *34*, 130-140.

[45]　Hodes, L. Clustering a large number of compounds. 1. Establishing the method on an initial sample. *J. Chem. Inf. Model.* **1989**, *29*, 66-71.

[46]　McGregor, M. J.; Pallai, P. V. Clustering of Large Databases of Compounds: Using the MDL "Keys" as Structural Descriptors. *J. Chem. Inf. Model.* **1997**, *37*, 443-448.

[47]　Turner, D. B.; Tyrrell, S. M.; Willett, P. Rapid Quantification of Molecular Diversity for Selective Database Acquisition. *J. Chem. Inf. Model.* **1997**, *37*, 18-22.

[48]　Durant, J. L.; Leland, B. A.; Henry, D. R.; Nourse, J. G. Reoptimization of MDL Keys for Use in Drug Discovery. *J. Chem. Inf. Comput. Sci.* **2002**, *42*, 1273-1280.

[49]　Tong, W.; Lowis, D. R.; Perkins, R.; Chen, Y.; Welsh, W. J.; Goddette, D. W.; Heritage, T. W.; Sheehan, D. M. Evaluation of Quantitative Structure-Activity Relationship Methods for Large-Scale Prediction of Chemicals Binding to the Estrogen Receptor. *J. Chem. Inf. Model.* **1998**, *38*, 669-677.

[50]　Varnek, A.; Baskin, I. Machine Learning Methods for Property Prediction in Chemoinformatics: Quo Vadis? *J. Chem. Inf. Mod.* **2012**, *52*, 1413-1437.

[51]　Petrauskas, A. A.; Kolovanov, E. A. ACD/Log P method description. *Perspectives in Drug Discovery and Design* **2000**, *19*, 99-116.

[52]　Artemenko, N. V.; Baskin, I. I.; Palyulin, V. A.; Zefirov, N. S. Prediction of Physical Properties of Organic Compounds Using Artificial Neural Networks within the Substructure Approach. *Dokl. Chem.* **2001**, *381*, 317-320.

[53]　Adamson, G. W.; Lynch, M. F.; Town, W. G. Analysis of Structural Characteristics of Chemical Compounds in a Large Computer-based File. Part II. Atom-Centered Fragments. *J. Chem. Soc. C* **1971**, 3702-3706.

[54]　Hodes, L. Selection of molecular fragment features for structure-activity studies in antitumor screening. *J. Chem. Inf. Comput. Sci.* **1981**, *21*, 132-136.

[55]　Varnek, A.; Fourches, D.; Horvath, D.; Klimchuk, O.; Gaudin, C.; Vayer, P.; Solov'ev, V.; Hoonakker, F.; Tetko, I. V.; Marcou, G. ISIDA - Platform for virtual screening based on fragment and pharmacophoric descriptors. *Curr. Comput.-Aided Drug Des.* **2008**, *4*, 191-198.

[56]　Poroikov, V. V.; Filimonov, D. A.; Borodina, Y. V.; Lagunin, A. A.; Kos, A. Robustness of biological activity spectra predicting by computer program PASS for noncongeneric sets of chemical compounds. *J. Chem. Inf. Comput. Sci.* **2000**, *40*, 1349-1355.

[57]　Filimonov, D.; Poroikov, V.; Borodina, Y.; Gloriozova, T. Chemical Similarity Assessment through Multilevel Neighborhoods of Atoms: Definition and Comparison with the Other Descriptors. *J. Chem. Inf. Comput. Sci.* **1999**, *39*, 666-670.

[58]　Xing, L.; Glen, R. C. Novel methods for the prediction of logP, pKa, and logD. *J. Chem. Inf. Comput. Sci.* **2002**, *42*, 796-805.

[59]　Bender, A.; Mussa, H. Y.; Glen, R. C.; Reiling, S. Molecular Similarity Searching Using Atom Environments, Information-Based Feature Selection, and a Naive Bayesian Classifier. *J. Chem. Inf. Comput. Sci.* **2004**, *44*, 170-178.

[60]　Bender, A.; Mussa, H. Y.; Glen, R. C.; Reiling, S. Similarity Searching of Chemical Databases Using Atom Environment Descriptors (MOLPRINT 2D): Evaluation of Performance. *J. Chem. Inf. Comput. Sci.* **2004**, *44*, 1708-1718.

[61] Glen, R. C.; Bender, A.; Arnby, C. H.; Carlsson, L.; Boyer, S.; Smith, J. Circular fingerprints: Flexible molecular descriptors with applications from physical chemistry to ADME. *IDrugs* **2006**, *9*, 199-204.

[62] Rodgers, S.; Glen, R. C.; Bender, A. Characterizing bitterness: Identification of key structural features and development of a classification model. *J. Chem. Inf. Mod.* **2006**, *46*, 569-576.

[63] Cannon, E. O.; Amini, A.; Bender, A.; Sternberg, M. J. E.; Muggleton, S. H.; Glen, R. C.; Mitchell, J. B. O. Support vector inductive logic programming outperforms the Naive Bayes Classifier and inductive logic programming for the classification of bioactive chemical compounds. *J. Comput.-Aided Mol. Des.* **2007**, *21*, 269-280.

[64] Faulon, J.-L.; Visco, D. P., Jr.; Pophale, R. S. The Signature Molecular Descriptor. 1. Using Extended Valence Sequences in QSAR and QSPR Studies. *J. Chem. Inf. Comput. Sci.* **2003**, *43*, 707-720.

[65] Faulon, J.-L.; Churchwell, C. J.; Visco, D. P., Jr. The Signature Molecular Descriptor. 2. Enumerating Molecules from Their Extended Valence Sequences. *J. Chem. Inf. Comput. Sci.* **2003**, *43*, 721-734.

[66] Churchwell, C. J.; Rintoul, M. D.; Martin, S.; Visco, D. P., Jr.; Kotu, A.; Larson, R. S.; Sillerud, L. O.; Brown, D. C.; Faulon, J. L. The signature molecular descriptor. 3. Inverse-quantitative structure-activity relationship of ICAM-1 inhibitory peptides. *J. Mol. Graph. Modell.* **2004**, *22*, 263-273.

[67] Avidon, V. V.; Leksina, L. A. Descriptor Language for the Analysis of Structural Similarity of Organic Compounds. *Nauchno.-Tekhn. Inf., Ser. 2* **1974**, 22-25.

[68] Carhart, R. E.; Smith, D. H.; Venkataraghavan, R. Atom Pairs as Molecular Features in Structure-Activity Studies: Definition and Applications. *J. Chem. Inf. Comput. Sci.* **1985**, *25*, 64-73.

[69] Karpov, P. V.; Baskin, I. I.; Palyulin, V. A.; Zefirov, N. S. Virtual screening based on one-class classification. *Dokl. Chem.* **2011**, *437*, 107-111.

[70] Horvath, D., High Throughput Conformational Sampling & Fuzzy Similarity Metrics: A Novel Approach to Similarity Searching and Focused Combinatorial Library Design and its Role in the Drug Discovery Laboratory. In *Combinatorial Library Design and Evaluation: Principles, Software Tools and Applications*, Ghose, A.; Viswanadhan, V., Eds. Marcel Dekker: New York, 2001; pp 429-472.

[71] Horvath, D.; Jeandenans, C. Neighborhood Behavior of *in Silico* Structural Spaces with Respect to *in Vitro* Activity Spaces-A Novel Understanding of the Molecular Similarity Principle in the Context of Multiple Receptor Binding Profiles. *J. Chem. Inf. Comput. Sci.* **2003**, *43*, 680-690.

[72] Bonachera, F.; Parent, B.; Barbosa, F.; Froloff, N.; Horvath, D. Fuzzy Tricentric Pharmacophore Fingerprints. 1. Topological Fuzzy Pharmacophore Triplets and Adapted Molecular Similarity Scoring Schemes. *J. Chem. Inf. Model.* **2006**, *46*, 2457-2477.

[73] Schuffenhauer, A.; Floersheim, P.; Acklin, P.; Jacoby, E. Similarity Metrics for Ligands Reflecting the Similarity of the Target Proteins. *J. Chem. Inf. Comput. Sci.* **2003**, *43*, 391-405.

[74] MOE, Molecular Operating Environment, Chemical Computing Group Inc., Montreal, Canada. *MOE, Molecular Operating Environment, Chemical Computing Group Inc., Montreal, Canada www.chemcomp.com.*

[75]   Franke, L.; Byvatov, E.; Werz, O.; Steinhilber, D.; Schneider, P.; Schneider, G. Extraction and visualization of potential pharmacophore points using support vector machines: application to ligand-based virtual screening for COX-2 inhibitors. *J. Med. Chem.* **2005**, *48*, 6997-7004.

[76]   Byvatov, E.; Sasse, B. C.; Stark, H.; Schneider, G. From virtual to real screening for D3 dopamine receptor ligands. *ChemBioChem* **2005**, *6*, 997-999.

[77]   Nilakantan, R.; Bauman, N.; Dixon, J. S.; Venkataraghavan, R. Topological Torsion: A New Molecular Descriptor for SAR Applications. Comparison with Other Descriptors. *J. Chem. Inf. Comput. Sci.* **1987**, *27*, 82-85.

[78]   Kearsley, S. K.; Sallamack, S.; Fluder, E. M.; Andose, J. D.; Mosley, R. T.; Sheridan, R. P. Chemical Similarity Using Physiochemical Property Descriptors. *J. Chem. Inf. Comput. Sci.* **1996**, *36*, 118-127.

[79]   Kuz'min, V.; Artemenko, A.; Muratov, E. Hierarchical QSAR technology based on the Simplex representation of molecular structure. *J. Comput.-Aided Mol. Des.* **2008**, *22*, 403-421.

[80]   Kuz'min, V. E.; Artemenko, A. G.; Muratov, E. N.; Volineckaya, I. L.; Makarov, V. A.; Riabova, O. B.; Wutzler, P.; Schmidtke, M. Quantitative Structure-Activity Relationship Studies of [(Biphenyloxy)propyl]isoxazole Derivatives. Inhibitors of Human Rhinovirus 2 Replication. *J. Med. Chem.* **2007**, *50*, 4205-4213.

[81]   Ruggiu, F.; Marcou, G.; Varnek, A.; Horvath, D. ISIDA Property-Labelled Fragment Descriptors. *Mol. Inf.* **2010**, *29*, 855-868.

[82]   Kireeva, N.; Baskin, I. I.; Gaspar, H. A.; Horvath, D.; Marcou, G.; Varnek, A. Generative Topographic Mapping (GTM): Universal Tool for Data Visualization, Structure-Activity Modeling and Dataset Comparison. *Mol. Inf.* **2012**, *31*, 301-312.

[83]   Guener, O. F. *Pharmacophore Perception, Development, and Use in Drug Design.* Wiley-VCH Publishers: Weinheim, 2000.

[84]   Langer, T.; Hoffman, R. D. *Pharmacophores and Pharmacophore Searches.* Wiley-VCH Publishers: Weinheim, 2000.

[85]   Wang, J.; Lai, L.; Tang, Y. Structural Features of Toxic Chemicals for Specific Toxicity. *J. Chem. Inf. Comput. Sci.* **1999**, *39*, 1173-1189.

[86]   Kazius, J.; McGuire, R.; Bursi, R. Derivation and validation of toxicophores for mutagenicity prediction. *J. Med. Chem.* **2005**, *48*, 312-320.

[87]   Cunningham, A. R.; Rosenkranz, H. S.; Zhang, Y. P.; Klopman, G. Identification of 'genotoxic' and 'non-genotoxic' alerts for cancer in mice: the carcinogenic potency database. *Mutat. Res.* **1998**, *398*, 1-17.

[88]   Klopman, G. Artificial intelligence approach to structure-activity studies. Computer automated structure evaluation of biological activity of organic molecules. *J. Am. Chem. Soc.* **1984**, *106*, 7315-7321.

[89]   Klopman, G.; Rosenkranz, H. S. Structural requirements for the mutagenicity of environmental nitroarenes. *Mutat. Res.* **1984**, *126*, 227-238.

[90]   Rosenkranz, H. S.; Mitchell, C. S.; Klopman, G. Artificial intelligence and Bayesian decision theory in the prediction of chemical carcinogens. *Mutat. Res.* **1985**, *150*, 1-11.

[91]   Klopman, G. MULTICASE. 1. A Hierarchical computer automated structure evaluation program. *Quant. Struct.-Act. Relat.* **1992**, *11*, 176-184.

[92]   Klopman, G.; Rosenkranz, H. S. Approaches to SAR in carcinogenesis and mutagenesis. Prediction of carcinogenicity/mutagenicity using MULTI-CASE. *Mutat. Res.* **1994**, *305*, 33-46.

[93]   Klopman, G.; Chakravarti, S. K.; Harris, N.; Ivanov, J.; Saiakhov, R. D. *In Silico* Screening of High Production Volume Chemicals for Mutagenicity using the MCASE QSAR Expert System. *SAR QSAR Environ. Res.* **2003**, *14*, 165-180.

[94]   Sanderson, D. M.; Earnshaw, C. G. Computer prediction of possible toxic action from chemical structure; the DEREK system. *Hum. Exp. Toxicol.* **1991**, *10*, 261-273.

[95]   Gombar, V. K.; Enslein, K.; Hart, J. B.; Blake, B. W.; Borgstedt, H. H. Estimation of maximum tolerated dose for long-term bioassays from acute lethal dose and structure by QSAR. *Risk Anal.* **1991**, *11*, 509-517.

[96]   Judson, P. N. QSAR and Expert Systems in the Prediction of Biological Activity. *Pestic. Sci.* **1992**, *36*, 155-160.

[97]   Judson, P. N. Rule Induction for Systems Predicting Biological Activity. *J. Chem. Inf. Comput. Sci.* **1994**, *34*, 148-153.

[98]   Barratt, M. D.; Rodford, R. A. The computational prediction of toxicity. *Curr. Opin. Chem. Biol.* **2001**, *5*, 383-388.

[99]   Lipinski, C. A.; Lombardo, F.; Dominy, B. W.; Feeney, P. J. Experimental and computational approaches to estimate solubility and permeability in drug discovery and development settings. *Adv. Drug Deliv. Rev.* **2001**, *46*, 3-26.

[100] Oprea, T. I. Property distribution of drug-related chemical databases. *J. Comput. Aided Mol. Des.* **2000**, *14*, 251-264.

[101] Veber, D. F.; Johnson, S. R.; Cheng, H. Y.; Smith, B. R.; Ward, K. W.; Kopple, K. D. Molecular properties that influence the oral bioavailability of drug candidates. *J. Med. Chem.* **2002**, *45*, 2615-2623.

[102] Hann, M. M.; Oprea, T. I. Pursuing the leadlikeness concept in pharmaceutical research. *Curr. Opin. Chem. Biol.* **2004**, *8*, 255-263.

[103] Leo, A. J. Calculating log Poct from structures. *Chem. Rev.* **1993**, *93*, 1281-1306.

[104] Tetko, I. V.; Livingstone, D. J., Rule-based systems to predict lipophilicity. In *Comprehensive Medicinal Chemistry II: In silico tools in ADMET*, Testa, B.; van de Waterbeemd, H., Eds. Elsevier: 2006; Vol. 5, pp 649-668.

[105] Kubinyi, H. Similarity and Dissimilarity: A Medicinal Chemist's View. *Persp. Drug Discov. Design* **1998**, *9-11*, 225–252.

[106] Martin, Y. C.; Kofron, J. L.; Traphagen, L. M. Do structurally similar molecules have similar biological activity? *J. Med. Chem.* **2002**, *45*, 4350-4358.

[107] Daylight Chemical Information Systems Inc. *Daylight Chemical Information Systems Inc.* *http://www.daylight.com.*

[108] Barnard Chemical Information Ltd. *Barnard Chemical Information Ltd.* *http://www.bci.gb.com/.*

[109] Tripos Inc. *Tripos Inc. http://www.tripos.com.*

[110] Jaccard, P. Distribution de la flore alpine dans le Bassin des Dranses et dans quelque regions voisins. *Bull. Soc. Vaud. Sci. Nat.* **1901**, *37*, 241-272.

[111] Taylor, R. Simulation Analysis of Experimental Design Strategies for Screening Random Compounds as Potential New Drugs and Agrochemicals. *J. Chem. Inf. Comput. Sci.* **1995**, *35*, 59-67.

[112] Delaney, J. S. Assessing the ability of chemical similarity measures to discriminate between active and inactive compounds. *Mol. Divers.* **1996**, *1*, 217-222.

[113] Schneider, G.; Neidhart, W.; Giller, T.; Schmid, G. "Scaffold-Hopping" by Topological Pharmacophore Search: A Contribution to Virtual Screening. *Angew. Chem. Int. Ed.* **1999**, *38*, 2894-2896.

[114] Hull, R. D.; Singh, S. B.; Nachbar, R. B.; Sheridan, R. P.; Kearsley, S. K.; Fluder, E. M. Latent semantic structure indexing (LaSSI) for defining chemical similarity. *J. Med. Chem.* **2001**, *44*, 1177-1184.

[115] Hull, R. D.; Fluder, E. M.; Singh, S. B.; Nachbar, R. B.; Kearsley, S. K.; Sheridan, R. P. Chemical similarity searches using latent semantic structural indexing (LaSSI) and comparison to TOPOSIM. *J. Med. Chem.* **2001**, *44*, 1185-1191.

[116] Willett, P.; Barnard, J. M.; Downs, G. M. Chemical similarity searching. *J. Chem. Inf. Comput. Sci.* **1998**, *38*, 983-996.

[117] Hert, J.; Willett, P.; Wilton, D. J.; Acklin, P.; Azzaoui, K.; Jacoby, E.; Schuffenhauer, A. Comparison of Fingerprint-Based Methods for Virtual Screening Using Multiple Bioactive Reference Structures. *J. Chem. Inf. Comput. Sci.* **2004**, *44*, 1177-1185.

[118] Ormerod, A.; Willett, P.; Bawden, D. Comparison of fragment weighting schemes for substructural analysis. *Quant. Struct.-Act. Relat.* **1989**, *8*, 115-129.

[119] Baskin, I. I.; Kireeva, N.; Varnek, A. The One-Class Classification Approach to Data Description and to Models Applicability Domain. *Mol. Inf.* **2010**, *29*, 581-587.

[120] Markou, M.; Singh, S. Novelty detection: A review - Part 1: Statistical approaches. *Signal Process.* **2003**, *83*, 2481-2497.

[121] Markou, M.; Singh, S. Novelty detection: A review - Part 2:: Neural network based approaches. *Signal Process.* **2003**, *83*, 2499-2521.

[122] Karpov, P. V.; Osolodkin, D. I.; Baskin, I. I.; Palyulin, V. A.; Zefirov, N. S. One-class classification as a novel method of ligand-based virtual screening: The case of glycogen synthase kinase 3OI inhibitors. *Bioorg. Med. Chem. Lett.* **2011**, *21*, 6728-6731.

[123] Godden, J. W.; Furr, J. R.; Xue, L.; Stahura, F. L.; Bajorath, J. Molecular Similarity Analysis and Virtual Screening by Mapping of Consensus Positions in Binary-Transformed Chemical Descriptor Spaces with Variable Dimensionality. *J. Chem. Inf. Comput. Sci.* **2004**, *44*, 21-29.

[124] Godden, J. W.; Stahura, F. L.; Bajorath, J. POT-DMC: A virtual screening method for the identification of potent hits. *J. Med. Chem.* **2004**, *47*, 5608-5611.

[125] Batista, J.; Godden, J. W.; Bajorath, J. Assessment of molecular similarity from the analysis of randomly generated structural fragment populations. *J. Chem. Inf. Model.* **2006**, *46*, 1937-1944.

[126] Godden, J. W.; Bajorath, J. Differential Shannon Entropy as a sensitive measure of differences in database variability of molecular descriptors. *J. Chem. Inf. Comput. Sci.* **2001**, *41*, 1060-1066.

[127] Batista, J.; Bajorath, J. Chemical database mining through entropy-based molecular similarity assessment of randomly generated structural fragment populations. *J. Chem. Inf. Model* **2007**, *47*, 59-68.

[128] Maldonado, A. G.; Doucet, J. P.; Petitjean, M.; Fan, B. T. Molecular similarity and diversity in chemoinformatics: from theory to applications. *Mol. Divers.* **2006**, *10*, 39-79.

[129] Bajorath, J. Chemoinformatics methods for systematic comparison of molecules from natural and synthetic sources and design of hybrid libraries. *Mol. Divers.* **2002**, *5*, 305-313.

[130] Waller, C. L. Recent advances in molecular diversity. *Mol. Divers.* **2002**, *5*, 173-174.

[131] Agrafiotis, D. K.; Myslik, J. C.; Salemme, F. R. Advances in diversity profiling and combinatorial series design. *Mol. Divers.* **1998**, *4*, 1-22.

[132] Trepalin, S. V.; Gerasimenko, V. A.; Kozyukov, A. V.; Savchuk, N. P.; Ivaschenko, A. A. New diversity calculations algorithms used for compound selection. *J. Chem. Inf. Comput. Sci.* **2002**, *42*, 249-258.

[133] Harper, G.; Bradshaw, J.; Gittins, J. C.; Green, D. V. S.; Leach, A. R. The Prediction of Biological Activity for High-Throughput Screening Using Binary Kernel Discrimination. *J. Chem. Inf. Comput. Sci.* **2001**, *41*, 1295-1300.

[134] Geronikaki, A. A.; Dearden, J. C.; Filimonov, D.; Galaeva, I.; Garibova, T. L.; Gloriozova, T.; Krajneva, V.; Lagunin, A.; Macaev, F. Z.; Molodavkin, G.; Poroikov, V. V.; Pogrebnoi, S. I.; Shepeli, F.; Voronina, T. A.; Tsitlakidou, M.; Vlad, L. Design of new cognition enhancers: from computer prediction to synthesis and biological evaluation. *J. Med. Chem.* **2004**, *47*, 2870-2876.

[135] Katritzky, A. R.; Kuanar, M.; Slavov, S.; Dobchev, D. A.; Fara, D. C.; Karelson, M.; Acree, W. E., Jr.; Solov'ev, V. P.; Varnek, A. Correlation of blood-brain penetration using structural descriptors. *Bioorg. Med. Chem.* **2006**, *14*, 4888-4917.

[136] Katritzky, A. R.; Dobchev, D. A.; Fara, D. C.; Hur, E.; Tamm, K.; Kurunczi, L.; Karelson, M.; Varnek, A.; Solov'ev, V. P. Skin permeation rate as a function of chemical structure. *J. Med. Chem.* **2006**, *49*, 3305-3314.

[137] Katritzky, A. R.; Kuanar, M.; Fara, D. C.; Karelson, M.; Acree, W. E., Jr.; Solov'ev, V. P.; Varnek, A. QSAR modeling of blood:air and tissue:air partition coefficients using theoretical descriptors. *Bioorg. Med. Chem.* **2005**, *13*, 6450-6463.

[138] Mannhold, R.; Rekker, R. F.; Sonntag, C.; ter Laak, A. M.; Dross, K.; Polymeropoulos, E. E. Comparative evaluation of the predictive power of calculation procedures for molecular lipophilicity. *J. Pharm. Sci.* **1995**, *84*, 1410-1419.

[139] Nys, G. G.; Rekker, R. F. Statistical Analysis of a Series of Partition Coefficients with Special Reference to the Predictability of Folding of Drug Molecules. Introduction of Hydrophobic Fragmental Constants (f-Values). *Eur. J. Med. Chem.* **1973**, *8*, 521-535.

[140] Leo, A.; Jow, P. Y. C.; Silipo, C.; Hansch, C. Calculation of hydrophobic constant (log P) from .pi. and f constants. *J. Med. Chem.* **1975**, *18*, 865-868.

[141] Ghose, A. K.; Crippen, G. M. Atomic Physicochemical Parameters for Three-Dimensional-Structure-Directed Quantitative Structure-Activity Relationships. 2. Modeling Dispersive and Hydrophobic Interactions. *J. Chem. Inf. Comput. Sci.* **1987**, *27*, 21-35.

[142] Ghose, A. K.; Crippen, G. M. Atomic Physicochemical Parameters for Three-Dimensional Structure-Directed Quantitative Structure-Activity Relationships I. Partition Coefficients as a Measure of Hydrophobicity. *J. Comput. Chem.* **1986**, *7*, 565-577.

[143] Ghose, A. K.; Pritchett, A.; Crippen, G. M. Atomic physicochemical parameters for three dimensional structure directed quantitative structure-activity relationships III: Modeling hydrophobic interactions. *J. Comput. Chem.* **1988**, *9*, 80-90.

[144] Wildman, S. A.; Crippen, G. M. Prediction of Physicochemical Parameters by Atomic Contributions. *J. Chem. Inf. Comput. Sci.* **1999**, *39*, 868-873.

[145] Suzuki, T.; Kudo, Y. Automatic log P estimation based on combined additive modeling methods. *J. Comput. Aided. Mol. Des.* **1990**, *4*, 155-198.

[146] Convard, T.; Dubost, J.-P.; Le Solleu, H.; Kummer, E. SMILOGP: A Program for a fast evaluation of theoretical log-p from the smiles code of a molecule. *Quant. Struct.-Act. Relat.* **1994**, *13*, 34-37.

[147] Wang, R.; Gao, Y.; Lai, L. Calculating partition coefficient by atom-additive method. *Persp. Drug Discov. Design* **2000**, *19*, 47-66.

[148] Wang, R.; Fu, Y.; Lai, L. A New Atom-Additive Method for Calculating Partition Coefficients. *J. Chem. Inf. Comput. Sci.* **1997**, *37*, 615-621.

[149] Balakin, K. V.; Savchuk, N. P.; Tetko, I. V. *In silico* approaches to prediction of aqueous and DMSO solubility of drug-like compounds: trends, problems and solutions. *Curr Med Chem* **2006**, *13*, 223-241.

[150] Varnek, A.; Kireeva, N.; Tetko, I. V.; Baskin, I. I.; Solov'ev, V. P. Exhaustive QSPR studies of a large diverse set of ionic liquids: How accurately can we predict melting points? *J. Chem. Inf. Mod.* **2007**, *47*, 1111-1122.

[151] Feuston, B. P.; Chakravorty, S. J.; Conway, J. F.; Culberson, J. C.; Forbes, J.; Kraker, B.; Lennon, P. A.; Lindsley, C.; McGaughey, G. B.; Mosley, R.; Sheridan, R. P.; Valenciano, M.; Kearsley, S. K. Web enabling technology for the design, enumeration, optimization and tracking of compound libraries. *Curr. Top. Med. Chem.* **2005**, *5*, 773-783.

[152] Green, D. V.; Pickett, S. D. Methods for library design and optimisation. *Mini Rev. Med. Chem.* **2004**, *4*, 1067-1076.

[153] Green, D. V. Virtual screening of virtual libraries. *Prog. Med. Chem.* **2003**, *41*, 61-97.

[154] Varnek, A.; Fourches, D.; Solov'ev, V. P.; Baulin, V. E.; Turanov, A. N.; Karandashev, V. K.; Fara, D.; Katritzky, A. R. \"*In Silico*\" Design of New Uranyl Extractants Based on Phosphoryl-Containing Podands: QSPR Studies, Generation and Screening of Virtual Combinatorial Library, and Experimental Tests. *J. Chem. Inf. Comput. Sci.* **2004**, *44*, 1365-1382.

[155] Varnek, A.; Fourches, D.; Solov'ev, V.; Klimchuk, O.; Ouadi, A.; Billard, I. Successful "*In Silico*" Design of New Efficient Uranyl Binders. *Solvent Extraction and Ion Exchange* **2007**, *25*, 433-462.

[156] Tetko, I. V. Neural Network Studies. 4. Introduction to Associative Neural Networks. *J. Chem. Inf. Comput. Sci.* **2002**, *42*, 717-728.

[157] Vapnik, V. N. *The Nature of Statistical Learning Theory*. Springer: 1995.

[158] Solov'ev, V. P.; Varnek, A. Anti-HIV Activity of HEPT, TIBO, and Cyclic Urea Derivatives: Structure-Property Studies, Focused Combinatorial Library Generation, and Hits Selection Using Substructural Molecular Fragments Method. *J. Chem. Inf. Comput. Sci.* **2003**, *43*, 1703-1719.

Send Orders for Reprints to reprints@benthamscience.net

CHAPTER 3

Fluorescent Probes for Cellular Assays

George T. Hanson and Bonnie J. Hanson[*]

Life Technologies, 501 Charmany Drive, Madison, Wisconsin, 53719, USA

Abstract: A fluorescent probe is a fluorophore designed to localize within a specific region of a biological specimen or to respond to a specific stimulus. Fluorescent probes have been used for nearly a century to study cellular processes due to their exquisite sensitivity and selectivity. Fluorescent probes have also gained in popularity as safety and environmental concerns over the use of radioactive probes have grown. At the same time, cellular assays are being more widely used now than ever before. This review will give a broad overview of types of fluorescent probes, types of fluorescent assays, and their application in cellular assays for a number of pharmaceutically relevant target classes.

Keywords: Fluorescence, cell based assays, high throughput screening, GPCR, pathway, ion channel, kinase, multiplexing, caspase.

INTRODUCTION

Over the years, much emphasis has been placed on generating sensitive, high-throughput and cost-effective assays for screening pharmaceutically relevant targets. In addition, there has been an increasing acknowledgement that the performance of a drug in a biochemical assay using isolated enzymes may not accurately reflect its performance in the context of a whole cell where many more components and complex pathways are present. Therefore, the use of cellular assays has increased as researchers strive to study compounds in a more biologically relevant context. Many researchers have also turned to various fluorescence based assays as these assays are sensitive, compatible with many of the high throughput screening (HTS) or high content screening (HCS) instruments available, and in many cases, allow for live cell imaging and/or sorting by flow cytometry which can speed assay development.

*Address correspondence to Bonnie J. Hanson: Life Technologies, 501 Charmany Drive, Madison, Wisconsin, 53719, USA; Tel: 608-204-5044; Fax: 608-204-5200; E-mail: bonnie.hanson@lifetech.com

Fluorescent probes have been used for nearly a century to study cellular processes due to their exquisite sensitivity and selectivity. This sensitivity arises in part because unless the fluorophore is irreversibly destroyed in the excited state by photobleaching, the same fluorophore can be repeatedly excited and detected. This allows a single fluorophore to generate many thousands of detectable photons. Fluorescent probes have also gained in popularity as safety and environmental concerns over the use of radioactive probes have grown. Fluorescence based assays come in a large variety. The fluorescent probe itself may be genetically encoded (like fluorescent proteins), may be a fluorescently labeled protein or peptide, or may be a small molecule. In addition, the fluorescent probe(s) may be designed for use in numerous assay modalities.

For instance, fluorogenic assays make use of a probe that undergoes an increase in fluorescence intensity upon a signaling event. Oftentimes, these types of assays involve some form of fluorescence quenching that is removed upon signaling. Intramolecular self-quenching is the quenching of one fluorophore by another and often occurs when high concentrations or labeling densities of a particular fluorophore are used. The EnzChek® assays from Life Technologies are an example of assays in which intramolecular self-quenching keeps a fluorescent substrate almost completely quenched until cleavage of the substrate occurs, leading to disruption of the quenching and an increase in fluorescence (Fig. **1**). The Fluo series of dyes, which will be discussed in more detail in the section on GPCR assays, are another example of a fluorogenic probe, in this case, one which becomes fluorescent upon binding to calcium.

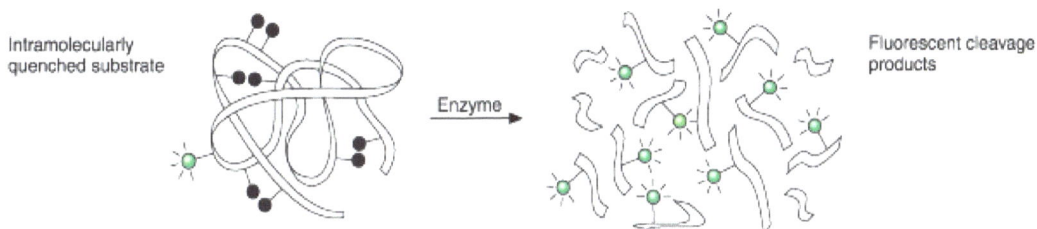

Fig. (1). Enzyme Detection *via* Disruption of Intramolecular Self-Quenching. Some fluorescent probes utilize intramolecular self quenching to keep the probe in a virtually non-fluorescent state. Upon enzyme catalyzed hydrolysis of the probe, the quenching is relieved, yielding a brightly fluorescent product.

Fluorescence polarization can also be used to develop cellular assays. Fluorescence polarization, first observed by Weigert in 1920 and later described by Perrin in 1926, is based upon the observation that fluorescent molecules in solution that are excited with plane-polarized light will emit light in a fixed plane [1,2]. However, as the molecules rotate and tumble, the plane into which the light is emitted may vary from the plane used for initial excitation. The rate at which a molecule rotates and tumbles is dependent upon its size. If a molecule is very large, little rotation will occur, whereas if a molecule is small, rotation will be faster. Therefore, a molecule can be excited with vertically polarized light and the intensity of light emitted can then be monitored in the vertical and horizontal planes to determine the relative mobility of the fluorescently labeled molecule. If the fluorescent probe is bound to a larger object such as an antibody or a protein, the emission light will remain highly polarized relative to the excitation plane. But if the probe is free in solution, it will rotate more quickly, and the light will be depolarized relative to the excitation plane.

There are also ratiometric fluorescent probes which are used to develop cellular assays. Ratiometric probes may be ratiometric by excitation or emission. With fluorescent probes that are ratiometric by emission, the probe is excited by one wavelength of light, and a ratio of the emission detected at two different wavelengths is calculated (Indo-1 is an example of a ratiometric by emission probe). Probes that are ratiometric by excitation, on the other hand, are excited at two different wavelengths, and the ratio of resulting emission at a single wavelength is calculated (Fura-2 is an example of a ratiometric by excitation probe). Ratiometric measurements can eliminate distortions in data caused by events such as variations in probe loading and retention, instrumental factors, and photobleaching. Ratiometric calculations are also usually used with fluorescent probes that undergo Förster resonance energy transfer (FRET). With a FRET assay, a donor fluorophore in its excited state can transfer energy non-radiatively to an acceptor fluorophore that is in close proximity and appropriately oriented if the emission spectra of the donor fluorophore overlaps with the excitation of the acceptor fluorophore. The end result being that a calculation of the emission of the acceptor fluorophore to that of the donor fluorophore can be made. When FRET is occurring, increased emission at the acceptor fluorophore wavelength will be observed and when FRET is not occurring, increased emission at the donor fluorophore wavelength will be observed. When FRET is utilized with species that undergo time-resolved fluorescence, so-called TR-FRET assays can be developed. With cellular assays, many components may have fluorescence including the compounds that are being screened or proteins present in the system that are naturally fluorescent. These background fluorescence signals can interfere with a fluorescent assay. However, by making use of long-lived fluorophores and combining them with time-resolved detection (allowing a delay between fluorophore excitation and emission detection) these fluorescence interferences can be avoided. Time-resolved fluorescence assays generally take advantage of the rare earth elements called lanthanides (such as terbium and europium). Lanthanide complexes exhibit large Stoke's shifts and also have extremely long emission lifetimes (on the order of micro- to milli- seconds) compared to traditional fluorophores (typically 1-4 nanoseconds). TR-FRET assays are then generally created by utilizing a lanthanide complex as the donor molecule which then transfers energy to an acceptor fluorophore. The fluorescent signal is read in a time-resolved manner which reduces assay interference while providing the benefits of a FRET assay.

To further delineate the spatial and temporal aspects of cellular processes, fluorescent probes may be used in this wide variety of modes to look at localization of components, the presence or absence of components (such as analytes or proteins of interest), kinetic events (such as the activation of enzymes over time), and can even be used in multiplexed assays to study multiple variables at the same time.

In this review, we will approach this complex picture by looking at various pharmaceutically relevant target classes and give a broad overview of fluorescent cellular assays that have been developed to study targets within each class. We will also cover a few multiplexed fluorescent assay approaches that have been developed and how these assays have been utilized.

KINASES

There are estimated to be 518 protein kinases within the human genome which comprises ~1.7% of all human genes [3]. Functionally, protein kinases transfer a phosphate group to specific tyrosine, threonine, or serine residues on target proteins and act to modify these proteins' activity, cellular location, and/or association with other proteins. Protein phosphorylation regulates most aspects of cell life including cell survival, growth, metabolism, and proliferation [4-7]. Perturbations in protein-kinase mediated cell signaling pathways have been linked to a number of different diseases including cancer, inflammation, diabetes, and heart failure (reviewed in [8,9]). In fact, of the >100 dominant oncogenes identified thus far, protein kinases comprise a large fraction [5,10]. Because of the central role played by kinases in cellular signaling and their involvement in human diseases, protein kinases have arisen as one of the major drug target classes being screened by pharmaceutical companies.

Many different biochemical kinase assays have been developed, but fewer cellular assays exist. Most of these assays are ELISA based assays that are functional with cell lysates. Many kinases are themselves regulated by phosphorylation, so measurements of phosphorylation status using phosphospecific antibodies to the kinase of interest can be used to approximate catalytic activity. In general, the way that these assays have been engineered to work in a plate based format is to use two antibodies—one directed towards the phosphorylated kinase of interest and the other to the total kinase of interest. Differential secondary antibody labeling is then used to distinguish the two signals, and the ratio of phosphorylated kinase to total kinase for a specific kinase can then be calculated [11]. Cell lysates prepared from cells treated with different potential activators or inhibitors are then used as the input. Several commercially available kits are available for measuring kinase activity in this context (LI-COR and R&D Systems).

A somewhat modified approach to this idea has also been used to produce a TR-FRET version of this type of assay (LanthaScreen® Cellular Assays from Life Technologies). With this assay system, a green fluorescent protein (GFP) fusion with the kinase of interest is created and transduced into a cell line. The cell line is then exposed to various potential activators or inhibitors of kinase activity and again a cell lysate is used with a phospho-specific kinase antibody labeled with a terbium tag. The GFP acts as the donor in this assay system to the terbium label.

The amount of TR-FRET signal observed is then used to determine kinase activity (Fig. **2**).

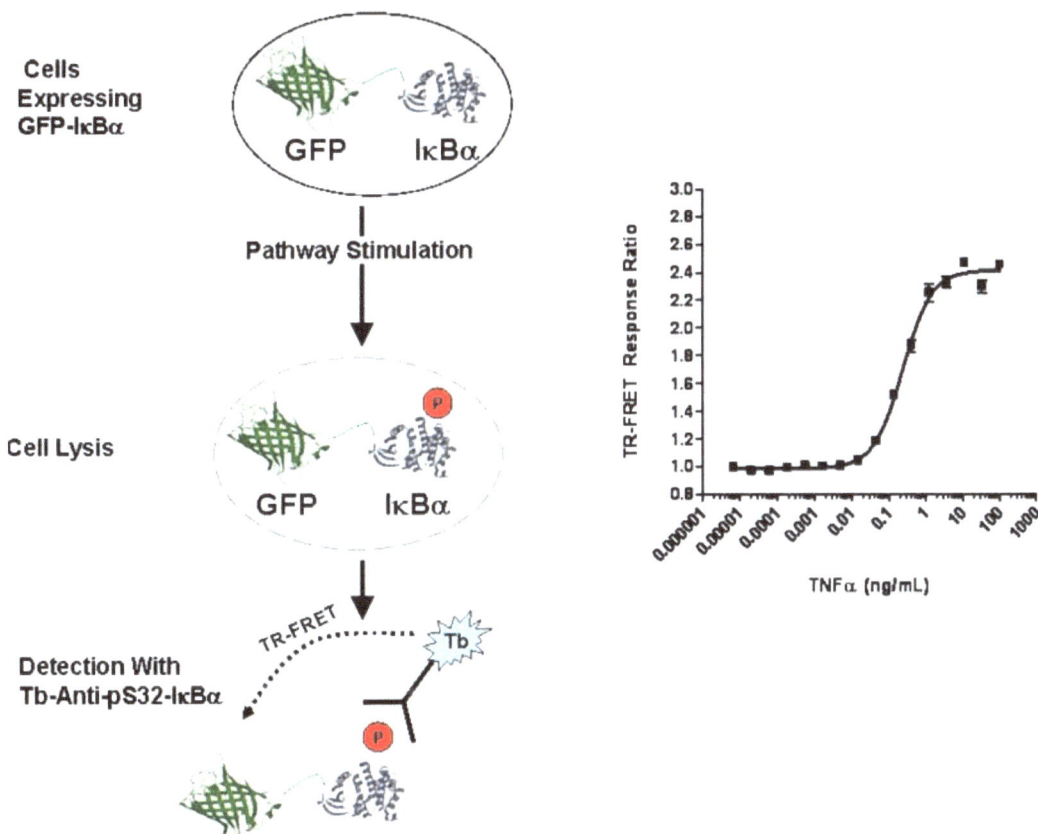

Fig. (2). Schematic Diagram of a Cell Based LanthaScreen Assay. The target of interest is fused to GFP and expressed in a cell line. Upon activation, the target of interest is phosphorylated. A phospho-specific antibody fused to terbium is then used to detect the phosphorylated target. The terbium is able to undergo TR-FRET with the GFP, leading to a fluorescent signal. Shown is a dose response of IκBα to TNFα.

The Omnia® Assay (Life Technologies) is another approach which has been used on cell lysates. In this assay, a fluorescence signal is generated when a non-natural Sox amino acid present in an optimized kinase peptide substrate undergoes chelation-enhanced fluorescence in the presence of divalent magnesium [12]. Although this assay is not dependent upon phospho-specific kinase antibodies, in order to use the assay with cell lysates, inhibitors of phosphatases and off-target kinases often need to be included, making this assay more difficult to use with cell lysates than with purified enzymes.

To look at kinase signaling within intact cells, reporter gene assays are often utilized. For these assays, a response element of interest is engineered in front of a reporter gene such as beta-lactamase or beta-galactosidase. A signaling pathway of interest can then be activated, and various kinases along this pathway may be activated or inhibited either with compounds or RNAi. The down-stream effect on the reporter gene can then be monitored. Although this assay format does not require the cells to be lysed, the response element of interest can often be activated by multiple convergent pathways so deconvolution of the contribution of specific kinases to this signal is often required through methods such as RNAi.

GPCRs

The G protein coupled receptor (GPCR) superfamily is comprised of ~800-1000 members and is one of the largest families of proteins identified in humans [13,14]. GPCRs contain seven transmembrane domains and reside on the cell surface where they respond to a diverse array of stimuli including light, odors, tastes, ions, small molecules, peptides, and hormones [15,16]. Excluding sensory GPCRs (those responding to signals such as odors or tastes which target sensory organs), endogenous ligands have been identified for ~230 different GPCRs, and there are another ~140 non-sensory orphan GPCRs [13]. The central importance of GPCRs to medicine can be seen in the large number of pharmaceuticals targeting these receptors. GPCRs account for ~27% of the currently marketed drugs [17]. In 2011 alone, five of the top seven selling pharmaceuticals targeted GPCRs accounting for sales exceeding US $25 billion (Table **1**).

The attractiveness of GPCRs as pharmaceutical targets stems from several avenues. The location of GPCRs on the cell surface makes them readily accessible to drugs, and small molecules have been shown to interact with GPCRs with a high degree of selectivity and sensitivity [18]. In addition, GPCRs are widely expressed throughout the body and play an essential role in physiology/ pathophysiology [19]. As such, they are involved in an enormous range of disease processes covering the majority of the current therapeutic markets including pain, asthma, allergies, ulcers, hypertension, migraines, schizophrenia, and depression [19, 20]. Despite their involvement in numerous cellular processes, current pharmaceuticals target only ~10% of the total non-sensory GPCRs [16], leaving numerous potential targets available for drug discovery well into the future.

Table 1.   Top Selling Drugs by Sales in the United States for 2011

Product	U.S. Sales (U.S. $ Billions)	Target	Therapeutic Indication
Lipitor	$7.43	HMG-CoA Reductase Inhibitor	High Cholesterol
Plavix	$6.56	P2RY12 Inhibitor	Preventing blood clots (*i.e.* after heart attacks or strokes)
Nexium	$5.96	Proton Pump Inhibitor	GERD
Abilify	$5.03	D2 Dopamine and 5HT1a Receptor Partial Agonist	Schizophrenia, Bi-polar Disorder, Depression
Advair Diskus	$4.49	β2-Adrenoceptor Agonist and Glucocorticoid Receptor Agonist	Asthma, COPD
Seroquel	$4.49	Dopamine and 5HT2 Serotonin Receptor Antagonist	Schizophrenia, Bi-polar Disorder
Singulair	$4.45	Cysteine Leukotiene Receptor Inhibitor	Allergies, Asthma

Source: http://www.drugs.com/top200.html.

Fluorescent cellular assays for GPCRs can be broken down into three main groups: those aimed at GPCR second messengers, reporter gene assays, and beta-arrestin/translocation based assays. Within the group of second messenger assays, the most common second messengers studied are calcium, cAMP, and IP(1 or 3). Many fluorescent probes, most of which are derivatives of the Ca^{2+} chelators EGTA, APTRA, and BAPTA [21], have been developed to look at calcium as a second messenger. The most common calcium indicators used for HTS in the GPCR field are based upon the Fluo dyes originally developed in Roger Tsien's lab [22]. These dyes undergo an increase in green fluorescence signal upon binding calcium (Fig. **3**). Several companies offer products within this area (such as Life Technologies' Fluo-4 Direct™ Kit and Molecular Devices' FLIPR® Calcium 5 Kit).

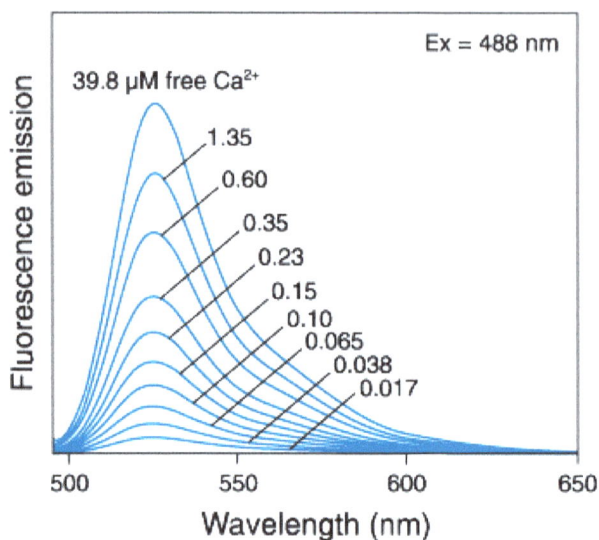

Fig. (3). Ca^{2+}-dependent fluorescence emission spectra of fluo-3. The spectrum for the Ca^{2+}-free solution is indistinguishable from the baseline.

Although calcium sensitive fluorescent dyes are often used in HTS, fluorescent protein based calcium sensors have also been developed. One such sensor based on the YC3.60 version of a GFP-based sensor family developed by Tsien, Miyawaki, and co-workers has been developed into a FRET assay [23,24]. This sensor utilized a yellow fluorescent protein (YFP) and a cyan fluorescent protein (CFP) separated by a calmodulin-M13 moiety. Binding of four calcium ions to the calmodulin-M13 moiety induces a conformational change which brings the CFP and YFP domains closer, allowing FRET to occur (Fig. **4**). This technology is now being commercialized by Life Technologies under the name Premo™ Cameleon Calcium Sensor. As this calcium sensor is genetically encoded, it is possible to use either transiently or stably transfected cells to study calcium mobilization within particular cells or cellular regions over time.

There are a large number of cAMP assays on the market, most of which are competition assays which utilize an anti-cAMP antibody and some form of a labeled cAMP employing technologies such as TR-FRET, FP, and enzyme complementation. Assays within the TR-FRET class are available from Perkin-Elmer and Cis-Bio. In general the assays utilize a labeled anti-cAMP antibody

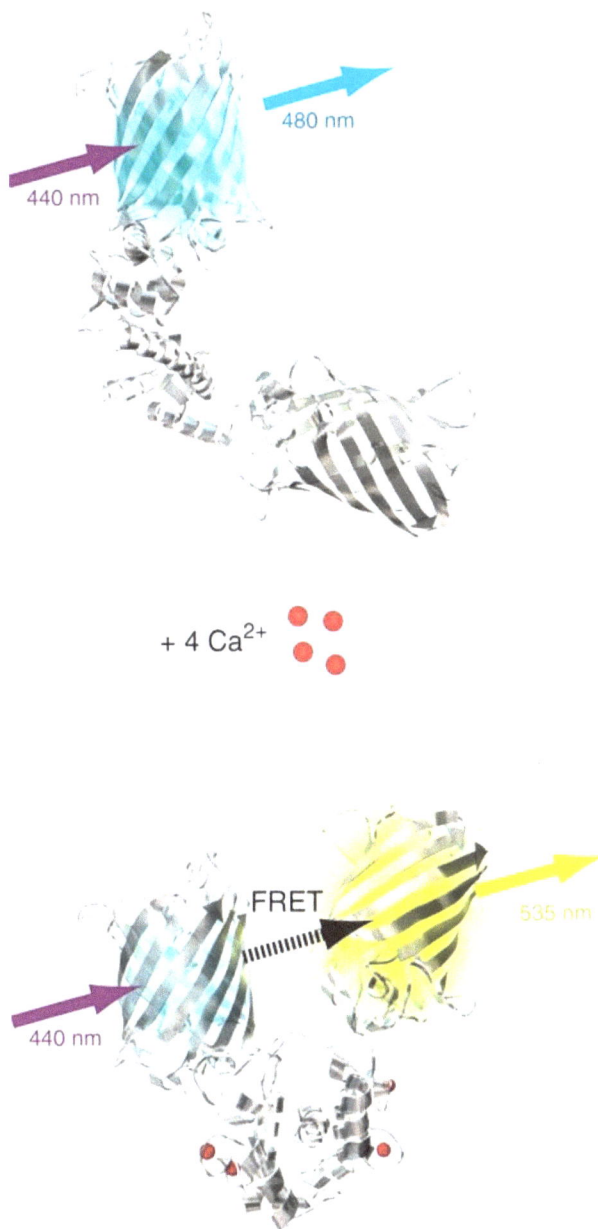

Fig. (4). Schematic of the Premo™ cameleon calcium sensor mechanism. The ECFP and cpVenus units of cameleon are connected by a calcium-sensitive moiety consisting of a calmodulin and a M13 domain. Upon binding calcium the calmodulin-M13 moiety undergoes a conformational change that brings the fluorescent protein-domains into close proximity. The resulting FRET signal allows ratiometric measurements of free intracellular calcium, independent of well-to-well (or cell-to-cell) differences in gene expression levels.

(labeled with either the lanthanide complex or the fluorescent molecule) and a labeled cAMP which contains the other TR-FRET partner. Free cAMP within the cell competes for binding to the labeled cAMP antibody disrupting TR-FRET and leading to a loss in signal. An FP based version of this assay is also available from Perkin-Elmer in which the fluorescence polarization of a fluorescently labeled cAMP is studied. When this labeled-cAMP is bound to an anti-cAMP antibody, the fluorescence will be more polarized than when cAMP within the cell is bound to the antibody leaving the labeled-cAMP free in solution. However, FP based assays are more susceptible to compound interference than the TR-FRET versions of the assays. The HitHunter® assay from DiscoveRx also uses a cAMP antibody, but in this case the labeled-cAMP contains an enzyme fragment of the beta-galactosidase enzyme. The enzyme-complementing partner of the beta-galactosidase enzyme is also present in the assay. Free cAMP now competes with the labeled cAMP for binding to the antibody. When free cAMP is bound, the labeled-cAMP is available for enzyme complementation and creation of a functional beta-galactosidase enzyme which can then be detected with a fluorescent substrate. All of these assays work with cell lysates and can be used for HTS [25]. cAMP assays are reviewed in much more detail in several recent reviews [26,27].

Although IP(1 or 3) accumulation assays were originally developed as radioactive displacement assays, there are now also several fluorescent based kits to look at IP accumulation which are amenable with HTS. Cis-Bio offers an IP-One assay which utilizes a europium-cryptate labeled IP_1 antibody donor and a d2-labeled IP_1 acceptor fluorophore to produce a TR-FRET signal. As intracellular levels of IP_1 increase there is a reduction of signal in this assay format. Alternatively, DiscoveRx offers a HitHunter® assay which is an FP assay dependent upon competitive binding between intracellular IP_3 and a fluorescently labeled tracer.

The reporter gene assay group for GPCRs include assays which again link a response element with a reporter gene of interest. In general, a cAMP response element is utilized to look at Gs coupled receptors and some Gi/o coupled receptors if first stimulated with forskolin, and an NFAT response element is used to look at Gq coupled receptors and Gi/o/s coupled receptors when used in conjunction with chimeric or promiscuous G proteins. The reporter genes utilized for fluorescent assays are usually beta-lactamase (which utilizes a ratiometric FRET substrate) or beta-galactosidase (which utilizes various fluorogenic substrates). These assays are highly sensitive as amplification of signal occurs through the signaling pathways monitored and can be used with live cells instead of cell lysates allowing further study of the cells after the cAMP assay. Several HTS campaigns for GPCRs have been enabled by these technologies [28,29].

Beta-arrestin/translocation based assays constitute a third group of assays for GPCRs. Upon GPCR activation, GPCRs will almost universally recruit beta-arrestin and in most cases cause internalization of the GPPCR, so various assay formats have been developed to capitalize on this phenomenon. These assays all have the advantage in that the natural signaling pathway of the GPCR of interest does not need to be known for the assay to be utilized, and a single assay format can be utilized for different classes of GPCRs (Gs, Gq, Gi/o). The first assay developed within this category was the Transfluor® assay now commercialized by Molecular Devices. The Transfluor® assay uses a GFP labeled beta-arrestin and was developed as an assay for high content imagers. With this assay, the GFP-labeled beta-arrestin starts out cytoplasmically localized, is recruited to the cell membrane upon GPCR activation and then becomes localized to cytoplasmic pits or vesicles upon GPCR desensitization. The location of the GFP is then utilized to determine the activation state of the GPCR. A similar assay called Redistribution® (available from Thermo-Fisher) instead labels the GPCR with the fluorescent protein and internalization of the receptor is studied. Again this assay works best with an imaging device. A bioluminescence resonance energy transfer (BRET) assay was developed as a modification of the Transfluor® assay. With the $BRET^2$/beta-arrestin assay commercialized by Perkin-Elmer, a GFP-tagged beta-arrestin is again utilized, but with this assay format the GPCR of interest is also tagged with a luciferase construct. In the presence of the luciferase substrate and when the luciferase and the GFP are in close proximity, BRET occurs from the luciferase to the GFP, resulting in a fluorescence signal. This fluorescence signal is then able to be monitored using a fluorescence plate reader, removing the necessity of having a high content imager to read out the assay. Two new assays have also recently hit the market within the beta-arrestin arena. These are the PathHunter® assay from DiscoverX and the Tango™ assay from Life

Technologies. The PathHunter® assay relies upon enzyme complementation of beta-galactosidase. For this assay a portion of the beta-galactosidase enzyme is linked to the GPCR of interest and the complementary portion is linked to the beta-arrestin. Upon GPCR activation, the beta-arrestin is recruited to the GPCR, enzyme complementation occurs and the now functional beta-galactosidase is available to act on its substrate. With the Tango assay, a non-native transcription factor is linked to the GPCR N terminus with a protease cleaveage site between the transcription factor and the GPCR. A protease is then attached to the beta-arrestin. When the GPCR is activated and beta-arrestin is recruited, this protease will now cleave the transcription factor and the transcription factor will activate a beta-lactamase reporter gene [30].

ION CHANNELS

Ion channels are a diverse group of membrane spanning, pore forming proteins. In general, ion channels control the flow of ions across biological membranes. Ligand-gated channels are controlled by small molecules whereas voltage-gated channels are dependant on membrane potential. Ion channels are not only validated targets for drug discovery for epilepsy and other neurological disorders, but are also safety targets. The human ether-a-go-go related gene (hERG), a cardiac potassium channel, if blocked can lead to lethal ventricular arrhythmias [31]. The 'gold standard' for evaluating ion channel activity is *via* electrophysiology measurements or "patch clamping." However traditional patch clamping is time consuming and laborious, which has lead to the development of fluorescent cell-based assays for ion channels including: direct ion detection, surrogate ion detection, and voltage sensitive dye assays.

Direct ion detection of ion channel activity depends on the ability of a fluorescent molecule to directly monitor the flow of ions through the channel and into the intracellular space. Fluo dyes (as described in the GPCR section) are the most commonly employed fluorescent dyes for direct ion detection of calcium flux through channels. Other calcium sensitive indicators such as the fluorescent protein based cameleon should also function to monitor calcium flux through ion channels. Since cameleon is a genetically encoded calcium sensor it can be targeted to subcellular locations such as the intracellular side of the plasma membrane leading to more proximal detection of calcium flux through ion channels.

Surrogate ion detection uses similar principles of the direct ion method with the exception that non-native or surrogate ions are measured. Weaver and coworkers noted the shortcomings of throughput, temporal resolution, sensitivity or selectivity of traditional electrophysiology and radioactive rubidium flux assays for potassium channels [32]. By exploiting the permeability of thallium through potassium channels, Weaver and coworkers demonstrated ion channel activity can be monitored by the fluorescent dye BTC-AM binding to thallium (Fig. **5**). Cells are loaded with thallium sensitive dye prior to compound addition. Compounds are added prior to or at the same time as a thallium containing solution. Thallium flux is monitored in kinetic mode analogous to a fluo-4 calcium assay on a fluorescence plate reader with liquid handling capabilities. Functional ion channel activity results in thallium influx. Thallium binding to BTC-AM results in an

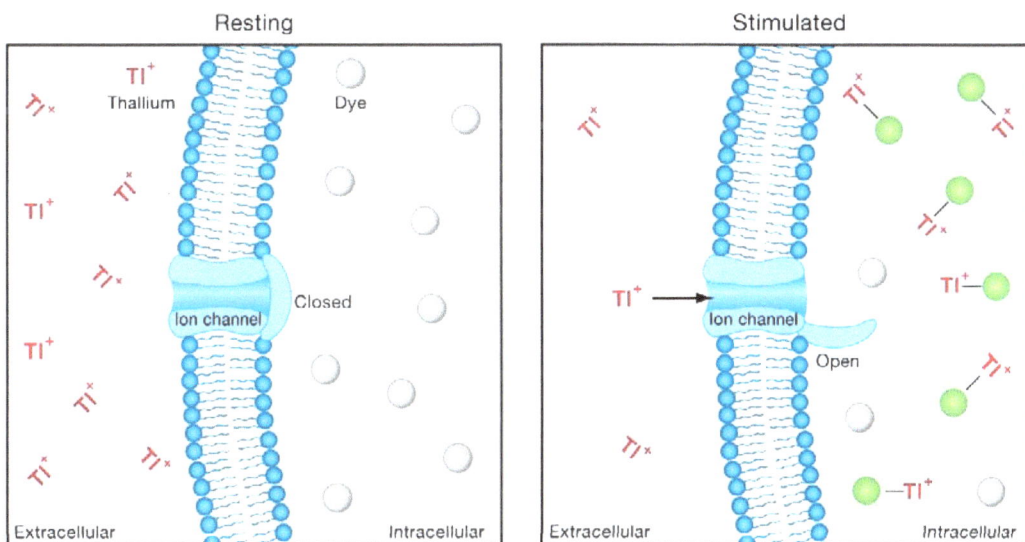

Fig. (5). Schematic diagram of the FluxOR™ surrogate ion assay for potassium channels. Cells are loaded with a fluorogenic thallium sensitive dye. Thallium is added to the extracellular media surrounding the cells. In the "resting" state or closed conformation of the channel, then the thallium remains segregated from the dye and little or no fluorescence is observed. Similarly if a channel is blocked by a specific compound, little fluorescence increase is observed. However if a potassium channel is open or "stimulated," thallium flows into the cell, binds to the dye and creates a highly green fluorescent complex.

increased fluorescence signal compared to the dye alone. The thallium surrogate ion detection assay is now commercially available as FluxOR™ from Life Technologies.

Voltage sensitive dyes indicate ion channel activity by monitoring changes in membrane potential. Voltage sensitive assay formats come in two flavors. The single dye approach, which is commercialized by Molecular Devices, uses an oxonol dye. This dye has the specialized property that it migrates within the lipid bilayer of the plasma membrane depending on membrane potential. In a resting state, the cell interior has a relatively negative potential and the oxonol dye is closely associated with the outer side of the plasma membrane. When the cell is depolarized, the oxonol dye migrates to the inner side of the plasma membrane where an increase in the fluorescence intensity of the dye takes place. Alternatively, Life Technologies offers a voltage sensing assay technology that is based on a pair of different wavelength dyes capable of FRET. The donor fluorophore is a blue coumarin-phospholipid dye that is anchored on the extracellular side of the plasma membrane. The acceptor dye is a red oxonol (DiSBAC) dye that, like the single dye approach, migrates within the lipid bilayer of the plasma membrane. In a resting state the cell interior has a relatively negative potential and the oxonol dye is in close proximity to the coumarin donor, resulting in more efficient FRET. When the cell is depolarized, the oxonol dye will migrate to the inner side of the plasma membrane, thus disrupting FRET. Voltage sensitive dyes have the advantage of being applicable to any ion channel target that changes the membrane potential and the assay can be performed in a high throughput setting.

CELLULAR PATHWAY ANALYSIS

Cell signaling pathways are complex networks by which intracellular or extracellular signals are detected and transduced into protein or gene expression changes. Oftentimes, the signals are detected by receptors, mediated by kinases through phosphorylation/dephosphorylation events, and terminated by transcription factors altering gene expression *via* specific promoter elements. Defects in cell signaling pathways can lead to a variety of disorders including autoimmune disease and cancer.

An important cell signaling pathway that is often studied is the apoptosis pathway. Apoptosis is a process of programmed cell death that is mediated by intracellular signaling pathways. There are numerous assayable apoptosis markers including mitochondrial function, plasma membrane integrity, and phosphotidylserine exposure. However, caspase activation remains one of the hallmark features of cells undergoing apoptosis. Caspases or cysteine-aspartic acid specific proteases can be monitored in cell lysates *via* peptides tagged with coumarin based fluorophores. For the detection of caspase-1 activity, the peptide Trp-Glu-His-Asp-AFC is often used. This peptide emits blue light (λ_{max} = 400 nm) until cleaved by caspase-1 or related caspases at which point free 7-Amino-Trifluoromethyl Coumarin (AFC) emits a yellow-green fluorescence (λ_{max} = 505 nm), which can be quantified using a fluorescence spectrophotometer or a fluorescence microtiter plate reader [33]. Comparison of the fluorescence of AFC from an apoptotic sample with an uninduced control allows determination of the fold increase in caspase activity. Millipore, R&D Systems, and Calbiochem provide commercially available fluorogenic assays for caspase detection from cell lysates.

Life Technologies uses a fluorescent inhibitor of caspases (FLICA™) methodology in their Vybrant® assay platform for studying caspase activity in live cells. The FLICA reagent is comprised of three parts: 1) a fluoromethyl ketone (FMK) moiety, 2) a carboxyfluorescein group (FAM) and 3) a valine-alanine-aspartic acid tripeptide linker. The FLICA reagent is cell permeant and noncytotoxic. The tripeptide region of FLICA is recognized by the caspase as a potential substrate. Upon close association of the reagent with the caspase, the FMK moiety covalently reacts with cysteines in the active site. Unbound FLICA molecules diffuse out of the cell and are washed away; the remaining green-fluorescent signal from the FAM group is a direct measure of the amount of active caspase present in the cell at the time the FLICA reagent was added. FLICA reagents are used widely to study apoptosis with flow cytometry and microscopy and have even been applied to high content screening [34].

Detection of caspase 3,7 activation can be performed in a cellular context *via* a fluorgenic substrate. The substrate contains the canonical caspase 3,7 peptide cleavage sequence Asp-Glu-Val-Asp (DEVD), connected to a non-fluorescent

DNA binding dye. Upon caspase activation, the peptide sequence is separated from the DNA binding dye. The dye is then free to migrate into the nucleus, bind DNA, and fluoresce. This assay methodology results in the illumination of apoptotic cell nuclei in a green fluorescent color. The assay is commercially available from the Life Technologies as CellEvent™ caspase 3,7 reagent or from Biotium as NucView™ 488 caspase-3 substrate.

Translocation or image based assays are another means to interrogate cell signaling pathways. GFP is often the work-horse for translocation assays. For example, fusion of GFP to a protein of interest, such as extracellular signal-regulated kinase (ERK), allows the mitogen-activated protein kinase (MAPK)/ERK signal transduction pathway to be studied. The basic pathway couples intracellular responses to the binding of growth factors to cell surface receptors. In many cell types, activation of this pathway promotes cell division. A key step in the pathway is the migration of ERK from cytoplasm to the nucleus of the cell. Therefore a GFP-ERK fusion results in diffuse cytoplasmic green fluorescence being concentrated in the nucleus, allowing the pathway to be followed by manual or automated fluorescence microscopy. Thermo Fisher commercializes the Redistribution® technology, which couples GFP to cellular targets for oncology, inflammation and other pathway analysis. Alternatively, the Tango™ assay commercialized by Life Technologies or enzyme fragment complementation (EFC) using beta-galactosidase and a fluorescent substrate as commercialized by DiscoveRx can be used to study cellular pathways [30,35].

Cellular pathway analysis is also enabled by the use of beta-lactamase as a reporter gene. Beta-lactamase is placed downstream of a particular promoter sequence or response element. Upon external stimuli a signaling cascade is initiated which terminates in beta-lactamase production. The extent of signaling is evaluated by using a fluorescent substrate specific for beta-lactamase. Alternatively, beta-galactosidase with an appropriate fluorescent substrate can be used, but without the ratiometric advantages of beta-lactamase.

The beta-lactamase approach commercialized by Life Technologies under the name CellSensors® has been used to study the combined affects of RNAi and small molecule inhibitors of epidermal growth factor receptor (EGFR) pathway components. O'Grady *et al.* showed 2.5-12 fold increased potency of small molecule inhibitors of EGFR and MEK-1 in combination with RNAi directed towards EGFR when stimulated with EGF [36]. The experiments were performed using beta-lactamase downstream of an AP-1 response element. They postulated that since the combination of these two targeted agents was shown to increase the efficacy of EGFR and MEK-1 kinase inhibitors, possible implications for overcoming or preventing drug resistance, lowering effective drug doses, and providing new strategies for interrogating cellular signaling pathways were all possible outcomes. This example nicely illustrates the ability of fluorescent reporter genes to aid in practical, clinically relevant experiments.

MULTIPLEXING

The key to successful multiplexing with fluorescent probes is to be able to distinguish the multiple parameters that are being studied. This could include looking at the same event in different populations of cells or different events occurring within the same cells. This can be done either through the use of fluorophores with different excitation and/or emission parameters, by individually tagging specific cells or populations of cells within a complex mixture, or by temporal or spatial differences in the signals being observed. We will look at several different versions of multiplexed assays within this review.

We can start by looking at localization assays. With these assays, specific cellular compartments or proteins are labeled with two different fluorophores. This can be done by using antibodies to specific proteins labeled with different fluorescent dyes, by using organic fluorophores known to concentrate in certain compartments (such as the MitoTracker® probes from Life Technologies), or by using fluorescent proteins to tag cellular compartments or specific proteins. Although antibodies allow for very specific labeling of a protein of interest, unless the protein is on the cell surface, live cell labeling is a challenge. With fluorescent proteins on the other hand, fusions of the protein of interest can be made with a fluorescent protein allowing the localization of that protein to be tracked in live cells. Fluorescent proteins now exist in a wide range of colors [37] enabling the differential labeling of multiple proteins or compartments without the need for antibodies. Life Technologies has commercialized fluorescent proteins that label specific cellular compartments with their series of Organelle Lights™ reagents [38]. These are baculovirus constructs which express in mammalian cell lines that deliver different colors of fluorescent proteins to various subcellular compartments to enable localization experiments.

A variation of localization type experiments comes with tracking populations of cells or cell lineages. For these types of experiments, different populations of cells are labeled with different fluorophores. The event of interest is then studied within specific populations of cells as shown by Krutzik *et al*. [39]. Reading the results of these types of multiplexed assays requires the use of a flow cytometer or an imaging device that scans for unique spectral signatures as well as for the fluorescence, luminescence or other labels that result from the assay of interest. Qdot® nanocrystals have been used for this application as they are extremely photostable for several generations and are not transferred to adjacent cells in the population. Qdots® are nanocrystals which can be excited at UV wavelengths and will emit at various wavelengths depending upon the size on the nanocrystal. Thus, different populations of cells can be labeled with different size Qdots® and different fluorescence emission wavelengths will be observed for each population while exciting with a single UV wavelength. Qtracker® products are being commercialized by Life Technologies for these types of applications.

Variations of this type of assay can also be made by attaching different biomolecules to the Qdot® nanocrystal or microsphere. The presence of different proteins or analytes within a complex mixture can then be studied. The Luminex assay is another example of this type of assay. With the Luminex assay, microspheres containing different ratios of red and infrared fluorophores are used to give the microsphere a unique spectral signature. Antibodies to proteins of interest can then be attached to the microspheres or nanocrystals, the microspheres/nanocrystals can be incubated with a sample such as a cell lysate or serum/plasma, a detection antibody can be added, and the sample can be analyzed *via* flow cytometry or a specialized imaging device.

Other types of multiplexed assays can be used with plate readers without the need for a flow cytometer or imaging device. These types of assays are generally looking at two different cellular events within the same population of cells and are dependent upon having two independent signals that have a large enough assay window to be distinguished within the population of cells. An example of this type of assay would be the ToxBLAzer™ assay from Life Technologies [40]. This assay simultaneously looks at the viability of cells using a red fluorescence marker and a GeneBLAzer® reporter gene assay that reports the results of a particular pathway of interest with a ratiometric blue and green fluorescent readout. Using the ToxBLAzer™ substrate, one can determine whether a compound activates or inhibits a receptor or pathway of interest and whether or not the compound is toxic to the cells. Another example would be the multiplexing of a membrane potential assay with a calcium mobilization assay as reported by Cassutt [41].

Another way to multiplex assays it to have the two signals of interest be separated temporally. An example of this would be the multiplexing of a Ca^{2+} mobilization assay with a reporter gene assay for GPCRs [42]. For Gq coupled GPCRs, activation of the receptor will lead to the release of calcium from intracellular stores and will also activate the NFAT transcription factor. With calcium mobilization assays, the calcium sensitive dye is loaded into the cells, and compounds are added to the cells while the fluorescence emission is simultaneously observed. An increase in the fluorescence emission spikes within one minute of compound addition but will return to baseline values within several minutes. The NFAT reporter gene on the other hand is dependent upon transcription and translation of the reporter gene and generally takes several hours. Therefore, it is possible to load cells with a calcium sensitive dye, add compounds and record the results from the calcium dye, return the cells to the incubator to allow the reporter gene to be activated, then detect the results from the reporter gene assay. In this way two different cellular assays can be performed on the same cells with the combined data having fewer false positives than either assay format individually.

CONCLUSION

Fluorescent cell-based assays have been developed for major drug targets such as kinases, GPCRs, ion channels, and cell signaling pathways. In many cases, the fluorescent assays replace time consuming radioactivity assays and have many advantages. Fluorescent cell-based assays share the attributes of being sensitive, multiplexible, selective, and high throughput.

As biological content expands, look for more genetically encoded fluorescent protein based sensors. The cameleon calcium sensor is the first commercially available sensor of its type; however sensors for cyclic AMP, IP_3, oxidation-reduction, and pH have all been described in literature. These sensors are well-suited for living cells and would be applicable to targets such as GPCRs that directly affect second messenger levels.

Surrogate ion detection *via* thallium for potassium channels is gaining popularity. The surrogate ion detection method as applied to potassium channels could also be applied to other channels. Trivedi and coworkers recently demonstrated the use of lithium as a surrogate ion for pharmacological characterization of sodium channel, Na_v 1.7 modulators [43]. The premise of this assay may form the basis for future fluorescent cell-based surrogate ion assays for sodium channels.

Finally as cell-based assays become the standard, primary cells and stem cells will be used as model systems. Immortal cell lines serve a purpose, but to move to a more disease relevant model, primary cells and stem cells will need to be employed. Techniques for the non-invasive delivery of fluorescent reporter molecules to these types of cells will likely see increased adoption. The BacMam approach is one such technology that has been heavily used in cell-based HTS screening by GlaxoSmithKline [44]. Overall, fluorescent technologies give a picture of the inner workings of cells and their important druggable targets.

ACKNOWLEDGEMENTS

The authors would like to recognize Lydia Jablonski for graphic design and Michael O'Grady for critical reading of the manuscript.

CONFLICTS OF INTEREST

George T. Hanson and Bonnie J. Hanson are employees of Life Technologies.

DISCLOSURE

This chapter is an update of our previous publication in CCHTS "Fluorescent Probes for Cellular Assays" published in *'Combinatorial Chemistry & High Throughput Screening'*, Volume 11, Issue 7, 2008, pp. 505 to 513.

REFERENCES

[1]   Weigert, F. *Vehr. d. Deutsch Phys. Ges.* **1920**, 3(*1*), 100-102.

[2]   Perrin, F. *J. Phys. Radium* **1926**, *7*, 390-401.

[3]   Manning, G.; Whyte, D.B.; Maritnez, R.; Hunter, T.; Sudarsanam, S. The protein kinase complement of the human genome. *Science*, **2002**, *298*(5600), 1912-1934.

[4]   Cohen, P. Protein kinases—the major drug targets of the twenty-first centruy? *Nat. Rev. Drug Discov.*, **2002**, *1*(4), 309-315.

[5]   Blume-Jensen, P.; Hunter, T. Oncogenic kinase signalling. *Nature*, **2001**, *411*(6835), 355-365.

[6]   Hanahan, D.; Weinberg, R.A. The hallmarks of cancer. *Cell*, **2000**, *100*(1), 57-70.

[7]   Yousefi, S.; Green D.R.; Blaser, K.; Simon, H.-U. Protein-tyrosine phosphorylation regulates apoptosis in human eosinophils and neutrophils. *Proc. Natl. Acad. Sci. USA,* **1994**, *91*(23), 10868-10872.

[8]   Noble, M.E.M.; Endicott, J.A., Johnson, L.N. Protein kinase inhibitors: insights into drug design from structure. *Science*, **2004**, *303*(5665), 1800-1805.

[9]   Vlahos, C.J.; McDowell, S.A.; Clerk, A. Kinases as therapeutic targets for heart failure. *Nat. Rev. Drug Discov.*, **2003**, *2*(2), 99-113.

[10] Futreal, P.A.; Kasprzyk, A.; Birney, E.; Mullikin, J.C.; Wooster, R.; Stratton, M.R. Cancer and genomics. *Nature*, **2001**, *409*(6822), 850-852.

[11] Chen, H.; Kovar, J.; Sissons, S.; Cox, K.; Matter, W.; Chadwell, F.; Luan, P.; Vlahos, C.J.; Scutz-Geschwender, A.; Olive, D.M. A cell-based immunocytochemical assay for monitoring kinase signaling pathways and drug efficacy. *Anal. Biochem.*, **2005**, *338*(1), 136-142..

[12] Shults, M.D.; Janes, K.A.; Lauffenburger, D.A.; Imperiali, B. A multiplexed homogeneous fluorescence-based assay for protein kinase activity in cell lysates. *Nat. Methods*, **2005**, *2*(4), 277-283.

[13] Foord, S.M.; Bonner, T.I.; Neubig, R.R.; Rosser, E.M.; Pin, J.P.; Davenport, A.P.; Spedding, M.; Harmar, A.J. International union of pharmacology. XLVI. G protein-coupled receptor list. *Pharmacol. Rev.*, **2005**, *57*(2), 279-288.

[14] Fredriksson, R.; Scioth, H.B. The repertoire of G-protein-coupled receptors in fully sequenced genomes. *Mol. Pharmacol.*, **2005**, *67*(5), 1414-1425.

[15] Jensen, A.A.; Spalding, T.A. Allosteric modulation of G-protein coupled receptors. *Eur. J. Pharm. Sci.*, **2004**, *21*(4), 407-420.

[16] Vassilatis, D.K.; Hohmann, J.G., Zeng, H.; Li, F.; Ranchalis, J.E.; Mortrud, M.T.; Brown, A.; Rodriguez, S.S.; Weller, J.R.; Wright, A.C.; Bergmann, J.E.; Gaitanaris, G.A. The G protein-coupled receptor repertoires of human and mouse. *Proc. Natl. Acad. Sci. USA*, **2003**, *100*(8), 4903-4908.

[17]   Overington, J.P.; Al-Lazikani, B.; Hopkins, A.L. How many drug targets are there ? *Nat. Rev. Drug Discov.* **2006**, *5*(12), 993-996.

[18]   Chalmers, D.T.; Behan, D.P. The use of constitutively active GPCRs in drug discovery and functional genomics. *Nat. Rev. Drug Discov.*, **2002**, *1*(8), 599-608.

[19]   Wise, A.; Gearing, K.; Rees, S. Target validation of G-protein coupled receptors. *Drug Discov. Today,* **2002**, *7*(4)*, 235-246.

[20]   Hill, S.J. G-protein-coupled receptors: past, present and future. *Br. J. Pharmacol.*, **2006**, *147 Suppl 1*, S27-37.

[21]   Tsien, R. New calcium indicators and buffers with high selectivity against magnesium and protons: design, synthesis, and properties of prototype structures. *Biochemistry* **1980**, *19*(11)*, 2396-2404.

[22]   Minta, A.; Kao, J.P.; Tsien, R.Y. Fluorescent indicators for cytosolic calcium based on rhodamine and fluorescein chromophores. *J. Biol. Chem.*, **1989**, *264*(14), 8171-8178.

[23]   Miyawaki, A.; Llopis, J.; Heim, R.; McCaffery, J.M.; Adams, J.A.; Ikura, M.; Tsien, R.Y. Fluorescent indicators for Ca2+ based on green fluorescent proteins and calmodulin. *Nature*, **1997**, *388*(6645), 882-887.

[24]   Nagai, T.; Yamada, S.; Tominaga, T.; Ichikawa, M.; Miyawaki, A. Expanded dynamic range of fluorescent indicators for Ca(2+) by circularly permuted yellow fluorescent proteins. *Proc. Natl. Acad. Sci. USA*, **2004**, *101*(29), 10554-10559.

[25]   Weber, M.; Ferrer, M.; Zheng, W.; Inglese, J.; Strolovici, B.; Kunapuli, P. A 1536-well cAMP assay for Gs- and Gi-coupled receptors using enzyme fragmentation complementation. *Assay & Drug Dev. Tech.* **2004**, *2*(1), 39-49.

[26]   Gabriel, D.; Vernier, M.; Pfeifer, M.J.; Dasen, B.; Tenaillon, L.; Bouhelal, R. High throughput screening technologies for direct cyclic AMP measurement. *Assay & Drug Dev. Tech.*, **2003**, *1*(2), 291-303.

[27]   Williams, C. cAMP detection methods in HTS: selecting the best from the rest. *Nat. Rev. Drug Discov.*, **2004**, *3*(2), 125-135.

[28]   Kunapuli, P.; Lee, S.; Zheng, W.; Alberts, M.; Kornienko, O.; Mull, R.; Kreamer, A.; Hwang, J.I.; Simon, M.I.; Strulovici, B. Identification of small molecule antagonists of the human mas-related gene-X1 receptor. *Anal. Biochem.*, **2006**, *351*(1)*, 50-61.

[29]   Oosterom, J.; van Doornmalen, E.J.; Lobregt, S.; Blmenrohr, M.; Zaman, G.J. High-throughput screening using beta-lactamase reporter-gene technology for identification of low-molecular-weight antagonists of the human gonadotropin releasing hormone receptor. *Assay Drug Dev. Technol.,* **2005**, *3*(2)*, 143-154.

[30] Barnea, G.; Strapps, W.; Herrada, G.; Berman, Y.; Ong, J .; Kloss, B.; Axel, R.; Lee, K.J. The genetic design of signaling cascades to record receptor activation. *Proc. Natl. Acad. Sci. USA.,* **2008**, *105*(1), 64-69.

[31] Sanguinetti, M.C.; Jiang, C.; Curran, M.E.; Keating, M.T. A mechanistic link between an inherited and an acquired cardiac arrhythmia: HERG encodes the IKr potassium channel. *Cell* **1995**, *81*(2), 299-307.

[32]   Weaver, C.D.; Harden, D.; Dworetzky, S.I.; Robertson, B.; Know, R.J. A thallium-sensitive, fluorescence-based assay for detecting and characterizing potassium channel modulators in mammalian cells. *J. Biomol. Screen.* **2004**, *9*(8), 671-677.

[33]   Thornberry, N.A.; Rano, T.A.; Peterson, E.P.; Rasperi, D.M.; Timkey, T.; Garcia-Calvo, M.; Houtzageri, V.M.; Nordstromi, P.A.; Royi, S.; Vaillancourti, J.P.; Chapman, K.T.; Nicholsoni, D.W. A combinatorial approach defines specificities of members of the caspase

family and granzyme B. Functional relationships established for key mediators of apoptosis. *J. Biol. Chem.* **1997**, *272*(29), 17907-17911.

[34] Lövborg, H.; Nygren, P.; Larsson, R. Multiparametric evaluation of apoptosis: effects of standard cytotoxic agents and the cyanoguanidine CHS828. *Mol Cancer Ther.* **2004**, *3*(5), 521-526.

[35] Eglen, R.M. Enzyme fragment complementation: a flexible high throughput screening assay technology. *Assay and Drug Dev. Technol.* **2002**, *1*(1Pt1), 97-104.

[36] O'Grady, M.; Raha, D.; Hanson, B.J.; Bunting, M.; Hanson, G.T. Combining RNA interference and kinase inhibitors against cell signaling components involved in cancer. *BMC Cancer*, **2005**, *5*, 125-134.

[37] Shaner, N.C.; Steinbach, P.A.; Tsien, R.Y. A guide to choosing fluorescent proteins. *Nature Methods* **2005**, *2*(12), 905-909.

[38] Ames, R.S.; Kost, T.A.; Condreay, J.P. BacMam technology and its application to drug discovery. *Exp Opin Drug Discovery* **2007**, *2*(12), 1669-1681.

[39] Krutzik, P.O.; Nolan, G.P. Fluorescent cell barcoding in flow cytometry allows high-throughput drug screening and signaling profiling. *Nat. Methods* **2006**, *3*(5), 361-368..

[40] Hallis, T.M.; Kopp, A.L.; Gibson, J.; Lebakken, C.S.; Hancock, M.; Vandenheuvel-Kramer, K.; Turek-Etienne, T. An improved beta-lactamase reporter assay: multiplexing with a cytotoxicity readout for enhanced accuracy of hit identification. *J. Biomol. Screen.* **2007**, *12*(5), 635-644.

[41] Cassutt, K.J. Multiplexing calcium mobilization and membrane potential assays. (Cell based assays supplement). *Bioscience Technology.* Advantage Business Media **2006**, *HighBeam Research.*

[42] Hanson, B.J. Multiplexing Fluo-4 NW and a GeneBLAzer transcriptional assay for high-throughput screening of G-protein-coupled receptors. *J. Biomol. Screen.* **2006**, *11*(6), 644-651.

[43] Trivedi, S.; Dekermendjian, K.; Julien, R.; Huang, J.; Lund, P.-E.; Krupp, J.; Kronqvist, R.; Larsson, O.; Bostwick, R. Cellular HTS assays for pharmacological characterization of Na(V)1.7 modulators. *Assay and Drug Dev. Technol.* **2008**, *6*(2), 167-179.

[44] Kost, T.A.; Condreay, J.P.; Jarvis, D.L. Baculovirus as versatile vectors for protein expression in insect and mammalian cells. *Nat. Biotech.* **2005**, *23*(5), 567-575.

Send Orders for Reprints to reprints@benthamscience.net

CHAPTER 4

Label-Free Cell Phenotypic GPCR Drug Discovery

Ye Fang[*]

Biochemical Technologies, Science and Technology Division, Corning Incorporated, Sullivan Park, Corning, NY 14831, USA

Abstract: G protein-coupled receptors (GPCRs) have been proven to be the largest family of druggable targets in the human genome. Given the importance of GPCRs as drug targets and the de-orphanization of novel targets, GPCRs are likely to remain the frequent targets of many drug discovery programs. Traditionally, molecular assays dominate in the process of GPCR drug discovery and development. With recent advances in instrumentation and pathway deconvolution of GPCR ligand-induced biosensor signatures, label-free biosensor-enabled cell-based assays have become a very active area for GPCR screening. This article reviews the principles, current status and future directions of leading label-free technology platforms for GPCR drug discovery.

Keywords: Dynamic mass redistribution, electrical biosensor, G protein-coupled receptor, high throughput screening, impedance, ligand-directed functional selectivity, optical biosensor, resonant waveguide grating biosensor.

G-protein-coupled receptors (GPCRs) are the largest family of cell surface receptors that share common structural motif - seven α-helical transmembrane-spanning domains joined by intra- and extracellular loops [1]. GPCRs are expressed in virtually all tissues, with distinct expression patterns in different cell systems [2]. The extracellular ligands for GPCRs are diverse, including biogenic amines, amino acids, ions, small peptides, proteins, and bioactive lipids [3]. This diversity of GPCR activators underscores the physiological importance of this receptor class - GPCRs control a wide variety of physiological processes, such as neurotransmission, chemotaxis, inflammation, and cell proliferation. It is no surprise that GPCRs have been implicated in almost every major disease class, including asthma, cancer, and inflammatory and cardiovascular diseases [4-6].

*Address correspondence to Ye Fang: Biochemical Technologies, Science and Technology Division, Corning Incorporated, Sullivan Park, Corning, NY 14831, USA; Tel: 607-9747203; Fax: 607-9745957; E-mail: fangy2@corning.com

Given their importance in health and disease together with their ability for therapeutic intervention by small molecule drugs, GPCRs represent the largest and most successful class of druggable targets in the human genome [7-10]. It is estimated that the human genome encodes as many as 1,000 GPCRs [11,12], of which ~360 are non-chemosensory receptors, predicted to bind endogenous ligands [13]. More than 100 of these receptors are classified as orphan GPCRs for which their cognate ligands or biological functions are unknown [14]. De-orphanization of these receptors could bring novel therapeutic targets and opportunities to the industry [15,16], in light of the enormous therapeutic success of this class of drug targets. The future potential becomes even much clearer when one considers that approximately 50% of all clinically available drug products are active, directly or indirectly, on this family of receptors [7], and ~27% of the 1357 unique drugs in the market before 2006 are GPCR drugs [10]. Yet, these drugs are active only on small percentages of all known GPCRs, which was initially estimated to be ~40 GPCRs [9] and now expanded to as many as 82 unique GPCRs [17].

Continued success in GPCR drug discovery and development has seen an evolution in pharmacological assays. These assays can be classified to phenotypic assays and target-based molecular assays [18]. Molecular assays can be performed in cell-free, living cells, tissues and organs, but are common in that they rely on the measurement of specific molecule(s) in the receptor signaling cascades to infer the molecular mode of action (MMOA) of drug molecules. Originating from the bias towards a specific MMOA, molecular assays are amenable to both screening and medicinal chemistry optimization, but the pharmacological profiles of drugs obtained often have limited predictive power for their therapeutic impacts [19]. In contrast, phenotypic assays directly measure the activity of a drug molecule linked to a specific physiological or cellular phenotype in physiologically relevant cells or cell systems, thus leading to a pharmacological activity of the drug that might be linked to its therapeutic impact for a given disease state [20]. However, the physiological or cellular phenotypes measured can be too complex to derive mechanistic descriptions of drug actions, thus making the phenotypic assays impracticable to optimize drug molecules [18]. Interestingly, in the era of target-centric drug discovery, first-in-class new medicines are discovered more by phenotypic screening than by target-based molecular assays. Swinney and Anthony recently found that among 259 new molecular entities and new biologics approved by the US Food and Drug Administration between 1999 and 2008, 75 were first-in-class drugs with new MMOAs, 164 were follower drugs and 20 were imaging agents [18]. Further analysis of 50 first-in-class small-molecule drugs showed that 28 were discovered by phenotypic screens, 17 by target-based approaches, and 5 as synthetic versions of natural substances.

Cell-based assays dominate GPCR drug discovery. This is partly due to the difficulty and high cost to obtain high quality proteins for ligand-binding screening, and partly due to the affinity and functional consequence of ligands binding to a GPCR that are dependent on the appropriate interaction of the receptor with its interacting proteins [21]. Furthermore, cell-based assays are amenable to high throughput screening (HTS), and offer far more useful information, such as the action, mode and mechanisms of compounds under conditions more closely resembling the physiological environment, compared to ligand-binding assays. In large measure, these benefits have helped drive the increasing use of whole cell systems for drug screening and testing in the last decade [20,22,23]. In the recent years, label-free cell-based assays have been emerging a powerful platform for GPCR drug discovery, due to their phenotypic measurement in nature, yet permitting molecular mechanistic deconvolution [24,25]. Given the availability of various comprehensive reviews of assay technologies for GPCR screening [26-34], this article first reviews recent progresses in GPCR biology, and is then focused primarily on recent advances in label-free cell-based assays that make use of biosensor technologies for GPCR screening.

GPCR SIGNALING

GPCRs participate in a wide array of cell signaling pathways, primarily mediated through their coupled G proteins [35,36]. GPCR signaling is encoded by the spatial and temporal flux of downstream signaling networks, which are tightly controlled by intracellular signaling and regulatory machineries [37]. The consensus model concerning GPCR signaling assumes a receptor as a functional monomeric entity interacting through its specific intracellular domains with a single G protein, once stabilized in its active conformation(s) by agonist binding. The agonist binding results in changes in the conformation of the receptor [38,39]. The receptor activation in turn leads to the activation of an associated G protein heterotrimer through the GTP-GDP exchange on G_α subunit [40]. The activated G protein then modulates the activity of several intracellular enzymes, which in turn control the production of several key intracellular second messengers such as cyclic AMP (cAMP), cGMP, Ca^{2+}, inositol triphosphate, and arachidonic acid. These second messengers then act on several downstream targets including ion channel and kinases that regulate gene transcription and cell functions [41]. For example, G_s-coupled receptor signaling proceeds through sequential activation of the receptor, G protein, and adenylyl cyclase (AC) at the plasma membrane, and increased accumulation of a diffusible second messenger cAMP, and activation of cAMP-activated protein kinase (protein kinase A, PKA). This pathway governs multiple cellular machines, including ion channels, transcription factors, cytoskeletal proteins, and metabolic enzymes. Furthermore, common to almost all GPCRs is rapid attenuation of the receptor responsiveness upon agonist stimulation (known as desensitization), followed by receptor resensitization after removal of agonist [42,43].

With the adoption of label-free phenotypic assays and the improvement in resolution and sensitivity of molecular assays, novel signaling pathways downstream the activation of GPCRs have been identified. Recently Verrier *et al.* have discovered using both label-free dynamic mass redistribution (DMR) assay and fluorescent imaging that the activation of G_i-coupled receptors including α_{2A}-adrenergic receptor in HeLa cells promotes the assembly of a multienzyme complex involved in regulating purine *de novo* biosynthesis [44]. Purine *de novo* biosynthesis, among other vital functions, provides important DNA building blocks. Mitogenic GPCR signaling has been known for many years, but is believed to mediate through the direct regulation of the nuclear expression of growth-promoting genes [45]. In another recent study, Lefkowitz and his colleagues have used the infusion of an adrenaline-like compounds for four weeks in the mice, as a model to mimic a hallmark of chronic stress, elevated adrenaline. Results suggest a specific molecular mechanism through which β-adrenergic catecholamines activate β_2-adrenoreceptors (β_2-AR) in body, acting through both G_s-PKA and β-arrestin-mediated signaling pathways to trigger DNA damage and suppress p53 levels, respectively, both of which, in turn, synergistically result in the accumulation of DNA damage [46].

GPCR OLIGOMERIZATION

The concept of GPCR oligomerization was proposed more than two decades ago, and it has been implicated in the regulation of GPCR physiological function. A growing body of evidence suggests the prevalence of GPCR dimers (or higher order oligomers), which can form among identical or different receptors [47-50]. It has been reported that oligomerization of some receptors is modulated by agonists [51,52]. However, many studies suggest that GPCR oligomerization is constitutive [53]. For example, some receptors may be synthesized as an oligomeric unit [54]; and oligomerization of some receptors is required for their appropriate cell surface expression [55]. Biophysical [56], bioinformatics [57] and evolutionary trace analysis [58] suggests that interactions between receptors in a dimeric assembly likely involve conformational changes at the dimer interface. Recent X-ray structure studies show that μ-opioid receptor crystallizes as a two-fold symmetrical dimer through a four-helix bundle motif formed by transmembrane segments V and VI [59], while human κ-opioid receptor also crystallizes as a parallel dimer with contacts involving helices I, II and VIII [60]. The receptor oligomerization has been shown to have effects on ligand binding, receptor activation, desensitization and trafficking, as well as receptor signaling [47-50]. However, much remains to be elucidated about the formation, regulation, and functional and physiological consequences of receptor oligomerization, particularly *in vivo* cell systems [61,62]. Nonetheless, the possibility of receptor oligomerization in the plasma membrane greatly diversifies their pharmacological and physiological properties, and has implications for pharmacological interventions.

Ligand-Directed Functional Selectivity

The past years have witnessed the solidification of the concept of ligand-directed functional selectivity or biased agonism, describing that many ligands display the ability to differentially activate some of the vectorial pathways over others mediated through a single receptor [63]. GPCRs are a family of shapeshifting and versatile signaling proteins, and known to mediate pleiotropic signaling through interacting with multiple signaling proteins [64]. Considering that a receptor may preferentially couple to different signaling proteins in different types of cells and tissues [65], biased agonism may be linked to the complex therapeutic profiles of drug molecules [66]. Experimentally, biased agonism is mostly observed in recombinant or heterologous systems using a variety of molecular assays, and is often inferred from the relative potency and efficacy to modify one signaling molecule/event over another [67]. Consequently, many ligands may display assay readout-dependent potency and efficacy, suggesting that the efficacy for many ligands is collateral [68]. The biased agonism may be due to the presence of a library of active receptor states, some of which could have their own signaling preference. This was postulated by protein ensemble theory [69,70], and has been experimentally confirmed by many studies [25]. Biased agonism was once thought to be an artifact of recombinant technology. In the recent years, label-free cellular assays have now amply demonstrated that biased agonism is a natural cellular control mechanism [71-73].

LABEL-DEPENDENT CELL-BASED ASSAYS

An important aspect of GPCR signaling is that it consists of a series of spatial and temporal events. Each discrete cellular event exhibits its own characteristics in terms of kinetics, dynamics, amplitude and location, depending on the cellular context of the cell system studied. For example, the kinetics of cellular events mediated through receptor activation greatly differs, ranging from milliseconds (*e.g.*, GPCR conformational changes) to tens of seconds (*e.g.*, Ca^{2+} flux), minutes (*e.g.*, cytoskeletal modulation, morphological changes), or hours (*e.g.*, change in gene transcription) [74]. A ligand-induced cellular event could also have distinct spatial and temporal dynamics (*e.g.*, cycling or oscillation) [75, 76]. In the context of assays, virtually every single GPCR signaling event has been utilized as the basis of various assay technologies. Since most of these cell-based assays require a certain degree of engineering, manipulation, or labeling, these assays are referred to as label-dependent cell-based assays.

Assays to Measure Changes in Second Messenger

GPCRs transmit signals mainly through their coupled G proteins, leading to changes in intracellular levels of second messengers such as cAMP, inositol triphosphate, and Ca^{2+}. Direct measurement of intracellular levels of these messengers upon stimulation is a popular functional assay format for GPCR screening in mammalian cells. The activation of $G_{\alpha q}$ or $G_{\alpha i}$ coupled-receptors promotes phospholipase C to hydrolyze phosphatidylinositol bisphosphate to form two second messengers, inositol 1,4,5-triphosphate (IP_3) and diacyl glycerol (DAG). DAG then activates protein kinase C, while IP_3 activates the IP_3 receptor on the endoplasmic reticulum, resulting in Ca^{2+} mobilization from the endoplasmic reticulum to the cytoplasm and consequently a transit elevation of intracellular Ca^{2+}. IP_3 is quickly hydrolyzed to IP_2, then to IP_1 and finally to inositol by a series of enzymatic reactions. For Ca^{2+} mobilization, there are highly sensitive calcium flux assays including FLIPR (fluorometric imaging plate reader) and Aequorin assays for receptors naturally coupled to G_q proteins or artificially coupled to a chimeric G_{qo5} or G_{qi5}, or a promiscuous $G_{15/16}$ proteins [77]. For IP_3 generation and hydrolysis, there are traditionally radiometric scintillation proximity assay (SPA) (GE Healthcare), and newer non-radiometric assays including AlphaScreen™ (PerkinElmer), HitHunter™ fluorescent polarization (DiscoveRx), and IP-One HTRF™ (Cisbio).

Today, most cAMP assays measure global changes in the concentration of intracellular cAMP in the presence of a broad spectrum phosphodiesterase inhibitor without and with the adenylyl cyclase activator forskolin for G_s- and G_i-coupled receptors, respectively. In these assays, changes in intracellular cAMP are generally detected by the competition between cellular cAMP and a labeled form of cAMP for binding to an anti-cAMP antibody [30]. A wide variety of assays are commercially available; these include SPA (GE Healthcare), FlashPlate™ and AlphaScreen™ amplified luminescent proximity homogeneous assay (PerkinElmer), HitHunter™ enzyme fragment complementation (DiscoveRx), and HTRF™ homogeneous time resolved fluorescence (Cisbio) cAMP assays. A gene reporter cAMP assay based on D-luciferin, Glosensor™ available from Promega, provides a non-lytic live-cell-based and facile kinetic measurement of intracellular cAMP dynamics [34]. Alternatively, a modified rat olfactory cyclic nucleotide gated (CNG) channel, which is engineered such that it enhances the cAMP binding affinity but reduces the cGMP binding affinity, is used as a cAMP biosensor for G_s and G_i GPCR screening [78] (BD ACTOne, BD Biosciences, Rockville, MD). Amassing evidence now suggests that instead of uniform alteration in cytosolic concentration, cAMP undergoes changes with unique spatial and temporal gradients upon the receptor activation [76]. The spatial and temporal compartmentalization of cAMP (and its target proteins) is central to the ability of this second messenger to govern cellular activity over timescales ranging from milliseconds to several hours. In the recent years, a variety of cAMP biosensors have been developed to directly monitor rapid subcellular cAMP dynamics. However, these single cell assays have not been fully developed for HTS.

Assays to Measure Protein-Protein Interaction

The recent realization that GPCRs potentially function as homo-oligomeric and hetero-oligomeric complexes [47-62] has given rise to rationale for developing assays based on protein-protein interactions. These assays are useful not only for elucidating the nature of receptor interaction within the oligomeric complexes, but also for monitoring ligand-induced reorientation of an existing oligomer, particularly relative to its bound G protein subunits. These assays have also been used to study distinct protein interactions during GPCR signaling cycle, including the interactions of ligand-activated receptor with its coupled G proteins, or other regulatory proteins such as G protein-coupled receptor kinases (GRKs) and β-arrestins.

These assays typically utilize either resonance energy transfer (RET) [62] or protein fragment complementation [79,80]. The RET is a non-radiative transfer of energy between a donor and an acceptor when they are in proximity, with an efficiency that varies inversely with the sixth power of the distance between the two molecules. The donor moiety can either be a fluorophore (FRET) or a bioluminescent enzyme that emits light upon oxidation of a substrate (BRET). In both cases, the acceptor is a fluorophore. The changes in RET in response to stimuli offers an effective means for real-time monitoring of dynamic protein-protein interactions that are involved in GPCR activation and regulation in cells.

Alternatively, protein fragment complementation assays utilize a pair of recombinantly engineered and inactive fragments of either an enzyme (e.g, β-galactosidase [79]) or a fluorescent protein (*e.g.*, green fluorescent protein (GFP) [80]), each being recombinantly attached to a protein. When the receptor activation leads to the interaction between the two proteins having the complementary fragments, this interaction drives the functional complementation of the protein mutant fragments. GPCR activation is measured directly by quantifying the restored activity of the enzyme or fluorescence of the fluorescent protein. Since protein fragment complementation is an amplified signal detection system, exquisite detection sensitivity is achieved, permitting analysis of protein interactions at concentrations approximating their normal levels of expression. Although protein interaction-driven functional complementation of 2 split fragments is reversible, the rate of association of the two yellow fluorescent protein (YFP) fragments fused to Fos and Jun has been reported to be much slower ($t_{1/2}$ of ~60sec) than the subunit exchange rate of Fos-Jun heterodimers ($t_{1/2}$ of ~10sec) [80]. Such a slow kinetics is mainly due to relatively weak affinity between the 2 split YFP fragments, and is typically slower than the reorientation of emitter/acceptor fluorescent proteins. Thus, cautious is warranted to interpret the kinetics of protein-protein interactions measured using protein fragment complementation assays. Recently, Javitch and his colleagues had combined protein complementation with resonance energy transfer to study conformational changes in response to activation of the putative dopamine D_1-D_2 receptor heteromer [81]. They found that the potency of the D_2 receptor agonist R-(-)-10,11-dihydroxy-N-n-propylnoraporphine to change the $G_{\alpha i}$ conformation *via* the D_2 protomer in the heteromer was enhanced ten-fold relative to its potency in the D_2 homomer, suggesting that GPCR dimerization can influence drug pharmacology.

Assays to Measure Protein Trafficking

GPCRs modulate diverse physiological signaling pathways by virtue of changes in receptor activation and inactivation states. Functional changes in receptor state lead to dynamic trafficking of the receptor [82] and many of its downstream signaling molecules (*e.g.*, arrestins [83], protein kinase C [75]) at various stages of GPCR signaling cycle. Direct visualization of protein trafficking provides a basis for high-content screening (HCS), which often provides a measure of receptor-ligand interactions in multiple dimensions with spatial and temporal dynamics [27]. Receptor internalization and β-arrestin recruitment assays are two of the most common protein trafficking assays. Receptor internalization is an essential component of receptor desensitization process, wherein GRKs phosphorylate agonist activated receptors, which, in turn, recruit cytosolic β-arrestins to the cell membrane. The β-arrestins then uncouple the activated receptors from their cognate G proteins and target the receptors to clathrin-coated pits for endocytosis. A wide array of techniques has been adopted for these protein trafficking assays [79,80,83].

These protein trafficking assays require no prior knowledge of the interacting G protein(s), enabling screening orphan GPCRs. Furthermore, these assays are often generic to all classes of GPCRs, thus overcoming the limitations associated with typical G protein-dependent functional assays. However, these trafficking assays suffer drawbacks associated with the fact that some agonists, particularly partial agonists, could stimulate receptor signaling without causing the trafficking event(s) seen when the receptor is stimulated by its native ligand [63,64]. For example, morphine, a partial agonist, exhibits relatively poor activity to cause internalization of the μ-opioid receptor [84].

Dissecting the biased agonism of GPCR ligands has led to a hypothesis that G protein and β-arrestin pathways are distinct and could be pharmacologically modulated independently; and both receptor internalization and β-arrestin recruitment assays are considered to be G protein-independent functional assays [34,63,66]. However, in a recent elegant study, Sauliere *et al.* used a set of new, highly sensitive and direct BRET-based G protein activation probes specific for all G protein isoforms to evaluate the G protein-coupling activity of $[^{1}Sar^{4}Ile^{8}Ile]$-angiotensin II (SII) [85]. SII has been described as a G protein-independent β-arrestin biased ligand for angiotensin AT_{1A} receptor. Results showed that SII promotes G_q and G_i protein-dependent signaling in both heterologous and primary cells, but stabilizes a specific AT_{1A} receptor conformation distinct from that of Ang II and triggers a different intracellular signaling pattern. The most notable finding is that the β-arrestin biased agonism of SII is not truly G protein independent, as evidenced by the partial G_q protein dependence on the β-arrestin 2 recruitment to AT_{1A} receptor upon SII stimulation in HEK293T cells, as well as a G_i- and G_q-dependent β-arrestin-mediated ERK1/2 activation in MEF cells. The observed G protein-dependent β-arrestin activation as a specific signature of SII may put G protein pathways back in the spotlight.

Assays to Measure Changes in Gene Reporter Activity

GPCR activation is well known to alter gene transcription. Several genes contain elements responsive to second messengers upon the activation of GPCRs; these elements include the cAMP response element (CRE), the nuclear factor of activated T-cells response element (NFAT-RE), the serum response element (SRE) and the serum response factor response element (SRF-RE), thus providing the basis for developing reporter gene assays [86]. Reporter gene constructs usually contain second messenger responsive elements upstream of a minimal promoter, which in turn regulate the expression of the reporter protein. Modulation of intracellular cAMP by G_s or G_i GPCRs is detected with a reporter gene, whose transcription is regulated by the transcription factor cAMP response-element binding protein (CREB) binding to upstream CREs. Calcium flux, mobilized by G_q-coupled receptors, is detected using a reporter gene, whose transcription is regulated by the calcium-sensitive AP1 (activator protein 1) or NFAT (nuclear factor of activated T cells) elements. Commonly used reporter genes include alkaline phosphatase, β-galactosidase, GFP, luciferase, and β-lactamase, all of which can lead to detectable colorimetric, fluorescent or luminescent signals. Different gene reporters exhibit distinct sensitivity and dynamic range. Owing to the high sensitivity, broad dynamic range and amenability to HTS, luciferase has become the most popular gene reporter. However, beside the requirement for long incubation periods and the gene expression signal quite distal from the receptor activation, these reporter assays also have disadvantages associated with the measurements far distinct from native systems due to the use of highly artificial systems. The substrates used may also introduce interference. Recently, D-luciferin, the most widely used luciferase substrate, has been found to pose partial agonist activity at the GPR35 [87].

Assays to Measure Changes in Phenotype

Receptor activation ultimately leads to changes in cell phenotype. One example is the melanophore technology (Arena Pharmaceuticals, San Diego, CA), which uses GPCR targets expressed in frog skin cells containing a pigment that is highly sensitive to changing levels of cAMP [3, 26]. The pigment disperses throughout the cell and causes the cell to appear black when the intracellular cAMP level increases; conversely, when the cAMP level decreases, the pigment aggregates at the center of the cell, causing the cell to appear clear. In a recent study, Marshall and his colleagues have used flagella of the green alga *Chlamydomonas reinhardtii* as a model to screen chemicals that may regulate signaling pathways critical for ciliogenesis and length regulation [88]. Based on the flagellar length, motility, and cell viability, they found that the most frequent target found to be involved in flagellar length regulation was the family of dopamine binding GPCRs, and cilium length was altered with expression of the dopamine D_1 receptor in mammalian cells.

LABEL-FREE CELL-BASED ASSAYS

Label-free cell-based assays generally employ a biosensor to monitor ligand-induced responses in living cells. A biosensor typically utilizes a transducer such as an optical, electrical, calorimetric, acoustic, or magnetic transducer, to convert a molecular recognition event or a ligand-induced change in living cells into a quantifiable signal. These label-free biosensors are commonly used for molecular interaction analysis, which involves characterizing how molecular complexes form and disassociate over time. Many comprehensive reviews are available for different aspects of these biosensors for bio-molecular interaction analysis [89-91]. Thus, this section only highlights the applications of these biosensors for whole cell sensing, particularly for functional GPCR screening. Fig. (**1**) highlight two types of biosensors that are currently used as the basis for label-free cell-based assays - resonant waveguide grating (RWG) biosensors [92] and electrical biosensors [93-95].

RWG Biosensors and Systems

An RWG biosensor consists of a substrate (*e.g.*, glass), a waveguide thin film with an embedded grating structure, and a cell layer (Fig. **1a**). The RWG biosensor utilizes the resonant coupling of light into a waveguide by means of a diffraction grating, leading to total internal reflection at the solution-surface interface, which in turn creates an electromagnetic field at the interface. This electromagnetic field is evanescent in nature [96], meaning that it decays exponentially from the sensor

a

b

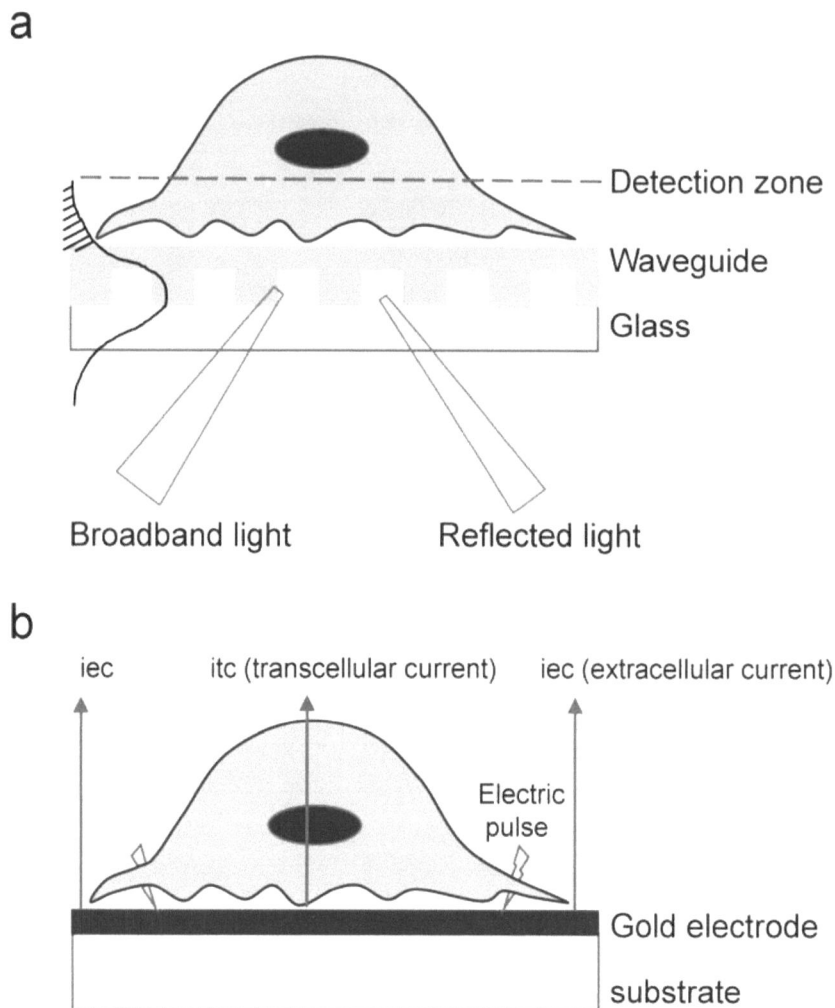

Fig. (1). Principles of two types of biosensors for living cell sensing. (**a**) A RWG biosensor for monitoring ligand-induced dynamic mass redistribution in living cells. Cells are directly cultured onto the surface of a biosensor. The biosensor consists of a glass substrate, a waveguide thin film within which a grating structure is embedded. Only the mass redistribution within the bottom portion of cells is directly measured. (**b**) An electric biosensor for monitoring the ionic environment surrounding the biosensor and the cells. Cells are cultured on the surface of a biosensor having arrayed gold microelectrodes. Both flows of extracellular (iec) and transcelular (itc) current are measured, and a low AC voltage at variable frequencies is applied to the cell layer.

surface; the distance at which it decays to $1/e$ (e is a numerical constant that is equal to 2.71828) of its initial value is known as the penetration [97] and is a function of the design of a particular RWG biosensor, but is typically on the order of 150 to 200nm [98]. Similar to biomolecular interaction analysis, RWG biosensor also exploits such an evanescent wave to characterize ligand-induced alterations of living cells at or near the sensor surface [99].

RWG instruments can be subdivided into systems based on angle- or wavelength-interrogation measurements. In a wavelength interrogation instruments, polarized light covering a range of incident wavelengths with a constant angle is used to illuminate the waveguide; light at specific wavelengths is coupled into and propagates along the waveguide [98]. Alternatively, in angle-interrogation instruments, the sensor is illuminated with monochromatic light and the angle at which light is resonantly coupled is measured [99]. In the recently developed swept wavelength interrogation RWG imagers, a light beam from a swept tunable light source is used to illuminate a partial or whole 384well biosensor microplate, resulting in distinct spatial resolution down to ~10 µm so single cell can be resolved [100,101]. A high speed digital camera is used to record the reflected resonant lights. During a single cycle of wavelength sweep from 825 to 840 nm total 150 spectral images are acquired every 3 sec, and the spectral image stack are then processed to generate signals. Since the wavelength sweeping is stepwise in 100 pm every 20 ms, this imaging system enables a time resolved resonance, leading to improved spatial resolution. However, for RWG biosensors the resonance conditions are influenced by the physical properties of the cell layer that contacts with the surface of the biosensor (*e.g.*, cell confluency, adhesion and status such as proliferating or quiescent states). When a ligand or an analyte interacts with a cellular target (*e.g.*, a GPCR, a kinase) in living cells, any change in local refractive index within the cell layer can be detected as a shift in resonant angle (or wavelength) [98,99].

The Epic® system (Corning Inc., Corning, NY) is a wavelength interrogation reader system tailored for RWG biosensors in microplate for both label-free biochemical and cell-based assays. This system consists of a RWG plate reader and Society for Biomolecular Screening (SBS) standard microtiter plates (Fig. **2a**) [102]. The detector system in the plate reader exploits integrated fibre optics to measure the shift in wavelength of the incident light, as a result of ligand-induced changes in cells. A series of illumination/detection heads are arranged in a linear fashion, so that reflection spectra are collected simultaneously from each well within a column of a 384-well microplate, leading to a temporal resolution of ~15 seconds. The whole plate is scanned so that each sensor can be addressed multiple times, and each column is addressed in sequence. The wavelengths of the incident light are collected and used for analysis. A temperature-controlling unit is built in the instrument to minimize spurious shifts in the incident wavelength due to the temperature fluctuations. The EnSpire system from PerkinElmer is a multimodal plate reader containing a label-free RWG module from Corning, which is based on a single light head to scan the whole plate, resulting in relatively low temporal resolution (~85 sec). The whole plate RWG imager from Corning can simultaneously image the whole plate with a temporal resolution of 3 seconds and a spatial resolution of 80 to 100 μm [100]. Both Epic and EnSpire are mostly operated at room temperature, whereas the imager can be placed inside standard cell culture incubators so assays can be performed under physiological conditions [101].

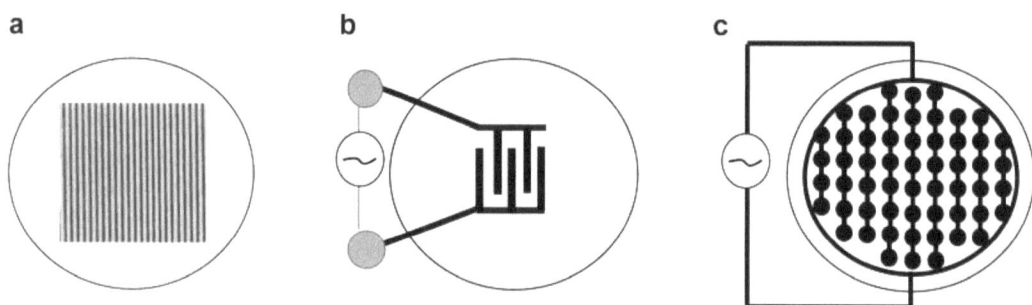

Fig. (2). Schematic drawings showing the surface configurations of three types of biosensor-embedded microplates used for label-free cell-based assays. (**a**) A well of a 384-well Epic® RWG biosensor microplate. The inset shows a scanning electronic microscopic graph of a portion of the biosensor. (**b**) A well of a 96-well CellKey gold electrode-embedded microplate. (**c**) A well of a 96-well RT-CES gold electrode-embedded microplate.

Electrical Biosensors and Systems

Electrical biosensors consist of a substrate (*e.g.*, plastic), an electrode, and a cell layer (Fig. **1b**). In this electrical detection method, cells are cultured on the substrate having an array of small gold electrodes. A small constant voltage at a fixed frequency or varied frequencies is applied to the electrode or electrode array, and the electrical current through the circuit is monitored over time. The ligand-induced change in electrical current provides a measure of cell response.

The first generation impedance systems suffer from high assay variability due to use of a small detection electrode and a large reference electrode [103]. To minimize this variability, the latest generation of systems, such as CellKey system (MDS Sciex, South San Francisco, CA) and RT-CES (ACEA Biosciences Inc., San Diego, CA), utilize an integrated circuit having a microelectrode array.

The CellKey system consists of an environmentally controlled impedance measurement system, a 96-well electrode-embedded microtiter plate, an onboard 96-well fluidics, and custom acquisition and analysis software [93]. The cells are seeded in the culture wells; each well has an integrated electrode array (Fig. **2b**). The system operates using a small-amplitude alternating voltage at 24 frequencies, from 1 KHz to 10 MHz. The resultant current is measured at an update rate of 2 sec. The system is thermally regulated and experiments can be conducted between 28°C and 37°C. A 96-well head fluid delivery device handles fluid additions and exchanges onboard.

The RT-CES system is composed of four main components: electronic microtiter plates (E-Plate™), E-Plate station, electronic analyzer, and a monitoring system for data acquisition and display [94]. The electronic analyser sends and receives the electronic signals. The E-Plate station is placed inside a tissue culture incubator. The E-Plate station comes in three throughput varieties: a 16x station for running six 16-well E-Plates at a time, a single 96-well E-Plate station, and the Mult-E-Plate™ station, which can accommodate up to six 96-well E-Plates at a time. The cells are seeded in E-Plates, which are integrated with microelectronic sensor arrays (Fig. **2c**). The system operates at a low-voltage (less than 20 mV) AC signal at multiple frequencies.

Optical Signals of GPCR Activation with RWG Biosensor

Cells are dynamic objects with relatively large dimensions - typically tens of microns. RWG biosensors enable detection of ligand-induced changes within the bottom portion of cells, determined by the penetration depth of the evanescent wave [99]. Furthermore, the spatial resolution of RWG biosensors is dependent on illumination scheme of the detection systems used. Generally, a confluent layer of cells is used in order to achieve optimal assay results, partly due to the requirement for effectively separate receptor signaling-mediated responses from the background, and partly due to the fact that current biosensors all measure an averaged response from a population of cells [25,98]. The background signal is sensitive to the adhesion and states of cells. Higher confluency generally leads to higher and more reproducible signal [100]. When the cells form a monolayer, the sensor configuration can be viewed as a three-layer waveguide composite, consisting of a substrate, waveguide thin film and a cell layer. Following a 3-layer waveguide biosensor theory [99] in combination with cellular biophysics [104], we found that for whole-cell sensing, a ligand-induced change in effective refractive index, the detected signal ΔN, is governed by [99]:

$$\Delta N = S(C)\Delta n_c$$

$$= S(C)\alpha d \sum_i \Delta C_i \left[e^{\frac{-z_i}{\Delta Z_c}} - e^{\frac{-z_{i+1}}{\Delta Z_c}} \right] \tag{1}$$

where $S(C)$ is the system sensitivity to the cell layer, and Δn_c is the ligand-induced change in local refractive index of the cell layer sensed by the biosensor. ΔZ_c is the penetration depth into the cell layer, α is the specific refractive index increment (about 0.18/mL/g for proteins), z_i is the distance where the mass redistribution occurs, and d is an imaginary thickness of a slice within the cell layer. Here the cell layer is divided into an equal-spaced slice in the vertical direction. We assumed that the detected signal is, to first order, directly proportional to the change in refractive index of the bottom portion of cell layer Δn_c. The Δn_c is directly proportional to changes in local concentration of cellular targets or molecular assemblies within the sensing volume, given that the refractive index of a given volume within cells is largely determined by the concentrations of bio-molecules, mainly proteins [104]. A weighted factor $\exp(-z_i/\Delta Z_c)$ is taken into account for a change in local protein concentration occurring, considering the exponentially decaying nature of the evanescent wave. Thus, the detected signal is a sum of mass redistribution occurring at distinct distances away from the sensor surface, each with unequal contribution to the overall response. Eq.1 suggests that the detected signal with an RWG biosensor is sensitive primarily to the vertical mass redistribution, as a result of any change in local protein concentration and where and when it occurs. The detected signal is often referred to as a dynamic mass redistribution (DMR) signal [105]. Thus, RWG-enabled cellular assay is also termed DMR assays.

GPCR activation leads to a series of spatial and temporal events, including ligand binding, receptor activation, protein recruitment, receptor internalization and recycling, second messenger alternation, cytoskeletal remodeling, gene expression, and cell adhesion changes, to name but a few. Each cellular event has its own characteristics regarding its kinetics, duration, amplitude, and mass movement. Thus it is reasonably assumed that these cellular events may contribute differently to the overall DMR signal, depending on the location where they occur. Using a panel of agonists targeting a variety of GPCRs, we identified three classes of DMR signals in human epidermoid carcinoma A431 cell, which reflect the signaling pathways mediated [106] (Fig. **3**). Since each is correlated with the activation of a class of GPCRs depending on the G protein with which the receptor is coupled, the DMR signals obtained were named G_q-, G_s- and G_i-DMR signals, respectively. Each class of DMR signals exhibits distinct kinetic and dynamic characteristics, reflecting the unique signaling integration mediated through different classes of GPCRs. Interestingly, G_q-type DMR signal appears to be rapid, whereas G_s-type DMR is comparatively slow. Our recent chemical biology and orthogonal fluorescence imaging studies suggest that G_q-DMR signal is downstream of Ca^{2+} mobilization [107], whereas G_s-DMR signal is downstream of cAMP accumulation [108].

Following classical receptor biology, a numerical analysis suggests that protein translocation and receptor internalization are two primary resources for the DMR signatures observed for G_q-coupled receptor signaling [99]. Unique to G_q-mediated signaling is the dramatic translocation of its signaling components, including several protein kinase C (PKC) isoforms, GRKs, β-arrestin, PIP-binding proteins, and DAG-binding proteins, to name but a few [75,109]. The protein trafficking mediated by the activation of G_q-coupled receptors is relatively rapid and become obvious shortly after agonist stimulation [75], leading to a rapid G_q-type DMR signal. On the other hand, we have found that an increase in cell adhesion is the major contributor to the positive-DMR (P-DMR) event in the β2-AR signal induced by epinephrine [unpublished data]. The occurrence of the initial negative-DMR (N-DMR) event reflects the fact that the majority of downstream signaling components directly involved in the β₂AR signaling complexes, with the exception (thus far) of AKAPs (A-kinase anchoring protein) and β-arrestins, are already compartmentalized at or near the cell membrane [110-112]. Thus, for G_s-coupled receptor the initial recruitment of intracellular targets to the activated receptors is much less pronounced than G_q-coupled receptor signaling, and is overwhelmed by other cellular events leading to the decrease in local mass density in cells. Our recent confocal fluorescence imaging studies suggest that the increase in cell adhesion appears to be only evident at least 10min after the activation of G_s-coupled receptors, consistent with the slow P-DMR event of the G_s-DMR signals [unpublished data]. Both G_q and G_s-DMR signals are quite universal across multiple cell lines tested, although their fine features differ, reflecting the important roles of cellular context in GPCR signaling [106,108,113].

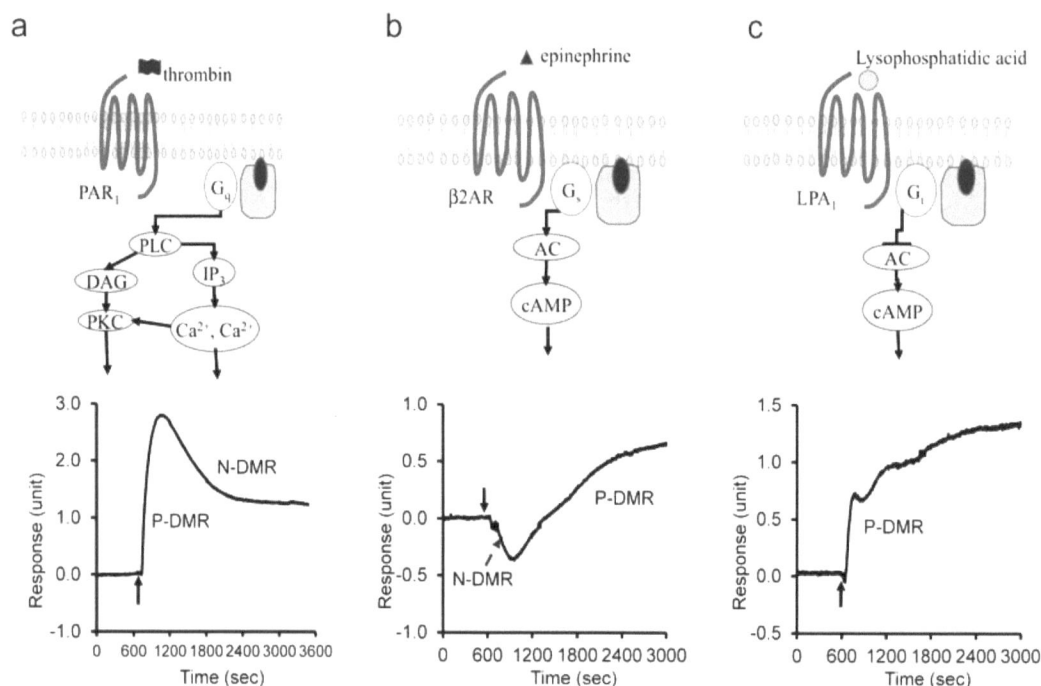

Fig. (3). The GPCR signaling and its DMR signal. (**a**) G_q-signaling and its DMR signal (CHO cells responding to 40unit/ml thrombin). (**b**) G_s-signaling and its DMR signal (A431 cells responding to 25 nM epinephrine). (**c**) G_i-signaling and its DMR signal (A431 responding to 200nM LPA). The solid arrows in the kinetic graphs indicate the time when the agonist solution is introduced (Reproduced with permission from *ref. 106,* Copyright Elsevier, 2007).

Bioimpedance Signals of GPCR Activation

The use of impedance measurements for whole cell sensing was first reported in 1984 [114]. These biosensors were later considered to a cell morphological sensor [115,116]. In a typical impedance-based cell assay, the total impedance of the sensor system is determined primarily by the ion environment surrounding the biosensor [95,117]. Under application of an electrical field, the ions undergo field-directed movement and concentration gradient-driven diffusion. For whole cell sensing, the total electrical impedance has four components: the resistance of the electrolyte solution, the impedance of the cell, the impedance at the electrode/solution interface, and the impedance at the electrode/cell interface. In addition, the impedance of a cell comprises two components - the resistance and the reactance. The conductive characteristics of cellular ionic strength provide the resistive component, whereas the cell membranes, acting as imperfect capacitors, contribute a frequency-dependent reactive component [118,119]. Thus, the total impedance is a function of many factors, including cell viability, cell confluency, cell numbers, cell morphology, degree of cell adhesion, ionic environment, the water content within the cells, and the detection frequency. Electric biosensor enabled cell assays can be viewed as bioimpedance assays.

In the RT-CES system, a percentage of the small voltage applied is coupled into the cell interior. Such signals applied to cells are believed to be much smaller than the resting membrane potential of a typical mammalian cell and thus present minimal invasiveness to cell function cell function. The RT-CES system measures these changes in impedance and displays it as a parameter called the cell index. The cell index is calculated according to the formula [120].

$$CI = \max_{i=1,\dots,N} \left(\frac{R_{cell}(f_i)}{R_0(f_i)} - 1 \right) \tag{2}$$

where N is the number of frequency points at which the impedance is measured (*e.g.*, N=3 for 10 kHz, 25 kHz, and 50 kHz), and $R_0(f)$ and $R_{cell}(f)$ are the frequency electrode resistance without cells or with cells present in the wells, respectively.

In the CellKey system, a change in sensor system's impedance is attributed to a change in complex impedance (delta Z or dZ) of a cell layer that occurs in response to receptor stimulation [93]. At low frequencies, the small voltage applied induces extracellular currents (iec) that pass around individual cells in the layer. However, the conduction currents through cell membrane due to ion channels may also be important at low measurement frequencies [119]. At high frequencies, the small voltage applied induces transcelular currents (itc) that penetrate the cellular membrane (Fig. **1b**). The ratio of the applied voltage to the measured current for each well is its impedance (Z) as described by Ohm's law.

When cells are exposed to a receptor ligand, signal transduction events are activated, leading to complex cellular events that cause changes in cell adherence, cell shape and volume, and cell-to-cell interaction. These cellular changes individually or collectively affect the flow of extracellular and transcelular current, thus resulting in the alteration in the magnitude and characteristics of the impedance. Fig. (**4**) shows three types of impedance signals mediated through the activation of three classes of GPCRs, depending on the G protein with which the receptor is coupled [93,121]. The profiles are obtained using CellKey system. Similar profiles were also recorded using the RT-CES system [122]. It is believed that these impedance signals are due to the different effects on the actin cytoskeleton that affect the cellular parameters measured by impedance, in

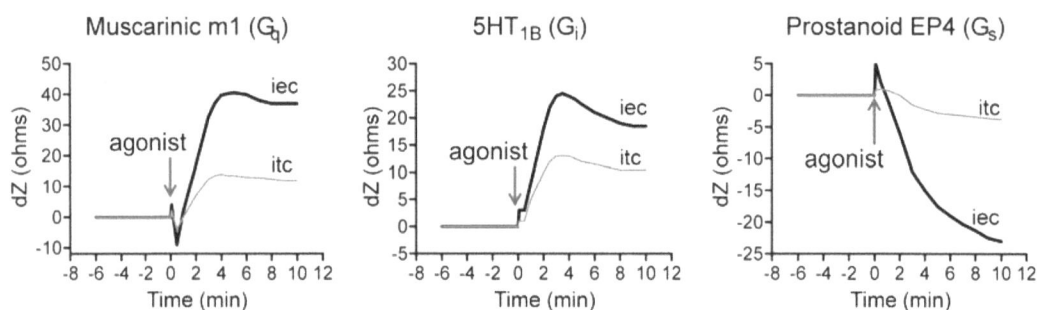

Fig. (4). The impedance signals of G_q, G_i and G_s GPCRs. The profiles are obtained in Chinese ovary hamster (CHO) cells stably expressing rat muscarinic receptor subtype 1 (CHOm1) using the CellKey system. The agonists are carbachol (M_1), 5-hydroxytryptamine ($5HT_{1B}$) and prostaglandin E2 (prostanoid EP_4). Both $5HT_{1B}$ and EP_4 receptors are endogenously expressed (Reproduced with permission from *ref. 95,* Copyright Elsevier, 2005).

response to the activation of different classes of GPCRs. It has been shown that activation of G_q [123,124] and G_i GPCRs [125,126] leads to increased actin polymerization, while stimulation of G_s GPCRs leads to actin depolymerization [127].

LABEL-FREE CELL-BASED ASSAYS FOR GPCR SCREENING

Receptor Panning

Label-free cell-based assays are capable of monitoring the activity of different classes of GPCRs with high sensitivity, offering a universal assay platform for assaying endogenously expressed and functionally active receptors in a cell system. Receptor panning with a GPCR agonist library, known to activate many families of GPCRs, would allow one to reliably map endogenous receptors in a cell system [106].

It is known that more than one family member of many GPCR families are endogenously expressed in a single cell system, such as purinergic P2Y receptors in A431 cells [106] as well as human embryonic kidney (HEK) cells [128]. In addition, a ligand including naturally occurring agonists could have cross-activity, but different efficacies, to activate distinct members of a receptor family [129]. Since the biosensor measures an integrated cellular response [24,25,99], a ligand-induced biosensor signal may represent the signaling mediated through the activation of multiple receptors. Determining the potency and efficacy of panels of non-selective and selective agonists against the receptor family should enable the assessment of which family member is dominated in the cell system, wherein multiple family members are co-endogenously expressed [128]. The cell system- or tissue-specific functional activity of a GPCR might indicate its physiological roles, or hold potential for the side-effect evaluation of a drug candidate. The ability of receptor panning at both the cell system and receptor family levels allows one to easily screen disease-relevant cell types, to determine the differentiation states of stem cells, and to permit more physiologically relevant selectivity and specificity screens. DMR profiling of endogenous receptors in A431 led to the identification of three types of DMR signatures (Fig. **3**), consistent with the primary signaling pathways of their cognate receptors [106]. Interestingly, the maximal amplitudes of the DMR signals arising from the activation of endogenous protease activated receptors by distinct agonists are found to be correlated well with their respective maximal increase in intracellular Ca^{2+} level [130].

DMR assay is a phenotypic whole cell assay with wide pathway coverage, yet permit molecular mechanistic deconvolution [24]. Thus, phenotypic profiling using DMR assays could be used to characterize the activity and specificity of a library of known agonists targeting known GPCRs in a native cell line or cell system including cancer cells and stem cells; and the resultant patterns of these agonists could be used to characterize the origin and states of the cell line or cell system [131]. Surveying functional receptors in the human epidermal squamous cancer cell line A431 with a library of agonists for most of all known GPCRs (~ 220) showed that a number of GPCR agonists led to robust DMR in A431, consistent with the expression pattern of their cognate receptor(s) [131]. For a subset of agonists tested, hierarchical Euclidean clustering analysis based on their real time kinetic responses led to a heat map consisting of two major classes at the low resolution, consistent with the primary pathways mediated through these receptors (Fig. **5**). The G_s-coupled receptor agonists are clustered together, and separated from the G_q-coupled receptor agonists. These results suggest that DMR assays faithfully report functional receptors in the cells.

The recent years have witnessed increasing interests in the regulated differentiation of stem cells into tissue specific cells for tissue repair, and to produce disease-relevant cells for drug discovery [132]. Essential to comprehend the potentials of stem cells for regenerative medicine and drug discovery is to determine the precise types and quality of differentiated cells derived from a specific stem cell or progenitor cell. However, challenges remain to quantitatively measure the functions of stem cells and their differentiated products. Recently, we applied DMR assays to characterize the differentiation process of ReNcell VM human neural progenitor stem cell [133]. The neuroprogenitor stem cell line can differentiate into dopaminergic neurons, astrocytes and oligodendrocytes after growth factor withdrawal, thus creating a neuronal cell system. Together with quantitative real time PCR, DMR profiling using an agonist library targeting over 100 known GPCRs revealed that a subset of receptors including dopamine D_1 and D_4 receptors underwent marked alterations in both receptor expression and signaling pathway during the differentiation process. These findings suggest that

DMR assays can decode the differentiation process of stem cells at the cell system level.

DMR responses at distinct time points

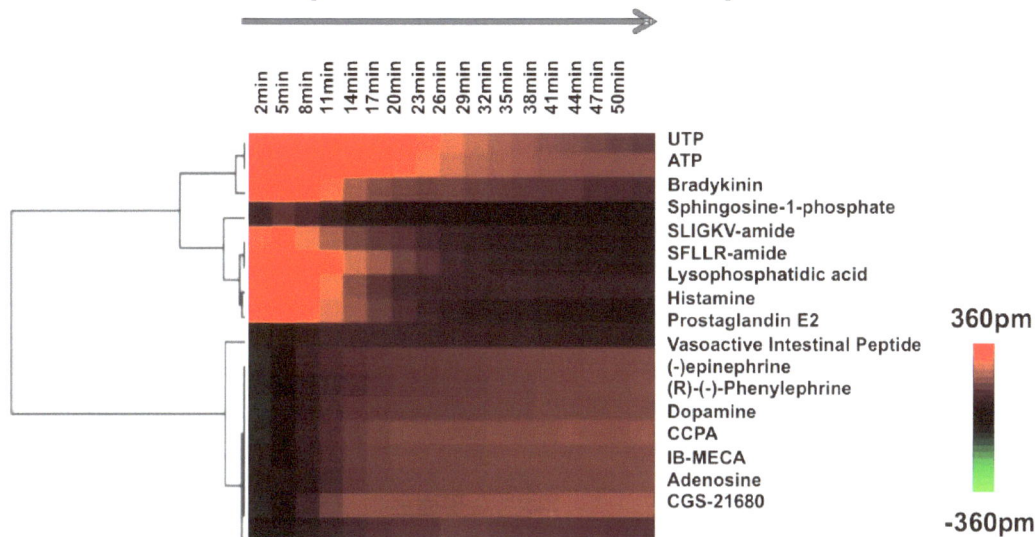

Fig. (5). The heat map classification of the DMR signals of quiescent A431 cells induced by diverse GPCR agonists. The heat map was generated using the Euclidean hierarchical cluster analysis. The real responses of all DMR signals at discrete time points, as indicated, were used as the basis for similarity analysis. The amplitude and direction are indicated by color: red indicates a positive value, green a negative value and black a value close to zero (Reproduced with permission from *ref. 131,* Copyright Informa Healthcare, 2010).

Bioimpedance assays have also been used to identify functional endogenous receptors in cells commonly used in drug discovery, such as HEK293, and U-2 OS cells [93,121]. Receptor panning has resulted in the identification of several functionally active, differently coupled endogenous receptors in these cell lines, some of which have not been previously documented in the literature, suggesting that these assays also have high sensitivity, enabling pharmacological profiling of endogenous receptors in native cells including primary cells.

Systems Cell Biology Studies of GPCR Signaling

Cells rely on highly dynamic network interactions in their decision-making processes. An exogenous signal typically affects the functionality of specific target(s), thus resulting in the execution of various cellular machineries in order for the cell to accommodate such a signal. GPCRs are competent to elicit a diverse array of cell signaling pathways [134]. Much information is available for the signaling of GPCRs *via* either G protein-dependent or independent mechanisms [35,36]. However, owing to the signaling complexity the network interactions of GPCR signaling cascades are still far from being completely understood [44,46]. Label-free cell-based assays are well positioned for dissecting the signaling pathways of GPCRs in distinct cellular backgrounds, because they oftenmeasure an integrated whole cell response, such as DMR signals obtained with RWG biosensors, or the bioimpedance signals obtained with electrical biosensors [24,25,44,99,105,107]. Schroder *et al.* demonstrated that a small set of well-established G protein modulators including the $G_{\alpha i}$ irreversible inhibitor pertussis toxin (PTx), the $G_{\alpha s}$ irreversible activator cholera toxin (CTx), the selective $G_{\alpha q/11}$ inhibitor GYM-254890, and the pan-G protein activator aluminum fluoride (AlF_4^-) can be used to deconvolute the signaling pathways of all four major classes of GPCRs [113]. The free fatty acid receptor FFA1, previously believed to signal only through the $G_{q/G11}$ pathway, was found to also activate the G_i pathway. In another study, Tran and Fang demonstrated that the adenylyl cyclase activator forskolin can be used as a probe molecule to dissect GPCR signaling, based on the experimental observations that the activation of adenylyl cyclase can partially desensitize $G_{\alpha q/11}$-mediated pathway, but cause complete heterologous desensitization of $G_{\alpha s}$-mediated pathway, and potentiate $G_{\alpha s}$-mediated signaling *via* heterologous sensitization [108].

Label-free biosensor cell-based assays can be used to delineate the molecular and cellular mechanisms of receptor signaling at system cell biology level [131]. This is possible for several reasons. First, the biosensor response of a receptor-ligand interaction is a whole cell phenotypic response, and consists of contributions of multiple signaling pathways and proteins. Second, HTS-compatible biosensor systems have made it feasible to probe the impact of a wide variety of probe molecules on the receptor-ligand biosensor signal at a unprecedented scale. Third, owing to its non-invasiveness in measurement, label-free can be easily integrated with chemical biology approaches. Chemical biology uses probe molecules including RNA interference to intervene cell-signaling components one at a time

[135]. Combined with chemical biology approaches, we have studied the systems cell biology of endogenous bradykinin B_2 receptor in A431 cells using RWG biosensors [107]. Stimulation with bradykinin of quiescent A431, obtained by continuous culturing using 0.1% fetal bovine serum for 20 hrs after the cells reach ~90% confluency, resulted in a dose-dependent and saturated DMR response, which can be inhibited by a B_2 specific antagonist HOE140 in a dose-dependent manner, suggested that the DMR signal is a direct result of B_2 receptor activation. The sensitivity of the DMR signal to cholesterol depletion by methyl-β-cyclodextrin argued that B_2 receptor signaling is dependent on the integrity of lipid rafts; disruption of these microdomains hinders the B_2 signaling. Modulations of several important intracellular targets with specific inhibitors suggested that B_2 receptor activation results in signaling *via* at least dual pathways - G_s- and G_q-mediated signaling, and the two signaling pathways counter-regulate each other (Fig. **6**). Several critical downstream targets including protein kinase C (PKC), protein kinase A (PKA), dynaminn, and epidermal growth factor receptor had been identified to involve in B_2 receptor signaling. These findings suggest that the RWG biosensor appears enable the analysis of GPCR signaling. Similarly, impedance-based biosensors have been used for studying complex signal transduction of GPCRs including human melanin-concentrating hormone receptor 1 (MCHR1) expressed heterologously in CHO and U2-OS cells and endogenously in MOLT4 cells. The biosensor detects both pertussis toxin-sensitive and Ca^{2+}-coupled events downstream the MCHR1 activation [93].

Label-free cell assays can be used to discover new pathways in GPCR signaling. Cellular metabolism is essential for cell proliferation, with abnormal metabolism a principal hallmark of human cancer. *De novo* nucleotide biosynthesis is often upregulated in cancer cells to provide building blocks for the synthesis of RNA, DNA and other effectors [136]. However, nothing is known about the role of GPCR signaling in modulating *de novo* purine biosynthesis. The two kinase inhibitors, TBB and DMAT, have been found to regulate in an opposite way a reversible multienzyme complex (*i.e.*, purinosome) participating in *de novo* purine biosynthesis in HeLa cells [137]. In conjunction with RNA interference, DMR profiling showed that the two kinase inhibitors triggered opposite types of DMR

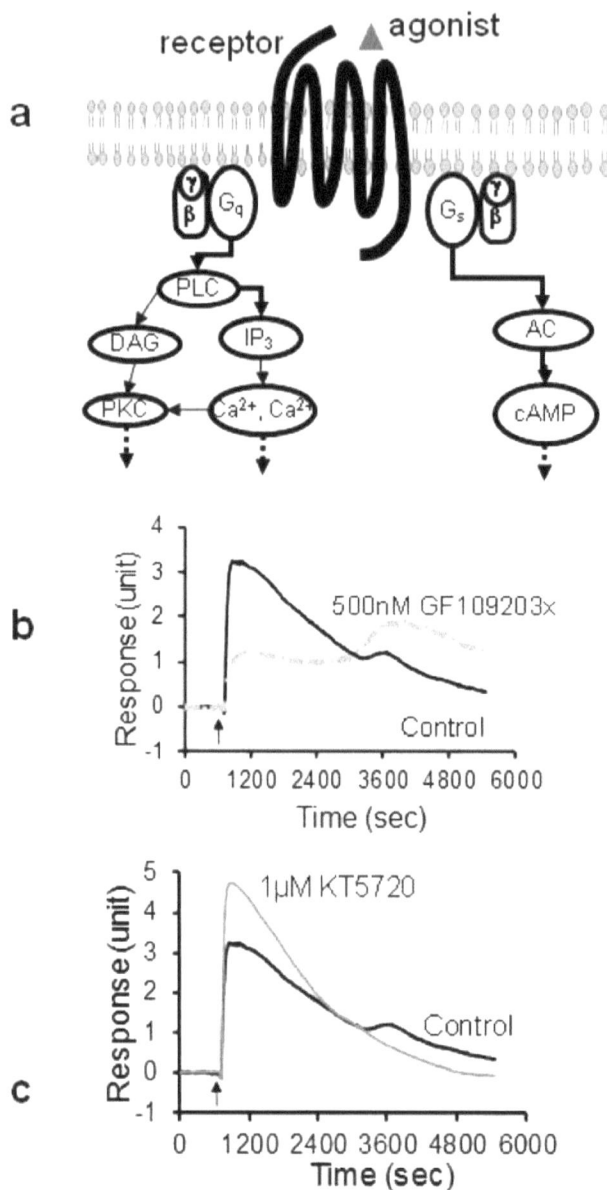

Fig. (6). RWG biosensors enable the analysis of complex GPCR signaling. (**a**) The dual signaling pathways of endogenous bradykinin B_2 receptors in A431 cells. (**b**) and (**c**) The modulation profiles of the bradykinin-induced DMR signal by GF109203x or KT5729, respectively. The quiescent A431 cells were pretreated with either compound for about 1 hr, before the stimulation with bradykinin at 16nM. GF 109203x is a PKC inhibitor, where KT 5720 is a PKA inhibitor. The black curves control in (**b**, **c**) represent the bradykinin response without any pretreatment. The solid arrows in (**b**, **c**) indicate the time when bradykinin is introduced (Reproduced with permission from *ref.* [107]).

signals in HeLa, correlated well with their opposite ability to regulate the purinosome [44]. Screening a library of GPCR agonists against the DMR arising from both TBB and DMAT led to identification of several agonists targeting endogenous $G_{\alpha i}$-coupled receptors including α_{2A}-AR that appears to promote the purinosome assembly. Interrogation of the DMR response with agents specific for the α_{2A}-AR or β_2-AR, coupled with fluorescent live-cell imaging supported that purinosome formation is associated with α_{2A}-AR, but not β_2-AR, activation. Pathway deconvolution further indicated the involvement of $G_{\alpha i}$ pathway in regulating the purinosome. These results were verified by fluorescent live-cell imaging. Collectively, this study suggests a novel pathway common to endogenous $G_{\alpha i}$-coupled receptors in HeLa cells.

Ligand Pharmacological Profiling

GPCR ligands display broad variations in their power to prompt receptor activation, leading to biased agonism [63-68]. The intrinsic property of distinct ligands for a specific receptor, known as efficacy, can be diverse, ranging from full agonism, partial agonism, natural antagonism, and inverse agonism [68]. The efficacy of a ligand is usually determined by the extent of the ligand-induced receptor configuration optimal for G protein activation [38,39,85]. GPCRs are a family of dynamic shapeshifting proteins [64]; and the ligand selects the most favored conformation for binding [138]. A receptor in cells may sample a vast ensemble of conformations, some of which are dominant for the binding of distinct classes of ligands [139], and could have their own signaling preference [64]. Thus, affinity and efficacy are often considered as properties that have independent and distinct structural requirements. Further, it is quite common that the efficacy of a specific agonist for a specific receptor can vary **greatly**, depending on which signaling event is measured [67]. For example, dopamine stabilizes an intermediate activation state of the β_2AR, acts as a strong partial agonist promoting interactions with $G_{\alpha s}$, but is only marginally effective at promoting the β_2AR internalization [140,141]. These findings suggest that instead of being linear (that is; controlling receptor activation also controls the full spectrum of GPCR behaviors such as desensitization, internalization, and phosphorylation) the efficacy of a ligand should be considered to be "collateral" [68].

Label-free cell assays are well-suited to characterize ligand-directed functional selectivity. This is because these assays have broad pathway coverage; yet, the biosensor responses are sensitive to pathways mediated through receptor-ligand interactions [25,106,113]. Thus, the biosensor signals of distinct ligands specific to a receptor in a cell line can be a direct readout of their biased agonism. Profiling of endogenous β_2-AR in A431 with a panel of β_2-AR-specific agonists revealed unique DMR patterns (Fig. 7) [71,72]. Full agonists such as epinephrine, formoterol, and isoproterenol behaved similarly - all gave rise to a DMR having relatively large amplitude and slow kinetics for its positive DMR event. In comparison, partial agonists such as catechol, pindolol, and halostachine led to a DMR with relatively smaller amplitudes but faster kinetics for its positive DMR event, while the known β-arrestin-biased agonist including carvedilol and propranolol triggered a DMR completely distinct from both full and partial agonists, and the antagonists including betaxolol and nadolol triggered no or little DMR. The distinct mechanisms of action of these agonists can be classified into different classes using similarity analysis based on either DMR kinetic parameters [24] or real time responses [72]. Similarity analysis is a powerful means to determine the relationships among different biological responses [142]. A biosensor signature can be simply considered a poly-dimensional coordinate at discrete time points; that is, a temporal series of biosensor output signals [24,72,73]. Thus, it can be directly used as a basis for similarity analysis.

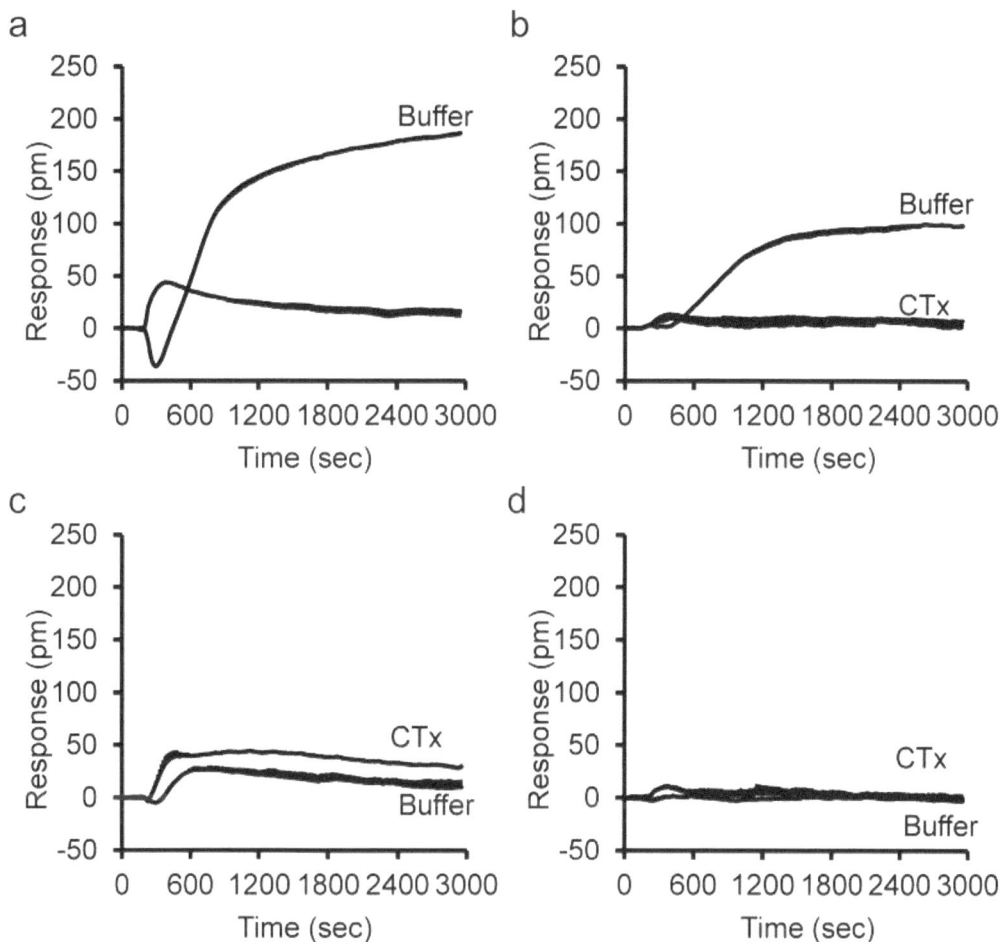

Fig. (7). Characteristic DMR and their sensitivity to the cholera toxin pretreatment of a panel of the β_2-AR ligands. (**a**) Formoterol; (**b**) pindolol; (**c**) S-(-)-propranolol; and (**d**) nadolol. The cells pretreated with the buffer vehicle (Buffer) and the cholera toxin (400 ng/ml, overnight) were stimulated with the ligands, each at 10 µM. Each graph represents mean±s.d. from four replicates.

Label-free cell assays can also be used to characterize the cellular context-dependent pharmacology (*i.e.*, phenotypic pharmacology) of ligands. Cell-specific parameters, such as the ratio of active to inactive receptor species, the rate constant of G protein activation, dynamics in receptor conformations and organizations, and expression levels and organization of receptors and their interactants, are known to influence the pharmacology of GPCR ligands, and to large extent, define drug actions [64, 65]. First, a ligand-receptor interaction may display distinct phenotypic pharmacology in different types of cells wherein the receptor preferentially couples to different signaling proteins. Peters and Scott [143] found, using bioimpedance assays, that melanocortin-4 receptor (MC4R) can trigger distinct signaling pathway in a cellular-context-dependent manner. In human embryonic kidney (HEK) cells stably expressing the MC4R, the receptor preferred $G_{\alpha s}$ coupling, while in Chinese hamster ovary (CHO) cells stably the MC4R, the receptor favored $G_{\alpha q}$ coupling. Second, a ligand-receptor interaction may display distinct phenotypic pharmacology in different cells even when the receptor is coupled with the same G protein. This is mostly due to the possibility that different cells may have different cell signaling machineries. DMR profiling showed that the activation of endogenous β_2-AR in A431 results in a biphasic DMR signal [71], but the activation of the β_2-AR stably expressed in CHO cells led to a single phase DMR [113]. Third, a ligand-receptor interaction may display distinct phenotypic pharmacology in the same type of cell, when the cellular background is preconditioned to distinct states. The DMR signals arising from distinct β_2-AR ligands were found to be sensitive to the pretreatment of A431 cells with different probe molecules including the $G_{\alpha s}$ permanent activator cholera toxin (Fig. **7**), the adenylyl cyclase activator forskolin, and the Casein kinase II (CKII) inhibitor TBB [72]. Finally, a ligand-receptor interaction may display distinct phenotypic pharmacology in the same cell but cultured on different surface chemistries. The surface chemistry is part of environmental cues that can influence the functional responses of receptors. DMR profiling of endogenous P2Y receptors in HEK293 cells showed that interactions of cells with the extracellular matrix coatings shape the signaling of endogenous P2Y receptors [128]. Compared to those on tissue culture treated surfaces, fibronectin coating increased the potency of all P2Y agonists examined, while gelatin coating had little impact.

Label-free cell assays can also differentiate both binding and functional selectivity of compounds. The selectivity profiling is critical to hit confirmation and lead selection. In an early study [92], bioimpedance profiling was used to examine the selectivity and potency of 48 compounds at the dopamine receptor D2 long isoform (D2L). Compound agonist activities were tested on four different cell lines, each over-expressing a different human dopamine receptor subtype, which is coupled to a different signaling pathway; that is, D1 ($G_{\alpha s}$), D2L ($G_{\alpha i}$), D4.4

($G_{\alpha i}$), and D5 ($G_{\alpha s}$). Results show that dopamine served as a non-selective agonist to all receptors, whereas quinpirole was identified as a D2L-selective agonist. Recently, a label-free integrative pharmacology on-target approach has been developed [72] and used to characterize a library of opioid ligands at the mu opioid receptor (MOR) [73]. Combining chemical biology with multiple assay formats was used to distinguish various ligands based on their specificity, relative potency and mechanisms of action at specific receptor sites. Results showed that this approach offers a global representation of ligand binding and functional selectivity at the MOR receptors over endogenous receptors. The opioid ligands BNTX and ICI 199441 were active in both parental and mu-expressing HEK-293 cells, suggesting that both ligands activate an endogenous target beside the MOR. Furthermore, distinct ligands displayed distinct functional selectivity towards G proteins and kinase pathways.

High Throughput Screening

Together with the adoption of microtiter plate formats with a density up to 1536-well, the ability of current generation biosensor systems to integrate with existing lab automation equipment and external liquid handling device makes HTS possible. Dodgson *et al.* screened 100K compounds using Corning Epic® label-free system for identifying antagonists of the muscarinic M_3 receptor [144]. The activation of the M_3 receptor triggered $G_{\alpha q}$-signaling pathway, leading to robust DMR signal. Based on the DMR kinetic profile of the M_3 receptor, they developed an endpoint assay, the DMR signal at 20min after the agonist stimulation, for HTS. The screen was found to be robust with a Z' value greater than 0.5. Using an inhibition threshold of > 40%, the hit rate was found to be 0.5%. In comparison, the hit rate obtained using FLIPR Ca^{2+} assays was 0.8% using the same inhibition threshold. Interestingly, comparing the hits identified using Epic® with those obtained using the FLIPR, they found that only 72 compounds were active in both assays, whereas 857 were specific to Epic® DMR screens, and 785 specific to the FLIPR. Almost all compounds (782) that are specific to the FLIPR were confirmed in follow-up studies to be false positives, and the remaining three compounds displayed low potency. Counter screening of the 392 compounds active in the Epic® screen showed that 149 compounds gave rise to a pIC_{50} 4.7 to >6.6.

Label-free cellular assays also permit multiplexed screening against multiple GPCRs including both endogenously and stably expressed receptors for both agonists and antagonists in a single screen. This is possible because GPCR signaling involves a series of highly regulated spatial and temporal events including the production and regulation of second messengers such as Ca^{2+} and cAMP [76,145], and label-free measurement is integrative in nature [25,44,92,99]. DMR profiling showed that the co-activation of both endogenous $G_{\alpha q}$-coupled histamine H_1 receptor and $G_{\alpha s}$-coupled β_2-AR in A431 resulted in a DMR that is close to the sum of both histamine and epinephrine DMR signals [146]. Histamine is the natural agonist for the H_1 receptor and epinephrine is the natural agonist for the β_2-AR. This result suggests that both $G_{\alpha q}$- and $G_{\alpha s}$-mediated signaling undergo spatially distinct routes at the complex pathway level. Furthermore, histamine was found to slightly attenuate the epinephrine response, while epinephrine partially attenuated the histamine response, suggesting the presence of cross-talks between the epinephrine- and histamine-mediated signaling. Based on these observations, a duplexed GPCR screening was performed using the Library of Pharmaceutically Active Compounds from Sigma which consists of 1280 compounds. This screen was performed in a sequential two step format. The first agonism step is to examine the ability of compounds to trigger DMR in A431. The second antagonism step is to determine the ability of compounds to modulate the DMR arising from the co-stimulation with both epinephrine and histamine. Given the wide pathway coverage of DMR assays, the agonism screen is multiplexed in nature. Results showed that the agonism screening has indeed correctly identified all known agonists from the library for several endogenous GPCRs including H_1 receptor, adenosine receptors, and β_2-AR (Fig. **8**). Using the DMR arising from the co-stimulation with epinephrine and histamine as a reporter, the antagonism screen identified seventy-seven apparent antagonists for the β_2-AR, and fifty-one apparent antagonists for the H_1 receptor. Correlating the agonism profile with the antagonism profile of each compound further segregated the hit compounds into both receptor-specific agonists and antagonists, and pathway modulators.

CHOICE OF ASSAY PLATFORMS FOR GPCR SCREENING

Because of the unpredictability in the intracellular signaling mediated by synthetic compounds, a single compound could behave as an agonist or antagonist, depending

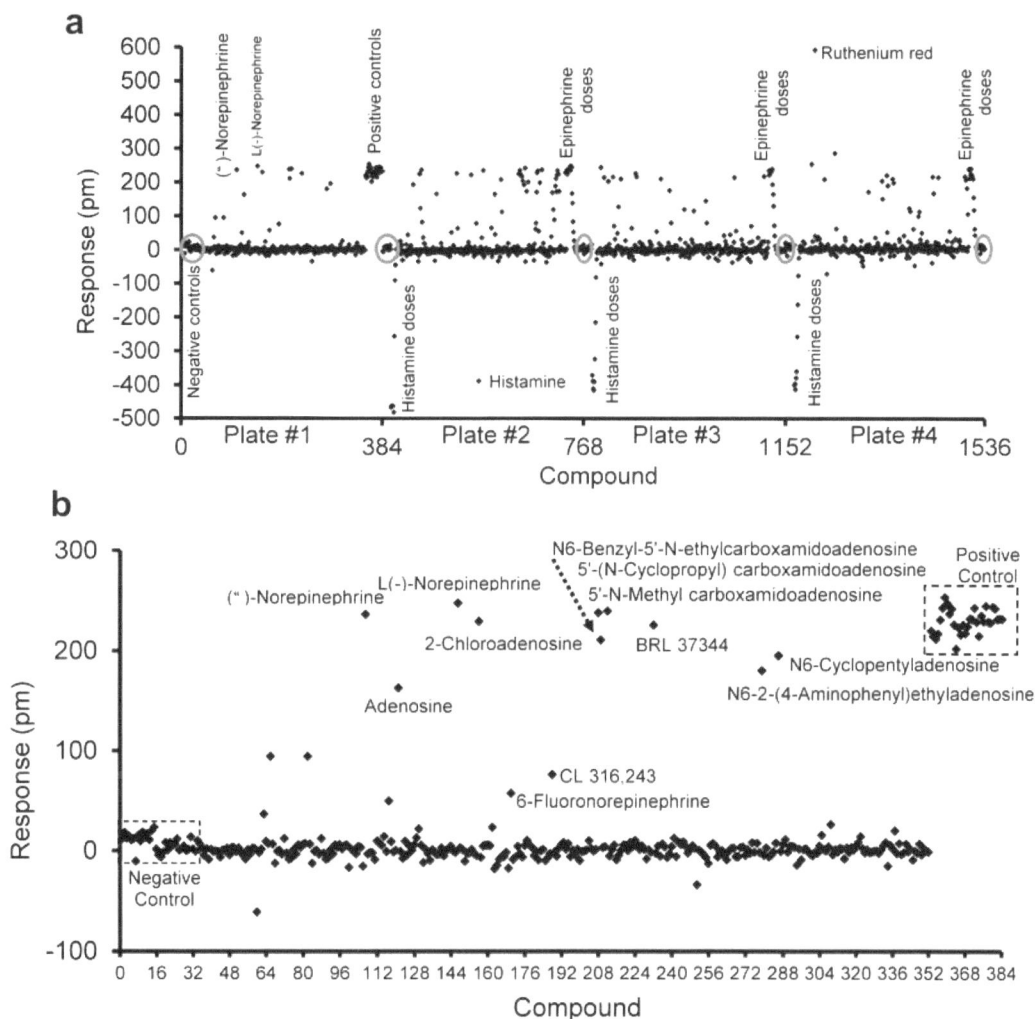

Fig. (8). DMR agonism screening results of the LOPAC compound library in A431. (**a**) The responses, the wavelength shifts between 2min and 50min after stimulation, were plotted as a function of compounds. In this set of LOPAC screen, the plate 1 contained 2 columns of the negative controls (*i.e.*, the assay buffer only), and 2 columns of the positive controls for 2 nM epinephrine. For the other three plates, there were also two columns of the negative controls (the column 1 and 24), a concentration series of epinephrine (the column 23) and of histamine (the column 2). (**b**) The representative hit map exampled in the plate 1. The positive hits for both the β_2AR and adenosine receptors were indicated in the hit plot (Reproduced with permission from *ref.* [146], Copyright SAGE, 2008).

on the signaling events measured; and the lack of efficacy on a given event measured does not guarantee a lack of receptor activation [16,65]. Thus, a rational

GPCR screen should not rely on a single assay; rather, an integrated approach should be employed to measure a multitude of signaling events. Technologies that are independent of cell signaling pathway(s) should be advantageous, given the recent findings that GPCR activation also leads to G protein-independent signaling under many circumstances [35,36].

Label-free cell-based assays measure a whole cell phenotypic response, thus offering a better ability to discriminate the efficacies of different agonists than conventional molecular assays. A receptor may couple to multiple signaling pathways in a given cell system. Molecular assays are often associated with a specific pathway or cellular event downstream the receptor signaling. Owing to the presence of ligand-directed functional selectivity, distinct molecular assays may give rise to distinct potency and efficacy for a specific ligand-receptor pair, a phenomenon termed pluridimensional efficacy [67]. Thus, the results obtained using distinct molecular assays may be contradictory to the global function of a ligand in a cell system or tissue. To this regard, the whole cell efficacy obtained using label-free biosensors should provide better discriminative power to assess ligand efficacy.

Label-free cell-based assays have high sensitivity, thus enabling phenotypic profiling of receptor-ligand interactions in native cells. This is in dramatic contrast to conventional GPCR cell assays which often require certain engineering or manipulations of cells in order to achieve the desired sensitivity for robust and reliable detection [147-149]. In these artificial systems, the potency or efficacy of a ligand might be altered, due to the high expression level of the receptor or the interference of certain manipulations (*e.g.*, GFP-tagging, or the act of transfection) on the cellular physiology of the targets. α_{2A}-adrenergic receptor ($\alpha_{2A}AR$) is commonly believed to couple to $G_{\alpha i}$ proteins, leading to the inhibition of adenylyl cyclase activity. In a native HEL 92.1.7 cells, $\alpha_{2A}AR$ ligand levomed is an inverse agonist, causing an increase in cAMP production [150]. However, in transfected PC10 fibroblasts levomed is found to be a positive agonist, causing an inhibition of cAMP production [151]. Besides the ability to detect the cell context dependent pharmacology in native cells, label-free phenotypic profiling also provides a unique means to characterize the origin and states of living cells including cancer cells and stem cells.

Label-free cell-based assays have broad pathway coverage, thus being well suited to screen ligands for orphan GPCRs. A receptor may couple to multiple pathways in a cell line, some of which may be unknown. The signaling pathways of orphan GPCRs are also mostly unknown. Recently, DMR profiling of compounds with unknown pharmacological activity in a native cancer cell line HT-29 had led to discovery of several novel chemical classes of GPR35 agonists including tyrphostin-51 and entacapone [152], thieno[3,2-b]thiophene-2-carboxylic acidderivatives [153,154], and the aspirin catabolite 2,3,5-trihydroxybenzoic acid [155]. The most notable finding is that multiple tyrosine metabolites including rosmarinic acid and 5,6-dihydroxyindole-2-carboxylic acid (DHICA) are natural agonists of the GPR35 [156,157]. Comparing the potency of efficacy of these agonists obtained using DMR assays with Tango beta-arrestin translocation gene reporter assays suggests that many of these agonists display biased agonism at the GPR35.

In addition, biosensor-based cell assays bypass the need of any sort of fluorescent labels, which are otherwise widely used for detection in many conventional assays. Such ability could significantly improve the date quality of GPCR screens, since the interference of cell components, labels or fluorescent compounds is minimized [87,150].

It is expected that like other functional assays [30], label-free cell-based assays are also prone to false positives - compounds that do not act on the receptor [92,144,146]. False positives that act on the downstream signaling cascade can easily be removed by further evaluation in an equivalent assay that uses different cell systems [144,146]. Those compounds that are found to act on the receptor also need further characterization for their ligand-binding properties to confirm whether they are competitive or allosteric modulators.

CONCLUSION

Biosensor-based cell assays enable the measurement of multiple classes of GPCRs simultaneously from a single assay, providing a universal yet highly sensitive platform for GPCR screens. Conversely, multiple conventional assays are required to evaluate all major GPCR classes; most of these assays require prior knowledge of the signaling transduction of the receptors of interest. Given the broad applicability for non-invasively assaying endogenous GPCRs in both kinetic and HTS formats, biosensor-based cell assays have a bright future and should further strengthen the role of GPCRs in drug development. The advent of HTS instruments and further understanding of the biosensor signals obtained will accelerate acceptance of label-free cell-based assays in new areas of drug discovery, for which physiologically relevant representation of GPCR drug candidates is important and valuable.

ACKNOWLEDGEMENTS

The author acknowledges the past contributions of Dr. Anthony G. Frutos and Mr. Ronald Verkleeren.

CONFLICT OF INTEREST

The author states that there is no conflict of interest.

DISCLOSURE

This chapter is an update of our previous publication in CCHTS:Fang, Y., Frutos, A.G., Verklereen, R. Label-free cell assays for GPCR screening. *Comb. Chem. High Throughput Screen.* **2008**, *11*, 357-369.

REFERENCES

[1]   Palczewski, K.; Kumasaka, T.; Hori, T.; Behnke, C.A.; Motoshima, H.; Fox, B.A.; Le Trong, I.; Teller, D.C.; Okada, T.; Stenkamp, R.E.; Yamamoto, M.; Miyano, M. Crystal structure of rhodopsin: A G protein-coupled receptor. *Science* **2000**, *289*, 739-45.

[2]   Chalmers, D.T.; Behan, D.P. The use of constitutively active GPCRs in drug discovery and functional genomics. *Nat. Rev. Drug Discov.* **2002**, *1*, 599-608.

[3]   Ji, T.H.; Grossmann, M.; Ji, I. G Protein-coupled receptors I. Diversity of receptor-ligand interactions. *J. Biol. Chem.* **1998**, 273, 17299-302.

[4]   Neubig, R.R.; Siderovski, D.R. Regulators of G-protein signaling as new central nervous system drug targets. *Nat. Rev. Drug Discov.* **2002**, *1*,187-97.

[5]   Rockman, H.A.; Koch, W.J.; Lefkowitz, R.J. Seven-transmembrane-spanning receptors and heart function. *Nature* **2002**, *415*, 206-12.

[6]   Schöneberg, T.; Schulz, A.; Gudermann, T. The structural basis of G-protein-coupled receptor function and dysfunction in human diseases. *Rev. Physiol. Biochem. Pharmacol.* **2002**, *144*, 143-227.

[7]   Drews, J. Drug discovery: a historical perspective. *Science* **2000**, *287*, 1960-4.

[8]   Zambrowicz, B.P.; Sands, A.T. Knockouts model the 100 best-selling drugs - will they model the next 100? *Nat. Rev. Drug Discov.* **2003**, *2*, 38-51.

[9]   Imming, P.; Sinning, C.; Meyer, A. Drugs, their targets and the nature and number of drug targets. *Nat. Rev. Drug Discov.* **2006**, *5*, 821-34.

[10]   Overington, J.P.; Al-Lazikani, B.; Hopkins, A.L. How many drug targets are there? *Nat. Rev. Drug Discov.* **2006**, *5*, 993-6.

[11]   Hopkins, A.L.; Groom, C.R. The druggable genome. *Nat. Rev. Drug Discov.* **2002**, *1*, 727-30.

[12]   Fredriksson, R.; Schioth, H.B. The Repertoire of G-protein-coupled receptors in fully sequenced genomes. *Mol. Pharmacol.* **2005**, *67*, 1414-25.

[13]   Chung, S.; Funakoshi, T.; Civelli O. Orphan GPCR research. *Br. J. Pharmacol.* **2008**, *153*, S339-46.

[14]   Levoye, A., Jockers, R. Alternative drug discovery approaches for orphan GPCRs. *Drug Discov Today.* **2008**, *13*, 52-8.

[15]   Williams, M. Target validation. *Curr. Opin. Pharmacol.* **2003**, 3, 571-7.

[16]   Sautel, M.; Milligan, G. Molecular manipulation of G-protein-coupled receptors: a new avenue into drug discovery. *Curr. Med. Chem.* **2000**, *7*, 889-96.

[17]   Rask-Andersen, M.; Almen, M.S.; Schioth, H.B. Trends in the exploitation of novel drug targets. *Nat. Rev. Drug Discov.* **2011**, *10*, 579-90.

[18]   Swinney, D.C.; Anthony, J. How were new medicines discovered? *Nat. Rev. Drug Discov.* **2011**, *10*, 507-519.

[19]   Gleeson, M., Hersey, A., Montanari, D., Overington, J. Probing the links between *in vitro* potency, ADMET and physicochemical parameters. *Nat. Rev. Drug Discov.* **2011**, *10*, 197-208.

[20]   Butcher, E. Can cell systems biology rescue drug discovery? *Nat. Rev. Drug Discov.* **2005**, *4*, 461-7.

[21]   Kenakin, T. Being mindful of seven-transmembrane receptor 'guests' when assessing agonist selectivity. *Br. J. Pharmacol.* **2010**, *160*, 1045-7.

[22]   Drews, J. Strategic trends in the drug industry. *Drug Discov. Today* **2003**, *8*, 411-20.

[23]   van der Greef, J.; McBurney, R.N. Innovation: Rescuing drug discovery: *in vivo* systems pathology and systems pharmacology. *Nat. Rev. Drug Discov.* **2005**, *4*, 961-7.

[24]   Fang, Y. Label-free receptor assays. *Drug Discov. Today Technol.* **2010**, *7*, e5-e11.

[25]   Fang, Y. The development of label-free cellular assays for drug discovery. *Exp. Opin. Drug Discov.* **2011**, *6*, 1285-98.

[26]   Lerner, M.R. Tools for investigating functional interactions between ligands and G-protein-coupled receptors. *Trends Neurosci.* **1994**, *17*, 142-6.

[27]   Milligan, G. High-content assays for ligand regulation of G-protein-coupled receptors. *Drug Discov. Today* **2003**, *8*, 579-85.

[28]   Fang, Y.; Lahiri, J.; Picard, L. G protein-coupled receptor microarrays for drug discovery. *Drug Discov. Today* **2003**, *8*, 755-61.

[29]   Cacace, A.; Banks, M.; Spicer, T.; Civoli, F.; Watson, J. An ultra-HTS process for the identification of small molecule modulators of orphan G-protein-coupled receptors. *Drug Discov. Today* **2003**, *8*, 785-92.

[30]   Williams, C. cAMP detection methods in HTS: selecting the best from the rest. *Nat. Rev. Drug Discov.* **2004**, *3*, 125-35.

[31]   Zaman, G.J. Cell-based screening. *Drug Discov. Today* **2004**, *9*, 828-30.

[32]   Leifert, W.R.; Aloia, A.L.; Bucco, O.; Glatz, R.V.; McMurchie, E.J. G-protein-coupled receptors in drug discovery: nanosizing using cell-free technologies and molecular biology approaches. *J. Biomol. Screen.* **2005**, *10*, 765-79.

[33]   McLoughlin, D.J.; Bertelli, F.; Williams, C. The A, B, Cs of G-protein-coupled receptor pharmacology in assay development for HTS. *Exp. Opin. Drug Discov.* **2007**, *2*, 603-19.

[34]   Zhang, R.; Xie, X. Tools for GPCR drug discovery. *Acta Pharmacol. Sin.* **2012**, *33*, 372-84.

[35]   Neves, S.R.; Ram, P.T.; Iyengar, R. G protein pathways. *Science* **2002**, *296*, 1636-9.

[36]   Holinstat, M.; Oldham, W.M.; Hamm, H.E. G-protein-coupled receptors: evolving views on physiological signaling. *EMBO Reports* **2006**, *7*, 866-9.

[37]   Kholodenko, B.N. Cell signaling dynamics in time and space. *Nat. Rev. Mol. Cell Biol.* **2006**, *7*, 165-76.

[38]   Kobilka, B. Agonist binding: a multistep process. *Mol. Pharmacol.* **2004**, *65*, 1060-2.

[39]   Swaminath, G.; Xiang, Y.; Lee, T.W.; Steenhuis, J.; Parnot, C.; Kobilka, B.K. Sequential binding of agonists to the β_2 adrenoceptor. Kinetic evidence for intermediate conformational states. *J. Biol. Chem.* **2004**, *279*, 686-91.

[40]   Oldham, W.M.; Hamm, H.E. Heterotrimeric G protein activation by G-protein-coupled receptors. *Nat. Rev. Mol. Cell Biol.* **2008**, *9*, 60-71.

[41]   Linderman, J.J. Modeling of G-protein-coupled receptor signaling pathways. *J. Biol. Chem.* **2009**, *284*, 5427-31.

[42]   Shenoy, S.K.; Lefkowitz, R.J. Multifaceted roles of beta-arrestins in the regulation of seven-membrane-spanning receptor trafficking and signaling. *Biochem. J.* **2003**, *375*, 503-15.

[43]   Kelly, E.; Bailey, C.P.; Henderson, G. Agonist-selective mechanisms of GPCR desensitization. *Br. J. Pharmacol.* **2008**, *153*, S379-88.

[44]   Verrier, F.; An, S.; Ferrie, A.M.; Sun, H.; Kyoung, M.; Fang, Y.; Benkovic, S.J. G protein-coupled receptor signaling regulates the dynamics of a metabolic multienzyme complex. *Nat. Chem.Biol.* **2010**, *7*, 909-15.

[45]   Rozengurt, E. Mitogenic signaling pathways induced by G protein-coupled receptors. *J. Cell. Physiol.* **2007**, *213*, 589-602.

[46]   Hara, M.R.; Kovacs, J.J.; Whalen, E.J.; Rajagopal, S.; Strachan, R.T.; Grant, W.; Towers, A.J.; Williams, B.; Lam, C.M.; Xiao, K.; Shenoy, S.K.; Gregory, S.G.; Ahn, S.; Duckett, D.R.; Lefkowitz, R.J. A stress response pathway regulates DNA damage through $\beta2$-adrenoreceptors and β-arrestin-1. *Nature* **2011**, 477, 349-53.

[47]   Bouvier, M. Oligomerization of G-protein-coupled transmitter receptors. *Nat. Rev. Neurosci.* **2001**, *2*, 274-86.

[48]   George, S.R.; O'Dowd, B.F.; Lee, S.P. G-protein-coupled receptor oligomerization and its potential for drug discovery. *Nat. Rev. Drug Discov.* **2002**, *1*, 808-20.

[49]   Prinster, S.C.; Hague, C.; Hall RA. Heterodimerization of G protein-coupled receptors: specificity and functional significance. *Pharmacol. Rev.* **2005**, *57*, 289-98.

[50]   Han, Y.; Moreira, I.S.; Urizar, E.; Weinstein, H.; Javitch, J.A. Allosteric communication between protomers of dopamine class A GPCR dimers modulates activation.*Nat. Chem. Biol.* **2009**, *5*, 688-95.

[51]   Berglund, M.M.; Schober, D.A.; Esterman, M.A.; Gehlert, D.R. Neuropeptide Y Y4 receptor homodimers dissociate upon agonist stimulation. *J. Pharmacol. Exp. Ther.* **2003**, *307*,1120-6.

[52]   Cheng, Z.J.; Miller, L.J. Agonist-dependent dissociation of oligomeric complexes of G protein-coupled cholecystokinin receptors demonstrated in living cells using bioluminescence resonance energy transfer. *J. Biol. Chem.* **2001**, *276*, 48040-7.

[53]   Roed, S.N.; Orgaard, A.; Jorgensen, R.; de Meyts, P. Receptor oligomerization in family B1 of G-protein-coupled receptors: focus on BRET investigations and the link between GPCR oligomerization and binding cooperativity.*Front. Endocrinol.* **2012**, *3*, 62.

[54]   Rocheville, M.; Lange, D.C.; Kumar, U.; Patel, S.C.; Patel, R.C.; Patel, Y.C. Receptors for dopamine and somatostatin: formation of hetero-oligomers with enhanced functional activity. *Science* **2000**, *288*, 154-7.

[55]   Mei, G.; Di Venere, A.; Rosato, N.; Finazzi-Agro, A. The importance of being dimeric. *FEBS J.* **2005**, *272*, 16-27.

[56]   Fotiadis, D.; Liang, Y.; Filipek, S.; Saperstein, D.A; Engel, A.; Palczewski, K. Atomic-force microscopy: Rhodopsin dimers in native disc membranes. *Nature* **2003**, *421*, 127-8.

[57]   Dean, M.K.; Higgs, C.; Smith, R.E.; Bywater, R.P.; Snell, C.R.; Scott, P.D.; Upton, G.J.G.; Howe, T.J.; Reynolds, C.A. Dimerization of G-protein-coupled receptors. *J. Med. Chem.* **2001**, *44*, 4595-614.

[58]   Lichtarge, O.; Bourne, H.R.; Cohen, F.E. Evolutionarily conserved $G_{\alpha\beta\gamma}$ binding surfaces support a model of the G protein-receptor complex. *Proc. Natl. Acad. Sci. USA* **1996**, *93*, 7507-11.

[59]   Manglik, A,; Kruse, A.C.; Kobilka, T.S.; Thian, F.S.; Mathiesen, J.M.; Sunahara, R.K.; Pardo, L.; Weis, W.I.; Kobilka, B.K.; Granier, S. Crystal structure of the μ-opioid receptor bound to a morphine antagonist. *Nature* **2012**, *485*,321-6.

[60]   Wu, H.; Wacker, D.; Mileni, M.; Katritch, V.; Han, G.W.; Vardy, E.; Liu, W.; Thompson, A.A.; Huang, X.P.; Carroll, F.I.; Mascarella, S.W.; Westkaemper, R.B.; Mosier P.D; Roth, B.L.; Cherezov, V.; Stevens, R.C. Structure of the human κ-opioid receptor in complex with JDTic. *Nature* **2012**, *485*, 327-32.

[61]   Breitwieser, G. E. G protein-coupled receptor oligomerization: Implications for G protein activation and cell signaling. *Circ. Res.* **2004**, *94*, 17-27.

[62]   Cottet, M.; Faklaris, O.; Maurel, D.; Scholler, P.; Doumazane, E.;Trinquet, E.; Pin, J.P.; Durroux, T. BRET and Time-resolved FRET strategy to study GPCR oligomerization: from cell lines toward native tissues. *Front. Endocrinol.* **2012**, *3*, 92.

[63]   Urban, J.D.; Clarke, W.P.; von Zastrow, M.; Nichols, D.E.; Kobilka, B.; Weinstein, H.; Javitch, J.A.; Roth, B.L.; Christopoulos, A.; Sexton, PM.; Miller, K.J.; Spedding, M.; Mailman, R.B. Functional selectivity and classical concepts of quantitative pharmacology. *J. Pharmacol. Exp. Ther.* **2007**, 320, 1-13.

[64]   Kenakin, T.; Miller, L.J. Seven transmembrane receptors as shapeshifting proteins: the impact of allosteric modulation and functional selectivity on new drug discovery. *Pharmacol. Rev.* **2010**, *62*, 265-304.

[65]   Kinzer-Ursem, T.L.; Linderman, J.J. Both ligand- and cell-specific parameters control ligand agonism in a kinetic model of G protein-coupled receptor signaling. *PLoS Comput. Biol.* **2007**, *3*, e6.

[66]   Mailman, R.B. Ligand-selective signaling and high content for GPCR drugs. *Trends Pharmacol. Sci.* **2007**, *8*, 390-6.

[67]   Galandrin S, Oligny-Longpre G, Bouvier M. The evasive nature of drug efficacy: implications for drug discovery. *Trends Pharmacol. Sci.* **2007**, *8*, 423-30.

[68] Kenakin, T. New concepts in drug discovery: collateral efficacy and permissive antagonism. *Nat. Rev. Drug Discov.* **2005**, *4*, 919-27.

[69] Kenakin, T.P.; Morgan, P.H. The theoretical effects of single and multiple transducer receptor coupling proteins on estimates of the relative potency of agonists. *Mol. Pharmacol.* **1989**, *35*, 214-22.

[70] Boehr, D.D.; Nussinov, R.; Wright, P.E. The role of dynamic conformational ensembles in biomolecular recognition. *Nat. Chem. Biol.* **2009**, *5*, 789-96.

[71] Fang, Y.; Ferrie, A.M. Label-free optical biosensor for ligand-directed functional selectivity acting on β_2-adrenoceptor in living cells. *FEBS Lett.* **2008**, *582*, 558-64.

[72] Ferrie, A.M.; Sun, H.; Fang, Y. Label-free integrative pharmacology on-target of drugs at the β_2-adrenergic receptor. *Sci. Rep.* **2011**, *1*, 33.

[73] Morse, M.; Tran, E.; Levension, R.L.; Fang, Y. Ligand-directed functional selectivity at the mu opioid receptor revealed by label-free on-target pharmacology. *PLoS One* **2011**, *6*, e25643.

[74] Lohse, M.J.; Nikolaev, V.O.; Hein, P.; Hoffmann, C.; Vilardaga, J.P.; Bünemann, M. Optical techniques to analyze real-time activation and signaling of G-protein-coupled receptors. *Trends Pharmacol. Sci.* **2008**, *29*, 159-65.

[75] van Baal, J.; de Widt, J.; Divecha, N.; van Blitterswijk, W.J. Translocation of diacylglycerol kinase theta from cytosol to plasma membrane in response to activation of G protein-coupled receptors and protein kinase C. *J. Biol. Chem.* **2005**, *280*, 9870-8.

[76] Willoughby, D.; Cooper, D.M. Live-cell imaging of cAMP dynamics. *Nat. Methods* **2008**, *5*, 29-36.

[77] Giannone, F.; Malpeli, G.; Lisi, V.; Grasso, S.; Shukla, P.; Ramarli, D.; Sartoris, S.; Monsurró, V.; Krampera, M.; Amato, E.; Tridente, G.; Colombatti, M.; Parenti, M.; Innamorati, G. The puzzling uniqueness of the heterotrimeric G15 protein and its potential beyond hematopoiesis. *J. Mol. Endocrinol.* **2010**, *44*, 259-69.

[78] Reinscheid, R.K.; Kim, J.; Zeng, J.; Civelli, O. High-throughput real-time monitoring of Gs-coupled receptor activation in intact cells using cyclic nucleotide-gated channels. *Eur. J. Pharmacol.* **2003**, *478*, 27-34.

[79] Yan, Y.X.; Boldt-Houle, D.M.; Tillotson, B.P.; Gee, M.A.; D'Eon, B.J.; Chang, X.J.; Olesen, C.E.; Palmer, M.A. Cell-based high-throughput screening assay system for monitoring G protein-coupled receptor activation using beta-galactosidase enzyme complementation technology. *J. Biomol. Screen.* **2002**, *7*, 451-9.

[80] Hu, C.-D.; Kerppola, T.K. Simultaneous visualization of multiple protein interactions in living cells using multicolor fluorescence complementation analysis. *Nat. Biotechnol.* **2003**, *21*, 539-45.

[81] Urizar, E.; Yano, H.; Kolster, R.; Galés, C.; Lambert, N.; Javitch, J.A. CODA-RET reveals functional selectivity as a result of GPCR heteromerization. *Nat. Chem. Biol.* **2011**, *7*, 624-30.

[82] Tan, C.M.; Brady, A.E.; Nickols, H.H.; Wang, Q.; Limbird, L.E. Membrane trafficking of G protein-coupled receptors. *Annu. Rev. Pharmacol. Toxicol.* **2004**, *44*, 559-609.

[83] Barak, L.S.; Ferguson, S.S.; Zhang, J.; Caron, M.G. A beta-arrestin/green fluorescent protein biosensor for detecting G protein-coupled receptor activation. *J. Biol. Chem.* **1997**, *272*, 27497-500.

[84]   Raehal, K.M,; Schmid, C.L.; Groer, C.E.; Bohn, L.M. Functional selectivity at the μ-opioid receptor: implications for understanding opioid analgesia and tolerance. *Pharmacol. Rev.* **2011**, *63*, 1001-19.

[85]   Saulière, A.; Bellot, M.; Paris, H.; Denis, C.; Finana, F.; Hansen, J.T.; Altié, M.F.; Seguelas, M.H.; Pathak, A.; Hansen, J.L.; Sénard, J.M.; Galés, C. Deciphering biased-agonism complexity reveals a new active AT_1 receptor entity. *Nat. Chem. Biol.* **2012**, doi: 10.1038/nchembio.961.

[86]   Cheng, Z.; Garvin, D.; Paguio, A.; Stecha, P.; Wood, K.; Fan, F. Luciferase reporter assay system for deciphering GPCR pathways. *Curr. Chem. Genomics* **2010**, *4*, 84-91.

[87]   Hu, H.; Deng, H.; Fang, Y. Label-free phenotypic profiling identified D-luciferin as a GPR35 agonist. *PLoS One* **2012**, *7*, e34934.

[88]   Avasthi, P.; Marley, A.; Lin, H.; Gregori-Puigjane, E.; Shoichet, B.K.; von Zastrow, M.; Marshall, W.F. A Chemical screen identifies class A G-protein coupled receptors as regulators of cilia. *ACS Chem. Biol.* **2012**, *7*, 911-9.

[89]   McDonnell, J.M. Surface plasmon resonance: towards an understanding of the mechanisms of biological molecular recognition. *Curr. Opin. Chem. Biol.* **2001**, *5*, 572-7.

[90]   Cooper, M.A. Label-Free Optical Biosensors in Drug Discovery. *Nat. Rev. Drug Discov.* **2002**, 1, 515-28.

[91]   Fang, Y. Ligand-receptor interaction platforms and their applications for drug discovery. *Exp. Opin. Drug Discov.* **2012**, *7*, 969-88.

[92]   Fang, Y. Label-free cell-based assays with optical biosensors in drug discovery.*Assays Drug Dev. Technol.* **2006**, *4*, 583-95.

[93]   Verdonk, E.; Johnson, K.; McGuinness, R.; Leung, G.; Chen, Y.-W.; Tang, H.R.; Michelotti, J.M.; Liu, V.F. Cellular dielectric spectroscopy: a label-free comprehensive platform for functional evaluation of endogenous receptors. *Assays Drug Dev. Technol.* **2006**, *4*, 609-19.

[94]   Atienza, J.M.; Yu, N.; Kirstein, S.L.; Xi, B.; Wang, X.; Xu, X.; Abassi, Y.A. Dynamic and label-free cell-based assays using the real-time cell electronic sensing system. *Assays Drug Dev. Technol.* **2006**, *4*, 597-607.

[95]   Leung, G.; Tang, H.R.; McGuinness, R.; Verdonk, E.; Michelotti, J.M.; Liu, V.F. Cellular dielectric spectroscopy: A label-free technology for drug discovery. *J. Assoc. Lab. Automat.* **2005**, *10*, 258-69.

[96]   Tiefenthaler, K.; Lukosz, W. Sensitivity of grating couplers as integrated-optical chemical sensors. *J. Opt. Soc. Am. B* **1989**, *6*, 209-20.

[97]   Steyer, J.A.; Almers, W. A real-time view of life within 100nm of the plasma membrane. *Nat. Rev. Mol. Cell Biol.* **2002**, *2*, 268-75.

[98]   Fang, Y. Non-invasive optical biosensor for probing cell signaling. *Sensors* **2007**, *7*, 2316-29.

[99]   Fang, Y.; Ferrie, A.M.; Fontaine, N.H.; Mauro, J.; Balakrishnan, J. Resonant waveguide grating biosensor for living cell sensing. *Biophys. J.* **2006**, *91*, 1925-40.

[100] Ferrie, A. M.; Wu, Q.; Fang, Y. Resonant waveguide grating imager for live cell sensing. *Appl. Phys. Lett.* **2010**, *97*, 223704.

[101] Ferrie, A.M.; Deichmann, O.D.; Wu, Q.; Fang, Y. High resolution resonant waveguide grating imager for cell cluster analysis under physiological condition. *Appl. Phys. Lett.* **2012**, *100*, 223701.

[102] Li, G.; Ferrie, A.M.; Fang, Y. Label-free profiling of endogenous G protein-coupled receptors using a cell-based high throughput screening technology. *J. Assoc. Lab. Automat.* **2006**, *11*, 181-7.

[103] Xiao, C.; Lachance, B.; Sunahara, G.; Luong, J.H. Assessment of cytotoxicity using electric cell-substrate impedance sensing: concentration and time response function approach. *Anal. Chem.* **2002** *74*, 5748-53.

[104] Barer, R.; Joseph, S. Refractometry of Living Cells. Part 1: basic principle. *Quarterly J. Microscopical Sci.*, **1954**, *95*, 399-423.

[105] Fang, Y.; Ferrie, A.M.; Fontaine, N.H.; Yuen, P.K. Characteristics of dynamic mass redistribution of EGF receptor signaling in living cells measured with label free optical biosensors. *Anal. Chem.* **2005**, *77*, 5720-5.

[106] Fang, Y.; Li, G.; Ferrie, A.M. Non-invasive optical biosensor for assaying endogenous G protein-coupled receptors in adherent cells. *J. Pharmacol. Toxicol. Methods* **2007**, *55*, 314-22.

[107] Fang, Y.; Li, G.; Peng, J. Optical biosensor provides insights for bradykinin B_2 receptor signaling in A431 cells. *FEBS Lett.* **2005**, *579*, 6365-74.

[108] Tran, E.; Fang, Y. Label-free optical biosensor for probing integrative role of adenylyl cyclase in G protein-coupled receptor signaling. *J. Recept. Signal Transduct. Res.* **2009**, *29*, 154-62.

[109] Sato, T.K.; Overduin, M.; Emr, S.D. Location, location, location: membrane targeting directed by PX domains. *Science* **2001**, *294*, 1881-5.

[110] Tao J, Wang HY, Malbon CC. AKAR2-AKAP12 fusion protein "biosenses" dynamic phosphorylation and localization of a GPCR-based scaffold. *J. Mol. Signal.* **2010**, *5*, 3.

[111] Ritter, S.L.; Hall, R.A. Fine-tuning of GPCR activity by receptor-interacting proteins. *Nat. Rev. Mol. Cell. Biol.* **2009**, 10, 819-30.

[112] Dai, S.; Hall, D.D.; Hell, J.W. Supramolecular assemblies and localized regulation of voltage-gated ion channels. *Physiol. Rev.* **2009**, *89*, 411-52.

[113] Schröder, R.; Janssen, N.; Schmidt, J.; Kebig, A.; Merten, N.; Hennen, S.; Müller, A.; Blättermann, S.; Mohr-Andrä, M.; Zahn, S.; Wenzel, J.; Smith, N.J. Gomeza, J.; Drewke, C.; Milligan, G.; Mohr, K.; Kostenis, E. Deconvolution of complex G protein-coupled receptor signaling in live cells using dynamic mass redistribution measurements. *Nat. Biotechnol.* **2010**, *28*, 943-9.

[114] Giaever, I.; Keese, C.R. Monitoring fibroblast behavior in tissue culture with an applied electric field. *Proc. Natl. Acad. Sci. U.S.A.* **1984**, *81*, 3761-4.

[115] Giaever, I.; Keese, C.R. Micromotion of mammalian cells measured electrically. *Proc. Natl. Acad. Sci. U.S.A.* **1991**, *88*, 7896-900.

[116] Giaever, I.; Keese, C.R. A morphological biosensor for mammalian cells. *Nature* **1993**, *366*, 591-2.

[117] Gheorghiu, E.; Balut, C.; Gheorghiu, M. Dielectric behaviour of gap junction connected cells: a microscopic approach. *Phys. Med. Biol.* **2002**, *47*, 341-8.

[118] Lo, C.M.; Keese, C.R.; Giaever, I. Impedance analysis of MDCK cells measured by electric cell-substrate impedance sensing. *Biophys. J.* **1995**, *69*, 2800-7.

[119] Miskian, M.; Kasianowicz, J.J.; Robertson, B.; Petersons, O. Frequency response of alternating currents through the Staphylococcus aureus alpha-hemolysin ion channel. *Bioelectromagnetics* **2001**, *22*, 487-93.

[120] Solly, K.; Wang, X.; Xu, X.; Strulovici, B.; Zheng, W. Application of real-time cell electronic sensing (RT-CES) technology to cell-based assays. *Assays Drug Dev. Technol.* **2004**, *2*, 363-72.

[121] Ciambrone, G.J.; Liu, V.F.; Lin, D.C.; McGuinness, R.P.; Leung, G.K.; Pitchford, S. Cellular dielectric spectroscopy: a powerful new approach to label-free cellular analysis. *J. Biomol. Screen.* **2004**, *9*, 467-80.

[122] Yu, N.; Atienza, J.M.; Bernard, J.; Blanc, S.; Zhu, J.; Wang, X.; Xu, X.; Abassi, Y.A. Real-time monitoring of morphological changes in living cells by electronic cell sensor arrays: an approach to study G protein-coupled receptors. *Anal. Chem.* **2006**, 78, 35-43.

[123] Kawanabe, Y.; Okamoto, Y.; Nozaki, K.; Hashimoto, N.; Miwa, S.; Masaki, T. Molecular mechanism for endothelin-1-induced stress-fiber formation: analysis of G proteins using a mutant endothelin(A) receptor. *Mol. Pharmacol.* **2002**, *61*, 277-84.

[124] Pierce, K.L.; Fujino, H.; Srinivasan, D.; Regan, J.W. Activation of FP prostanoid receptor isoforms leads to Rho-mediated changes in cell morphology and in the cell cytoskeleton *J. Biol. Chem.* **1999**, *274*, 35944-9.

[125] Schraufstatter, I.U.; Chung, J.; Burger, M. IL-8 activates endothelial cell CXCR1 and CXCR2 through Rho and Rac signaling pathways. *Am. J. Physiol. Lung Cell Mol. Physiol.* **2001**, *280*, L1094-103.

[126] Saltarelli, D. Heterotrimeric Gi/o proteins control cyclic AMP oscillations and cytoskeletal structure assembly in primary human granulosa-lutein cells. *Cell Signal.* **1999**, 11, 415-33.

[127] Perez, V.; Bouschet, T.; Fernandez, C.; Bockaert, J.; Journot, L. Dynamic reorganization of the astrocyte actin cytoskeleton elicited by cAMP and PACAP: a role for phosphatidylInositol 3-kinase inhibition. *Eur. J. Neurosci.* **2005**, *21*, 26-32.

[128] Tran, E.; Sun, H.; Fang, Y. Dynamic mass redistribution assays decodes surface influence on signaling of endogenous purinergic receptors. *Assays Drug Dev. Technol.* **2012**, *10*, 37-45.

[129] Fliri, A.F.; Loging, W.T.; Thadeio, P.F.; Volkmann, R.A. Biological spectra analysis: linking biological activity profiles to molecular structures. *Proc. Natl. Acad. Sci. USA* **2005**, *102*, 261-6.

[130] Fang, Y.; Ferrie, A.M. Optical biosensor differentiates signaling of endogenous PAR_1 and PAR_2 in A431 cells. *BMC Cell Biol.* **2007**, *8*, 24.

[131] Fang, Y. Probing cancer signaling with resonant waveguide grating biosensors. *Exp. Opin. Drug Discov.* **2010**, *5*, 1237-48.

[132] Murry, C.E.; Keller, G.Differentiation of embryonic stem cells to clinically relevant populations: lessons from embryonic development. *Cell* **2008**,*132*, 661-80.

[133] Pai, S.; Verrier, F.; Sun, H.; Hu, H.; Ferrie, A. M.; Eshraghi, A.; Fang, Y. Dynamic mass redistribution assay decodes differentiation of a neural progenitor stem cell. *J. Biomol. Screen.* **2012**, *17*, 1180-91.

[134] Hermans, E. Biochemical and pharmacological control of the multiplicity of coupling at G-protein-coupled receptors. *Pharmacol. Ther.* **2003**, *99*, 25-44.

[135] Lavis, L.D.; Raines, R.T. Bright ideas for chemical biology. *ACS Chem. Biol.* **2008**, *3*, 142-55.

[136] Kaelin Jr., W. G.; Thompson, C. B. Q&A: Cancer: clues from cell metabolism. *Nature* **2010**, *465,* 562-4.

[137] An, S.; Kyoung, M.; Allen, J.J.; Shokat, K.M.; Benkovic, S.J. Dynamic regulation of a metabolic multi-enzyme complex by protein kinase CK2. *J. Biol. Chem.* **2010**, *285*, 11093-9.

[138] Boehr, D.D.; Nussinov. R.; Wright. P.E. The role of dynamic conformational ensembles in biomolecular recognition. *Nat. Chem. Biol.* **2009**, *5*, 789-96.

[139] Dror, R.O.; Pan, A.C.; Arlow, D.H.; Borhani, D.W.; Maragakis, P.; Shan, Y.; Xu, H.; Shawa, D.E. Pathway and mechanism of drug binding to G-protein-coupled receptors. *Proc. Natl. Acad. Sci. USA* **2011**, *108*, 13118-23.

[140] Wang, H.; Berrios, M.; Malbon, C.C. Localization of beta-adrenergic receptors in A431 cells *in situ*. Effect of chronic exposure to agonist. *Biochem. J.* **1989**, *263*, 519-38.

[141] Desaphy, J.-F.; Pierno, S.; de Luca, A.; Didonna, P.; Camerino, D.C. Different ability of clenbuterol and salbutamol to block sodium channels predicts their therapeutic use in muscle excitability disorders. *Mol. Pharmacol.* **2003**, *63*, 659-70.

[142] Eisen, M.B.; Spellman, P.T.; Brown, P.O.; Botstein, D. Cluster analysis and display of genome-wide expression patterns. *Proc. Natl. Acad. Sci. USA* **1998**, *95*, 14863-8.

[143] Peters, M.F.; Scott, C.W. Evaluating cellular impedance assays for detection of GPCR pleiotropic signaling and functional selectivity. *J. Biomol. Screen.* **2009**, *14*, 246-55.

[144] Dodgson, K.; Gedge, L.; Murray, D.C.; Coldwell, M. A 100K well screen for a muscarinic receptor using the Epic label-free system - a reflection on the benefits of the label-free approach to screening seven-transmembrane receptors. *J. Recept. Signal. Transduct. Res.* **2009**, *29*, 163-72.

[145] Jordan, J.D.; Landau, E.M.; Iyengar, R. Signaling networks: the origins of cellular multitasking. *Cell* **2000**, *103*,193-200.

[146] Tran, E.; Fang, Y. Duplexed label-free G protein--coupled receptor assays for high-throughput screening. *J. Biomol. Screen.* **2008**, *13*, 975-85.

[147] Kenakin T. The classification of seven transmembrane receptors in recombinant expression systems. *Pharmacol Rev.* **1996**, *48*, 413-63.

[148] Kukkonen, J.P.; Nasman, J.; Akerman, K.E.O. Modelling of promiscuous receptor-Gi/Gs-protein coupling and effector response. *Trends Pharmacol. Sci.* **2001**, *22*, 616-22.

[149] Kenakin, T. Differences between natural and recombinant G protein-coupled receptor systems with varying receptor/G protein stoichiometry. *Trends Pharmacol. Sci.* **1997**, *18*, 456-64.

[150] Jansson, C.C.; Kukkonen, J.P.; Nasman, J.; Huifang, G.E.; Wurster, S.; Virtanen, R.; Savola, J. M.; Cockcroft, V.; Akerman, K. E. Protean agonism at alpha2A-adrenoceptors. *Mol. Pharmacol.* **1998**, *53*, 963-8.

[151] Jansson, C.C.; Savola, J.M.; Akerman, K.E.O. Different sensitivity of alpha 2A-C10 and alpha 2C-C4 receptor subtypes in coupling to inhibition of cAMP accumulation.*Biochem Biophys. Res. Commun.* **1994**, *199*, 869-75.

[152] Deng, H.; Hu, H.; Fang, Y. Tyrphostin analogs are GPR35 agonists. *FEBS Lett.* **2011**, *585*, 1957-62.

[153] Deng, H.; Hu, H.; He, M.; Hu, J.; Niu, W.; Ferrie, A.M.; Fang, Y. Discovery of 2-(4-methylfuran-2(5H)-ylidene)malononitrile and thieno[3,2-b]thiophene-2-carboxylic acid derivatives as G protein-coupled receptor-35 (GPR35) agonists. *J. Med. Chem.* **2011**, *54*, 7385-96.

[154] Deng, H.; Hu, J.; Hu, H.; He, M.; Fang, Y. Thieno[3,2-b]thiophene-2-carboxylic acid derivatives as GPR35 agonists. *Bioorg. Med. Chem. Lett.* **2012**, *22*, 4148-52.

[155] Deng, H.; Fang, Y. Aspirin metabolites are GPR35 agonists. *Naunyn-Schmiedeberg's Arch. Pharmacol.* **2012**, *385*,729-37.

[156] Deng, H.; Hu, H.; Fang, Y. Multiple tyrosine metabolites are GPR35 agonists. *Sci. Rep.* **2012**, *2*, 373.

[157] Deng, H.; Fang, Y. Synthesis and agonistic activity at the GPR35 of 5,6-dihydroxyindole-2-carboxylic acid analogs. *ACS Med. Chem. Lett.* **2012**, *3*, 550-4.

Send Orders for Reprints to reprints@benthamscience.net

CHAPTER 5

Immunological Assays: Biotools for High Throughput Screening and Characterisation of Combinatorial Libraries

Maria Â. Taipa*

Institute for Biotechnology and Bioengineering (IBB), Centro de Engenharia Biológica e Química, Instituto Superior Técnico, Lisboa, Portugal

Abstract: In the demanding field of proteomics, there is a great need for affinity-catcher molecules to implement effective and high throughput methods for analysing the human proteome or parts of it. Antibodies have an essential role in this endeavour, and selection, isolation and characterization of specific antibodies represent a key issue to meet success. Alternatively, it is expected that novel affinity reagents generated in fast, cost-effective manners will also be used to facilitate the deciphering of the function, location and interactions of the high number of encoded protein products. Combinatorial approaches combined with high throughput screening technologies have become essential for the generation and identification of robust affinity reagents from biological combinatorial libraries and the discovery of active/mimic molecules in large chemical libraries. Phage and yeast display provide the means for engineering a multitude of antibody-like molecules against any desired antigen. The construction of peptide libraries is commonly used for the identification and characterisation of ligand-receptor specific interactions, and the search for novel ligands for protein purification. Further improvement of chemical and biological resistance of affinity ligands encouraged the "intelligent" design and synthesis of chemical libraries of low-molecular-weight bio-inspired mimic compounds. No matter what the ligand source, selection and characterization of leads is a most relevant task. Immunoassays are a biological tool of inestimable value for the iterative screening of combinatorial ligand libraries for tailored specificities, and improved affinities. Particularly, enzyme-linked immunosorbent assays are frequently the method of choice in a large number of screening strategies, for both biological and chemical libraries.

Keywords: Antibody engineering, combinatorial chemistry, biological libraries, chemical synthesis, high throughput screening, immunoassays, ELISA.

*Address correspondence to Maria Â. Taipa: IBB – Institute for Biotechnology and Bioengineering, Centro de Engenharia Biológica e Química, Instituto Superior Técnico, Av. Rovisco Pais, 1049-001 Lisboa, Portugal; Tel: +351.21.841 9065; Fax + 351.21.841 9062; E-mail: angela.taipa@ist.utl.pt

INTRODUCTION

The explosion in genome sequencing, and in subsequent DNA array experiments, has provided extensive information on gene sequencing, organization and expression. This has resulted in the need to perform similarly broad experiments on all the proteins encoded by a genome, namely the human genome [1,2]. Taking into account that the human genome contains $\approx$ 25000 genes, and that their products are found in different splice variants and produce proteins with post-translaccional modifications, it is expected that there are at least 100 000 different protein products to be investigated [3]. The endeavour of this task requires wide panels of appropriate tools based on highly specific affinity ligands, such as antibodies and genetically engineered related molecules [4-6], that can provide high throughput, high sensitivity and resolution, as well as the possibility of identifying post-tranlacctional modifications in proteins. Alternatively, it is also expected that bioengineered/synthetic affinity ligands for protein/antibody separation and selection will gain increasing importance in the proteome analysis as affinity reagents for the enrichment or depletion of proteins or groups of proteins by affinity chromatography [7].

Due to their remarkable properties regarding affinity and specificity of molecular recognition, antibodies became invaluable tools in research, diagnostic and clinical applications. As so, this class of proteins is routinely used in biochemical and biological research as analytical reagents for qualitative or quantitative determination of molecules in a wide variety of assays [8]. Antibody-based (or immuno-) assays are the fastest technology for the analysis of biomolecules, and are the most used and potent reagents available for facing the challenging task of analysing of the human proteome [5,6,9].

The first milestone for the generalised use of immunoassays in biomedical research and clinical chemistry was the development of hybridoma technology, described in 1975 by Köhler and Milstein [10], that made possible an almost universal method to produce monoclonal antibodies against any desired antigen. Monoclonal antibodies are produced by fusion of spleen cells from an immunized animal with myeloma cells, to obtain hybridoma cells that are further selected and screened for the production of anti-target antigen antibodies. Hybridoma cells can be continuously grown, producing exactly the same antibody molecule for months or years. The mouse hybridoma technology has therefore paved the way for the exploitation of high affinity and specificity of antibody molecules in many different applications, and for the emergence of antibodies as effective therapeutics in inflammation, cancer, autoimmune, and infectious diseases [11-13].

During the 1980's, the integration of hybridoma technology with recombinant DNA technology and the development of new and valuable screening tools as the phage display technique enabled the design and construction of human-like antibodies with desirable affinity/specificity, and low human immunogenicity [14-17]. The immense possibilities offered by hybridoma technology combined with protein-engineering display techniques turned out possible to treat antibodies more like a 'chemical' instead of a variable biological product [8]. This has granted antibodies a distinguished position as therapeutic agents and as the most used ligand in industrial downstream processing of therapeutic proteins. For such applications antibodies have to be produced and purified to reach highly demanding purity criteria [17,18]. Affinity chromatography is traditionally the method of choice for the purification of antibodies. The need for antibodies with unique specificities and designed for single applications, has encouraged the search for novel purification methodologies and novel affinity ligands for antibody capture and purification [19]. Further advances over the 1990's in molecular modeling *in silico*, genetic engineering and biological and chemical combinatorial methods, combined with high throughput screening (HTS) technologies provided important means for generating and selecting new biologically active molecules, tailored to specific biotechnological needs. These molecules that often display enhanced robustness, resistance, stability and cost-efficient production as compared to their natural templates include engineered antibody-like molecules such as affibodies [20,21], novel antibody-mimic domains based on resistant protein scaffolds [22-24] small peptides [18,25], and triazine-based synthetic affinity ligands that mimic the interaction of natural specific receptors with their complementary proteins [26]. All these molecules have also a broad spectrum of potential applications as antibody substituents in diagnostic [27,28], separomics/proteomics [7] and therapy [29-32].

Immense progress in HTS strategies, which possess the "power of numbers" for the resolution, characterisation, and purification of target active molecules, have encompassed the advances in genetic engineering to allow the screening of large combinatorial libraries of either biological or chemical origin. The screening strategies are often based in the immunoafinity concept, relying on the specific, reversible and non-covalent interaction between an antibody and a biological target, or its analogue. ELISA (enzyme linked immunosorbent assays) assays in 96w microtitre plates have been so far the most used HTS methodology for the screening and characterization of combinatorial libraries. In the last decades, improvements in automation devices have resulted in reduced assays time without compromising specificity and sensitivity [3,33]. Furthermore, antibody engineering and phage-display antibodies has resulted in greater assay flexibility, allowing the design of antibody-like molecules with improved affinity and fine specificity, fused to useful molecular markers, which today hold also great promise within the area of microarray technology [8,9,34].

THE ANTIBODY MOLECULE

Antibodies or immunoglobulins are a group of bifunctional glycoproteins with unique structural features that play a central role in the functioning and regulation of the immune system in mammals. Structurally, immunoglobulins are glycoproteins with a common Y-shaped building block comprised of two identical light chains (L) (~25 kDa) and two identical heavy chains (H) (~50 kDa), associated by disulphide covalent forces and by non-covalent interactions. Each chain is composed of constant (C_L and C_H) and variable domains (V_L and V_H). These form two antigen-binding fragments (Fab) and a constant region involved in the effector function and biodistribution of the antibody (Fc), linked *via* the flexible hinge region (Fig. **1**).

All the antibody or immunoglobulin domains have the same conformation, consisting of a "sandwich" of two β pleated sheets, each containing anti-parallel β-strands of amino acids. The chains fold into a tertiary structure of compact globular domains (40x25x25 Å) that interact to form the quaternary structure, and thus create the biological activity of the molecule. Five distinct classes of immunoglobulin molecules (corresponding to heavy chain isotypes: α (IgA), δ

(IgD), ε (IgE), μ (IgM) and γ (IgG)) are recognised in higher mammals and differ in size, charge, biological properties, amino acid composition and carbohydrate content [35]. Immunoglobulin oligosaccharide chains are extremely hetero-geneous and differ in both composition and structure. Carbohydrate components can facilitate the maintenance of the structure of the C_H2 domains and are also crucial for the bioactivity of the molecule as they are involved in IgG interactions with ligands and receptors [36].

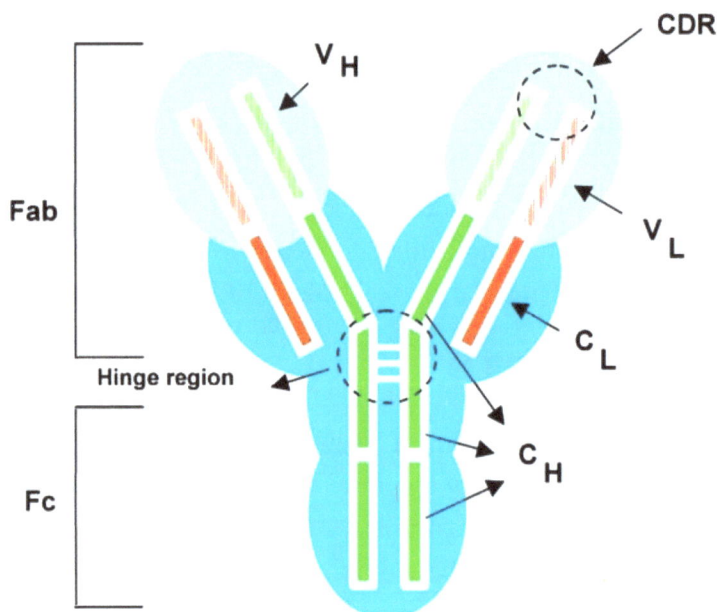

Fig. (1). Simplified schematic representation of the several domains of the Y-shaped building block of immunoglobulins, comprising two identical heavy chains (H) and two identical light chains (L). CDR represents the 'complementary determining region', a region of amino acid hypervariability within the variable domains of both light and heavy chains that is responsible for the specific molecular recognition of the antigen.

ANTIBODY ENGINEERING

Genetic and Combinatorial Tools

The increasing knowledge of antibody gene structure, *in vivo* expression, and regulation has made possible the generation of antibody-derived structures, which were denominated as "man-made" antibodies [37]. Genetic engineering of the basic immunoglobulin structure allowed the combination of target regions of antibodies into small functional monomeric or multimeric proteins such as Fab and F(ab')$_2$ fragments, single chain variable fragments (ScFv) which are heterodimers of V_H and V_L stabilised by an hydrophilic linker - and combination of these as diabodies and triabodies (reviewed in [17,21]).

The notion that immunoglobulins owe their function to the hypervariable composition of a conserved framework region has also attracted considerable attention in the context of generating novel types of ligand receptors mimicking antibody specificity and affinity [22,23,38,39]. Using combinatorial chemistry to create novel binding molecules based on protein frameworks or 'scaffolds' is a concept that has been strongly promoted during the past years in both academia and industry (reviewed in [40]). Non-antibody recognition proteins derive from different structural families and mimic the binding principle of immunoglobulins to varying degrees. For example, peptides with known affinity towards a certain target can be fused to, or inserted to, a carrier protein to combine its binding properties with desired favourable chacteristics (*e.g.* effective tissue penetration for the treatment of solid tumours, or intracellular activity for targeting cellular signal pathways) of the scaffold [40]. Combinatorial engineering of immnunoglobulin-binding domains such as the α helical bacterial receptor derived from staphylococcal Protein A, allowed the generation of novel binding proteins called affibodies, which present a secondary structure similar to the scaffold domain and specific affinity for a wide range of targets they are selected for [20,21]. The 10th human fibronectin type III (^{10}Fn3) domain is one of protein scaffolds used in recent years to design, and select, *in vitro*, a wide range of proteins that bind with high affinity and specificity to a variety of macromolecular targets [41-43]. The fibronectin III domain constitutes a small (94 amino acids), monomeric natural β-sandwich protein that is made up of seven strands with three connecting loops at one end of the β-sheet. Randomisation of three N terminal loops in the tenth of the 15 repeating units in human fibronectin showed that this domain can be used as a scaffold to obtain different protein binding variants with detectable biological activity [40]. High-affinity antibody mimics (Kd ≤ 350 pM) based on the ^{10}Fn3 scaffold have been generated and selected by evolution of an interloop disulphide bond, using a yeast surface display combinatorial approach [24].

BIOLOGICAL COMBINATORIAL LIBRARIES

Biological combinatorial libraries arose with the dawn of display techniques as a platform for the presentation of several biomolecules. Phage display has been applied in many fields of biological and medical sciences, and used to generate biological combinatorial libraries of different sizes. The use of filamentous phages that infect *Escherichia coli* has proven to be a very powerful technique to display millions or even billions of different peptides or proteins as coat-protein fusions. These include the expression and selection of peptides binding to antibody epitopes [44-46], the selection of peptides for protein purification [47,48], the expression and affinity maturation of proteins, as, for example, multivalent avimer proteins, a class of binding proteins evolved from a large family of human extracellular receptor domains (human A-domains) with high therapeutic potential [39], and the selection of specific antibodies from large naïve, immune libraries or synthetic libraries [49-52].

Numerous display methods have been developed to obtain recombinant antibodies with remarkable properties, some of which were never obtained with immunization (reviewed in [4,52,53]). Antigen-specific antibodies can be directly selected either from antibody gene repertoires expressed *in vivo* on the surface of filamentous bacteriophages [49], bacterial [54-56], and yeast cells [56-60] or *in vitro* by ribosome display [61-63] or puromycin display [64]. However, phage display is likely the most successful and commonly used method for combinatorial antibody display and screening [3-6,52,65].

Phage Display

Phage display can mimic the strategy used by the humoral immune system to produce fully human antibodies or antibody fragments *in vivo* and can exempt immunisation and hybridoma construction. A prerequisite for such an approach is the ability to isolate and clone the variable (V) region genes from the immunoglobulin heavy and light chains to display them functionally on a phage.

According to the source of the immunoglobulin genes, antibody phage libraries can be classified as immune (where the variable (V) genes are isolated from IgG-secreting plasma cells from immunised donors) or "single pot", that comprise non-imune naïve, semi-synthetic and fully synthetic libraries [9,52,66]. In the case of naïve libraries, the size of the library influences directly the affinity and specificity of the isolated antibodies, with Kd values ranging from 10^{-6} to 10^{-7} for smaller libraries to 10^{-9} for larger ones [67]. However, when selected antibodies are used as the basis for subsequent libraries and selection, affinities can be further increased to levels unobtainable in natural immune systems. Affinity and specificity can be improved by *in vitro* affinity maturation [65,68] or the use of semi-synthetic strategies by selecting one [69] or more antibody frameworks as a scaffold and randomising sequences within the CDR loops [14]. "Single pot" antibody libraries with a theoretical diversity of 10^{11} have been generated, with affinities ranging from 10^{-6} to 10^{-11}, being useful sources for selecting antibodies for proteome research [9].

Recombinant antibodies are generally displayed in non-lytic filamentous phages, fd or M13, that infect strains of *Escherichia coli* containing the F conjugative plasmid. Phage particles attach to the tip of F pillus that is encoded by genes on this plasmid, and the phage genome, a circular stranded DNA molecule, is translocated into the cytoplasm. The genome is replicated involving both phage and host derived proteins, and packaged by the infected cell into a rod-shaped particle which is released into the medium. Though fusion to different phage coat proteins can be used (pIII, PVI and PVIII) [49] most successful phage antibody libraries, against a wide variety of antigens, have been created by cloning antibody variable genes upstream of the gene of the pIII coat protein gene and using phage or phagemid as the display vehicles [4,66,70]. Phagemids are small plasmid vectors that carry gene III with appropriate cloning sites and a packaging signal, providing high transformation efficiencies and, therefore, are ideally suited for generating very large combinatorial repertoires [49,66]. Strategies as the addition of helper phage systems [71,72] and, engineering *E. coli* cell lines [73] have been developed to improve the enrichment of specific antibodies in phaghemid-based combinatorial libraries. A general scheme for the lead isolation of antibody fragments by phage display is shown in Fig. (**2**). Due to limitations of the *E.coli* folding machinery, complete IgG molecules can hardly be expressed in *E. coli* and displayed on the surface of phages. Therefore, smaller antibody fragments such as Fab, ScFv fragment or diabody fragments are mainly used for antibody phage display. However, once a Fab or scFv fragment with high affinity and specificity for a target antigen has been obtained, it is possible to genetically reconstruct these fragments into an intact fully human antibody for therapeutic application [53,74].

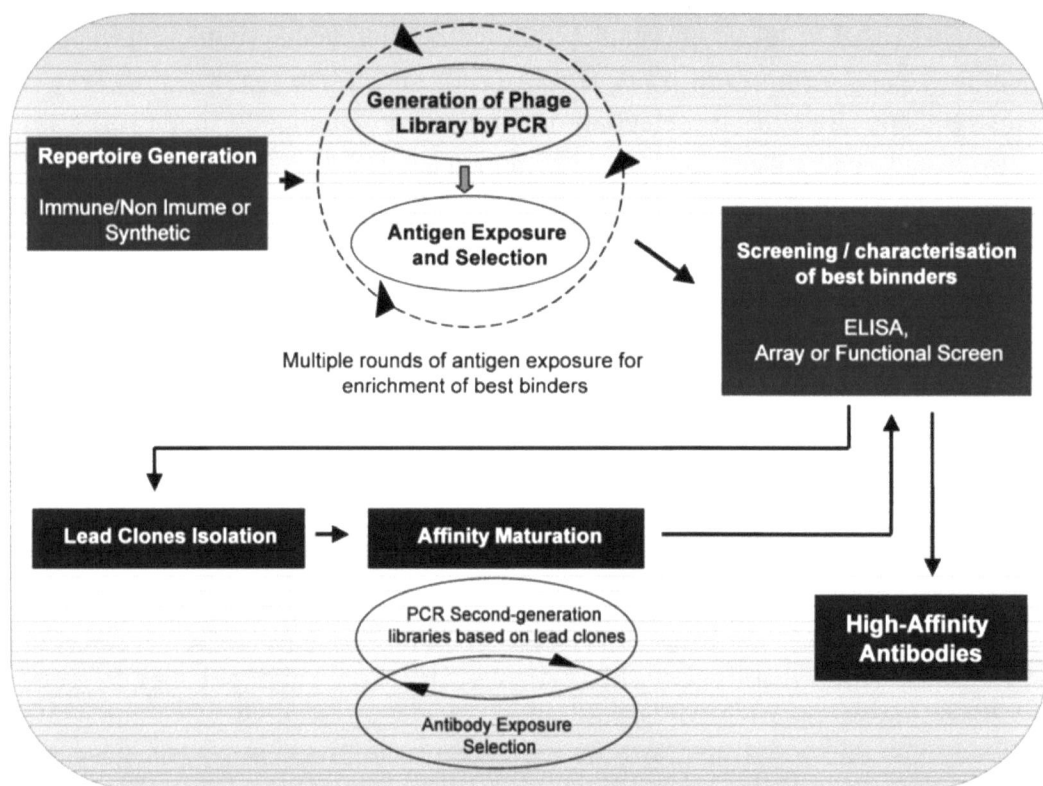

Fig. (2). Isolation of lead antibody fragments from a phage-display library involves several steps which include the generation of the library from a repertoire of antibody variable genes, the production of antibody-displaying phages, and the selection of high affinity binders by exposure to antigen. Lead antibodies can be identified by ELISA assays, microarrays or biological-activity screening in cell-based or receptor assays. Although screening thousands of clones can yield a greater diversity of "binders" after a single round, generally several (two-to-three) rounds of enrichment of the preliminary library and *in vitro* affinity maturation of the gene repertoire of lead clones are required to isolate antibodies that bind with high affinity and specificity to the target antigen (adapted from [23]).

Yeast Surface Display

Among the display techniques, yeast surface display has emerged as a powerful platform to the combinatorial engineering of a variety of proteins, including antibody fragments, for improved affinity, specificity, expression and stability (reviewed in [60]). The fundamentally different folding capacity of prokaryotic and eukaryotic cells restrict the presentation of some correctly folded proteins on bacteria, in particular if they are large and of mammalian origin. Typical examples are antibody fragments consisting of at least two domains, and sometimes two independent polypeptide chains (Fabs), requiring disulphide bond formation [56]. Yeast display system offers an alternative to phage display. It is based on a carboxyterminal fusion to the Aga 2p subunit of *Saccaromyces cerevisiae a*-agglutinin receptor. Through two disulphide bridges, the Aga 2p fusion protein is bound to the Aga 1p subunit of the same receptor that anchors the assembly to the yeast cell wall [57]. This display system has been used for the successful display and engineering (by directed evolution) of several proteins, including the single-chain T cell receptor (scTCR) [75], the epidermal growth factor receptor (EGFR) [60], and fibronectin-based high-affinity antibody mimics [24].

One of the most successful applications of yeast surface display has been the selection and affinity maturation of antibody fragments, both scFv fragments [57-59,76] and Fab [77,78]. Recently, yeast display has also been used for clinical applications such as the assessment of tumour-specific antibody responses in cancer patients and the screening of novel tumour antigens not previously detected from prokaryote-displayed libraries. In addition, several groups have isolated novel lead antibodies binding to a variety of targets from immune and non-imune human scFv libraries [60]. Bowley and co-workers have compared yeast surface display and phage display using identical immune ScFv antibody libraries and target antigens [79]. They have found that yeast display is able to sample a considerable more fully immune ScFv library repertoire, while also being less labour intensive. A main advantage of combinatorial yeast display over phage display advents from the employment of a eukaryotic host system possessing the secretory biosynthetic machinery for efficient protein folding and

N-linked glycosylation, and the ability to characterise isolated antibody clones directly in display format by flow-cytometry analysis [60,76].

CHEMICAL COMBINATORIAL LIBRARIES

Combinatorial chemistry allows nowadays the synthesis of a myriad of small molecules with biological applications, in a time and resource effective manner. Several formats can be used to undertake combinatorial synthesis, namely liquid-phase (in solution or using soluble polymers as supports), solid-phase or surfaces. Solid-phase chemistry, conjugated with "mix-and-split" strategies, continues to hold a dominant position in combinatorial chemistry. Generally, compounds being synthesized are attached, through a linker group, to insoluble, functionalised polymeric materials or beads, facilitating their separation from excess reagents, soluble reaction by-products or solvents [80].

Combinatorial methods have become important tools for several applications as in discovery of better catalysts [81] or active small molecules of agrochemical relevance [82]. A major hit of chemical combinatorial approaches has been in the area of medicinal chemistry, in the discovery and screening of novel active chemical compounds for binding specific cell receptors [83-85], the design of specific and potent bacterial endotoxin sequestrants [86] or the design, synthesis and evaluation of novel anticancer agents based on natural products-templated ligand libraries with cytotoxic activity [87]. Also, small-molecule microarrays emerged as a tool for identifying proteins and/or drugs with pharmacological interest [88]. In the last decade, the integration of molecular modeling with combinatorial chemistry has also allowed the generation of synthetic affinity ligands for protein and antibody purification [18,20,26,89]. Many of these novel ligands may advantageously replace naturally occurring receptor-binding proteins in biotechnological, clinical or diagnostic applications as they possess high selectivity, stability and low-cost of production.

Peptides

Peptide libraries have assumed an important role in the identification and characterization of ligand-receptor interactions. Large synthetic combinatorial libraries (SCL) of peptides may be chemically synthesised by different techniques ('split synthesis', 'peptides on beads', 'synthetic peptide combinatorial libraries' and photolithography) which can be used separately or combined [68,90,91]. Application of SCL of peptides is a relevant approach in both proteome and cancer research context [92]. The identification of synthetic mimics that are recognized by an antibody, regardless of whether the mimic retains apparent resemblance to immunogen, is of major interest in the development of more effective immunoassays and synthetic vaccine candidates [93]. Peptide SCLs have been extensively used for epitope and mimotope mapping of monoclonal antibodies against clinical relevant proteins and cell receptors [94-96]. Combinatorial libraries of peptides are also collections of synthetic compounds of utmost importance to screen for diagnostic agents that bind to disease-specific antibodies [27,28], for inhibitors of protein-protein interactions that regulate important mechanisms in diseases such as type 2 diabetes [32] and for affinity ligands for protein purification [97]. A major application in this area has been the synthesis of mimic peptides for antibody purification [19,98,99]. Difficulties in mimicking discontinuous epitopes with linear peptides has led to specific approaches (*e.g.* cyclization) for generating peptides with better mimic properties, increased resistance to enzymatic degradation and constrained flexibility [68]. As an example, Verdoliva and co-workers reported the selection of a disulfide-bridged cyclic peptide with the general formula $(NH_2-Cys_1-Phe-His-His-)_2-Lys-Gly-OH$ (named Fc-receptor mimetic – (FcRM) as an efficient ligand for the removal of antibodies from biological fluids to purities up to 90% [30]. Another relevant example has been the search for staphyloccocal IgG-binding Protein A mimetic by the design of a tetrameric peptide library, where four identical peptide chains were assembled starting from a tetradentate lysine core [100]. A multimeric library has been produced by solid-phase peptide synthesis. After three screening cycles, the multimer $(Arg-Thr-Tyr)_4-K_2-K-G$, denoted PAM (protein A mimetic) or TG19318, was proven to be an efficient ligand for one-step purification of antibodies directly from crude sera up to 95% purity.

Biomimetic Ligands

Increased knowledge of the mechanisms involved in molecular recognition have allowed the design of new and powerful bio-inspired ligands, for protein and antibody purification. Bio-inspired or 'biomimetic' ligands are synthetic compounds of full chemical nature, with similar molecular recognition properties of their biological model (template). The general concept behind this class of ligands - often referred as 'artificial proteins'- relies on the combination of molecular modeling, through the evaluation and study of available X-ray crystallographic structures of the target proteins or complexes of proteins, solid-phase combinatorial synthesis and high throughput screening techniques. A most effective way for obtaining these compounds was pioneered by Lowe and co-workers [101], by integrating rational design with combinatorial synthesis and screening on the same support. Such integration avoids numerous indefinite factors that can be introduced upon immobilisation of solution-phase synthesised ligands on a matrix. Apart from the ligand properties and its ability to bind the target protein, the nature of the support and the coupling chemistry can affect the overall binding process. A well-established procedure is used to construct a solid-phase combinatorial library of triazine-based near-neighbour ligands interacting in a complementary affinity-like mode with protein surface-exposed key residues (Fig. **3**). The solid support utilised, agarose, has proven to satisfy both the exigencies for solid-phase synthesis and the properties required for ligand screening and application in affinity chromatography [26]. The triazine scaffold serves as the spatial framework for the display of attached functional groups and has shown to deliver effective protein binding ligands.

Using this strategy, a series of triazine scaffolded molecules presenting affinity to different proteins or families of proteins have been developed [102-107]. From a proteomics point of view, these synthetic ligands can represent a good alternative to biological ligands for depletion of complex biological mixtures to be further analysed by specific techniques (*e.g.* two-dimensional electrophoresis or mass spectrometry). For example, affinity chromatography has been recently reported for the elimination of immunoglobulin content in crude samples [108]. Synthetic affinity ligands mimicking the interaction of immunoglobulins with bacterial IgG-binding receptors have been synthesised and characterised, and can surmount problems associated with biological ligands, while preserving affinity and specificity to target antibodies. Ligand 22/8 (4-[4-Chloro-6-(3-hydroxy-phenylamino)-[1,3,5]triazin-2-ylamino]-naphthalen-1-ol), - named as 'artificial

Protein A' -mimics the natural *Staphylococcus* aureus Protein A receptor in binding IgG from various sources, separating IgG from human plasma to purities of 98-99% [103,104].

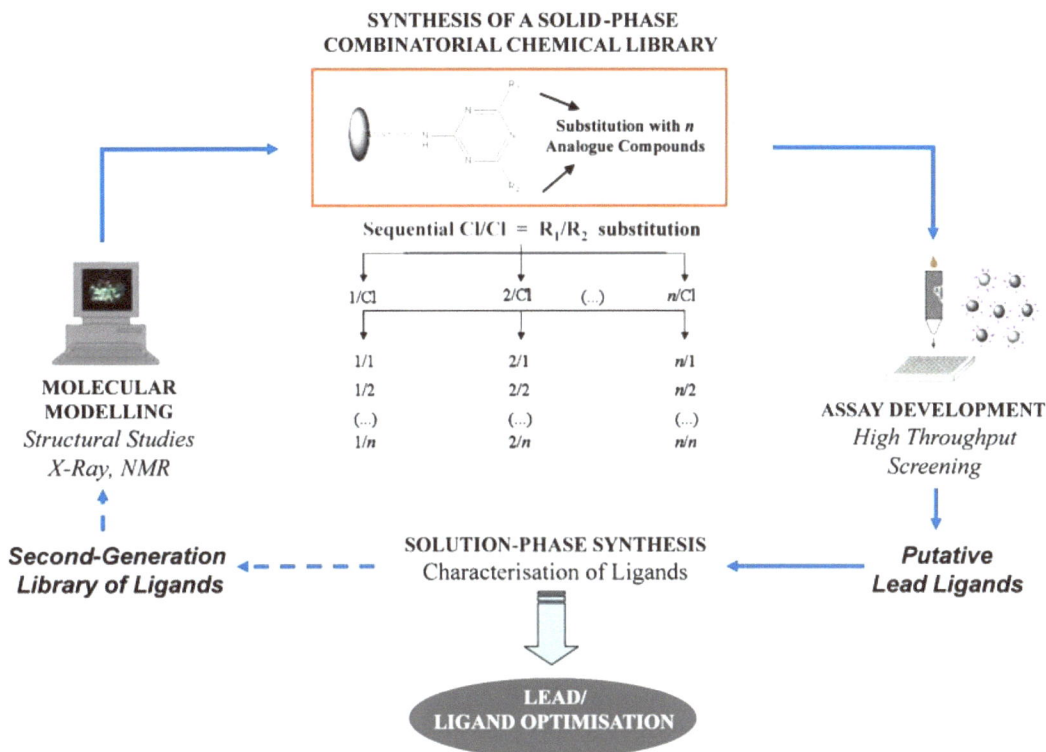

Fig. (3). General research strategy for the generation of effective *de novo* designed triazine-based synthetic mimic ligands is based on *i)* molecular modeling based on structural information available on the interaction of target biomolecules and *in silico* design of *n* chemical analogues that mimic the key residues involved in the specific molecular recognition; *ii)* synthesis of a *nxn* member solid-phase combinatorial library in agarose; *iv)* development of high throughput assays for ligand screening; *v)* solution-phase synthesis and further characterisation/optimisation of lead ligands. In some cases a second-generation library may be designed to improve the affinity/selectivity of selected putative lead ligands, allowing the synthesis of better mimics. Cl/Cl corresponds to dichlorotriazinyl agarose, R_1/Cl corresponds to a mono-substituted ligand and R_1/R_2 represents final bi-substituted ligands, with R_1 and R_2 being analogue compunds 1 to *n*.

Ligand 8/7 ('artificial Protein L') has shown to mimic Protein L in terms of antibody separation performance, while binding to IgG_1 with κ and λ isotypes [107]. Protein L is a bacterial receptor from *Peptostreptococcus magnus* strains, which binds to the Fab portion of immunoglobulins. The artificial Protein L has shown to compare well with the natural receptor in the isolation of immunoglobulins from different classes (human IgG, IgA and IgM) and species (rabbit, goat), and of a recombinant scFv-based protein from crude extracts [109,110]. More recently, the multicomponent Ugi-reaction was explored as an alternative to the well-defined triazine chemistry for generating synthetic ligands for immunoglobulin purification. A lead compound, selected from a combinatorial library, exhibited high dynamic binding capacity and ability to isolate highly pure IgG and Fab fragments from both eukaryotic and prokaryotic crude expression systems [89].

IMMUNOLOGICAL METHODS

HTS Tools for the Selection of Lead Ligands

Within the proteomic context, the development of HTS technologies is essential for the selection and screening of robust affinity reagents, either from biological [111] or chemical [112,113] combinatorial libraries. Immunological methods have proven to be a biological tool of inestimable value to reach this goal. Antibody-based assays are known for their versatility, sensitivity, specificity, high throughput and ease of automation. The interaction between of an antibody with its target (whether this a cell, tissue, a peptide or synthetic mimic ligand) is easily and rapidly monitored by classical assays involving the detection and measure of a complex formation by using a fluorescent-, radio- or enzyme- labelled antibody. Other biophysical techniques, such as NMR (Nuclear Magnetic Resonance) and optical spectroscopy (fluorescence, circular dichroism) combined with computer-assisted molecular modeling (CAAM) are also available to assist investigators to gain a better understanding of how antibodies can be used as tools for ligand library screening, and help to improve prediction, modeling and design of novel antigen-binding sites [25,114] and the construction of antibody microarrays.

Despite the efforts and progress in developing efficient microarrays [8] and adapting selection methods to robotics [9] screening and characterization of lead ligands of combinatorial libraries from both biological and chemical are frequently based on ELISA assays, in a classical or automated mode [111]. Given the ubiquity of the microtiter plate and its derivatives, it is probable that this format will continue as the standard in HTS of antibody and synthetic libraries. An automated, cost-effective, high throughput method, based on a 96-well plate filtration assay for the screening of affinity ligands has been reported by Sarawast and co-workers [115]. This methodology has shown to be effective for producing lead compounds, and may find utility as a generic tool for the functional characterization of novel proteins emerging from proteomics work.

Several methods have been reported and can be applied, alone or combined, for antibody phage display screening including biopanning, ELISA, antigen-coated magnetic particles, BIAcore sensor chips and selection by sorting procedures such as FACS [3,4,67,70,116]. However, most commonly, 96 phage antibodies (*i.e.*, a single microtitre plate) are screened by ELISA using relevant and non-relevant antigens [4,49]. If the antigen is not available, alternative strategies may involve the use of natural antigen sources (*e.g.* tissue sections, cell or tissue extracts) or small peptides (epitopes) [116]. Lou and co-workers have developed a successful and high throughput method for antibody-phage libraries screening based on the use of multipins in 96-well microtiter plates [70]. The method enables single aliquots of the library to be incubated with 96 different antigens simultaneously, using minimum volumes, with the advantage that fully library diversity can be assessed for each antigen. After several rounds of selection, a polyclonal ELISA is performed for each pool of amplified phage in order to identify positives that need to be further analysed at a monoclonal level [116]. Further approaches and developments in antibody library screening are reviewed in [52,111].

Regarding peptide mimetics or synyhetic mimic triazine-based affinity ligands, the iterative, high throughput screening of large solid-phase combinatorial libraries is commonly performed by combination of several methods as "on-bead" fluorescence microscopy binding assessment, affinity chromatography, and quantitative/competitive ELISA assays for ligand selection and characterization [30,103,104,107].

ELISA-Based Methods

ELISA or enzyme linked immunosorbent assays involve the adsorption of an antigen, (protein, peptide or analogue) or an antibody on the bottom of a 96 well microtiter plate. ELISA tests can be developed in different formats, based either on competitive or non-competitive principles, depending on whether or not the antigen/analogue to be identified/quantified competes with a different labelled antigen for antibody binding sites [117]. Four common ELISA formats are represented in Fig. (**4**).

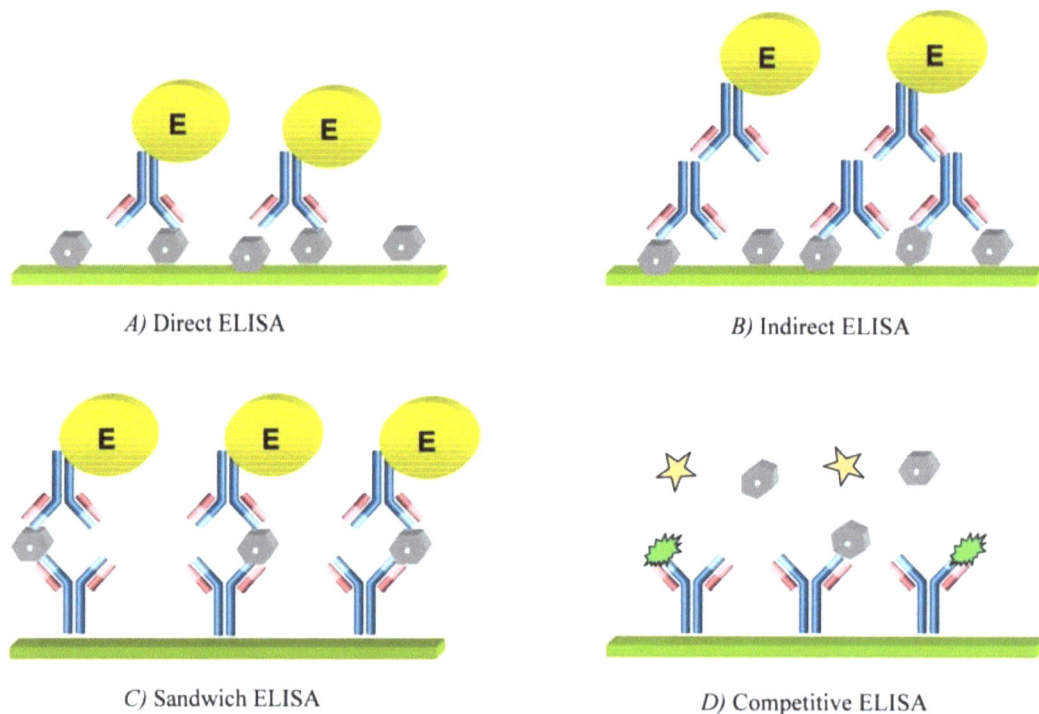

A) Direct ELISA

B) Indirect ELISA

C) Sandwich ELISA

D) Competitive ELISA

Fig. (4). Common formats for ELISA assays; **A**) a direct assay is possible if the antigen is pure and a primary labelled antibody is available; **B**) indirect assays are performed when not a primary but a secondary enzyme-labelled anti-antibody is available for detection; **C**) if the antigen is impure, a sandwich format is the best option. This assay requires two specific monoclonal antibodies that bind the antigen through two different epitopes, one that ensures antigen capture, and a second that acts as a detector; **D**) Competition ELISA assays allow the monitoring of the strength of binding of antibodies to different antigens or to their synthetic mimic analogues.

Non-competitive assays are simple and based on the direct or indirect detection of a target, using primary or secondary antibody reagents labelled with an enzyme (ELISA) - or alternatively, in a more late sense, with a fluorescent or a radioactive compound [117]. Competitive assays are frequently used to characterise the interaction of synthetic/analogue affinity reagents with target molecules or the competition of lead peptides and natural antigens in epitope mapping [29]. Competitive ELISA (C-ELISA) assays allow the relative comparison of the strength of binding to target proteins, and give evidence on epitope recognition by different antibody molecules. In the case of synthetic/antibody mimics, C-ELISA assays can monitor if bioengineered or synthetic molecules compete with their natural biological template in binding to the same *loccus* on the target protein surface [103,104,109]. Enzymes commonly used for antibody labelling are horseradish peroxidase (HRP) and alkaline phosphatase (AP), and a panoply of HRP- or AP- labelled primary and secondary antibodies and compatible substrate kits for ELISA assay development are commercially available. Table **1** summarizes examples of different types of combinatorial libraries where ELISA-based assays were used for lead-ligands selection and characterisation.

An interesting work from Vicennati and co-workers has demonstrated that, aside from spectroscopic and chemosensor-based methods, techniques that exploit the specific binding of antibodies might be highly valuable for the high throughput screening of combinatorial libraries of enantioselective catalysts. As so, a general-analytical tool based on a sandwich immunoassay was developed and successfully used for the screening of catalysts for cross-coupling reactions [81].

A representative example of a screening strategy based on a high throughput ELISA assay for the lead discovery of a bio-inspired synthetic ligand is illustrated in Fig. (**5**).

The development of an artificial Protein L binding specifically to the Fab domain of human IgG involved the iterative screening of a 169-membered solid-phase combinatorial library of triazine-scaffolded ligands [107]. All the synthesised ligands were firstly assessed for binding to human IgG by a new "on-bead" FITC-based method [118].

"Scale-down" and high throughput analysis for selectivity against human IgG fragments was driven by their high cost and limited availability. Positive ligands for human IgG were assessed for binding to human Fab by a quantitative ELISA assay, combined with a micro-scale affinity chromatography. Putative leads binding to Fab were screened against human Fc using a similar methodology

[107]. Ligand 8/7 which emerged as the lead from this screening strategy has proved to be highly selective to the Fab moiety of human IgG [109] and represents a synthetic mimic of Protein L with affinity and specificity for small antibody fragments (scFv, Fab or F(ab')$_2$).

Table 1. **Examples of Application of ELISA-Based Assays for the Screening and Characterisation of Diverse Biological and Chemical Combinatorial Ligand Libraries**

Target	Application	ELISA Selection Methods	References
Phage-displayed peptides	Identification of peptide affinity ligand for Anti-tenascin –C Ab	Competitive ELISA	[46]
Phage-Displayed Multivalent Avimer Proteins	Antibody mimics with potential therapeutic application	Indirect ELISA (for specificity analysis)	[39]
ScFv antibody phage-library based on chicken V genes	Selection of ScFv recognising haptens, proteins and virus	Indirect, Sandwich and Competitive ELISA assays	[51]
Phage-displayed Fab fragments	Selection of high-affinity Abs against human kallikrei tissue 1	Fab- automated Indirect ELISA Assay (Primary Screening)	[119]
Synthetic Peptides	Epitope Mapping: Peptides recognised by an Anti-carbohydrate Ab	Competitive ELISA	[29]
Synthetic Peptides (mimotopes)	Binding human IgM with an unusual combining site	Indirect ELISA	[120]
Synthetic Peptides (Antagonists)	Inhibitors of protein-protein interactions mediating molecular mechanisms in type 2 diabetes	Direct and Competitive ELISA Assays	[32]
Palladium-based Catalysts for Sonogashira Reaction	Selection of enantioselective catalysts for cross-coupled reactions	Sandwich ELISA	[81]
PAM (Protein A mimetic peptide)	Purification of Abs by targeting Fc region	Competitive ELISA	[100]
FcRM Mimetic Peptide	Purification of Abs	Indirect and Competitive ELISA	[30]
Artificial Protein A	Fc receptor synthetic mimic for Ab purification	Competitive ELISA	[103,104]
Artificial Protein L	Synthetic mimic for purification of Ab fragments (Fab, ScFv (Fab)$_2$	Quantitative Direct ELISA Competitive ELISA	[107,109]

Abs.; Antibodies.

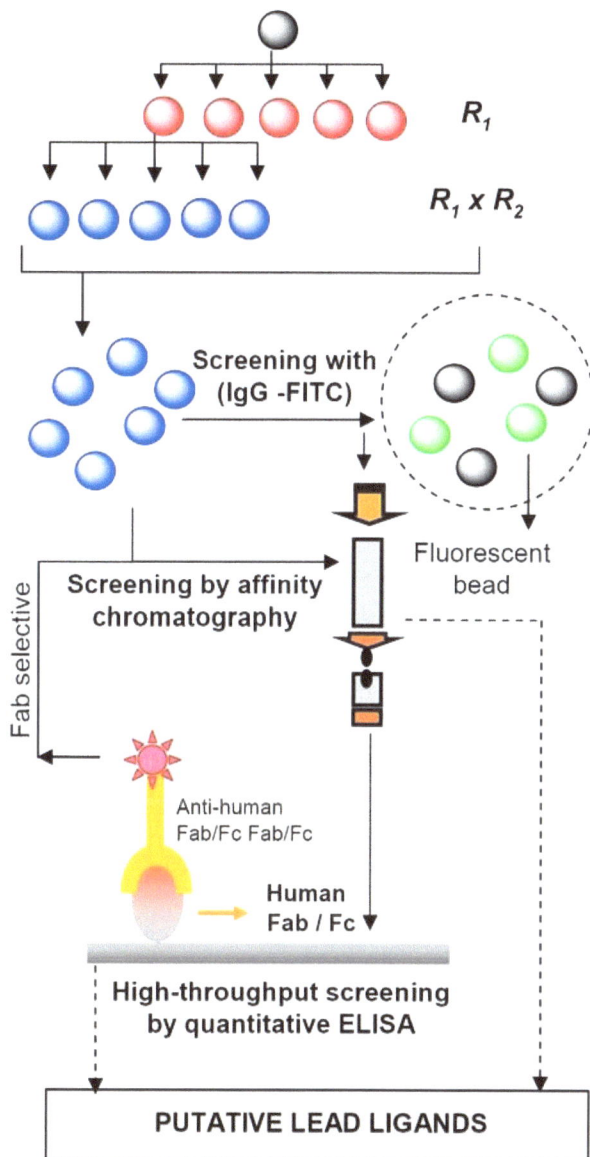

Fig. (5). Screening strategy for the lead discovery of a synthetic affinity ligand (artificial Protein L) mimicking the interaction between Protein L and the Fab portion of human IgG.

CONCLUSIONS

Combinatorial molecular technology and display systems combined with high throughput screening methodologies provide the means to generate variability in a diverse set of molecules such as engineered antibody-like molecules, novel antibody-mimic domains based on resistant protein scaffolds, small peptides, and synthetic affinity ligands that mimic the interaction of natural specific receptors with their complementary proteins. Antibody-based proteomics is nowadays the most powerful approach for the study of the human proteome. Nonetheless, it is desirable and expectable that novel robust protein-specific affinity reagents, generated by rapid and cost-effective manners, will gain increasing importance in the proteome analysis. Discovery of most potent and selective binders from combinatorial libraries depends on the development and use of suitable and efficient screening strategies. Immunological methods have proven to be a biological tool of inestimable value to reach this goal. Antibody-based assays are known for their versatility, sensitivity, specificity, high throughput and ease of automation. Particularly, ELISA assays have been extensively used as HTS methodology on the screening and characterization of lead ligands from both biological and chemical combinatorial libraries. Given the ubiquity of the 96 well microtiter plate and its derivatives, this format will likely continue to be a 'standard' in high throughput screening and selection strategies.

ABBREVIATIONS

Ab   = Antibodies

AP   = Alkaline phosphatase

CAAM = Computer-assisted molecular modeling

CDR   = The complementary determining region

C_H = Heavy chain constant domain

C_L = Light chain constant domain

ELISA = Enzyme-linked immunosorbent assay

Fab   = Antigen-binding fragment

Fc   = Cristallizable fragment

FcRM   = Fc-receptor mimetic

FITC   = Fluorescein isothiocyanate

HRP   = Horseradish peroxidase

HTS   = High throughput screening

IgA   = Immunoglobulin of class A

IgD   = Immunoglobulin of class D

IgE   = Immunoglobulin of class E

IgG   = Immunoglobulin of class G

IgM   = Immunoglobulin of class M

PAM   = Protein A mimetic

ScFv   = Single chain variable fragments

SCL   = Synthetic combinatorial libraries

V_H = Heavy chain variable domain

V_L = Light chain variable domain

ACKNOWLEDGMENTS

Dr. Isabel Teixeira de Sousa is gratefully acknowledged for her precious help in editing the updated references of this paper.

CONFLICT OF INTEREST

The author states that there is no conflict of interest.

DISCLOSURES

This chapter has been updated from its original publication: Taipa, M.A. (2008) Immunoassays: Biological tools for high throughput screening and characterisation of combinatorial libraries. *Comb. Chem. High Throughput Screen. 11*, 325-335.

REFERENCES

[1]   Uhlen, M.; Szigyarto, A. C.; Ottosson, H. J.; Nilsson, E. P.; Andersson, K. A. C.; Caroline, K.; Fredrik, P. A human protein atlas for normal and disease tissue. *Mol Cell Proteomics,* **2005**, *4*, S15-S15.

[2]   Uhlen, M. Mapping the human proteome using antibodies. *Mol Cell Proteomics,* **2007**, *6*, 1455-1456.

[3]   Konthur, Z.; Hust, M.; Dubel, S. Perspectives for systematic *in vitro* antibody generation. *Gene,* **2005**, *364*, 19-29.

[4]   Bradbury, A.; Velappan, N.; Verzillo, V.; Ovecka, M.; Chasteen, L.; Sblattero, D.; Marzari, O.; Lou, J. L.; Siegel, R.; Pavlik, P. Antibodies in proteomics I: generating antibodies. *Trends Biotechnol,* **2003**, *21*, 275-281.

[5]   Dubel, S.; Stoevesandt, O.; Taussig, M. J.; Hust, M. Generating recombinant antibodies to the complete human proteome. *Trends Biotechnol,* **2010**, *28*, 333-339.

[6]   Hust, M.; Meyer, T.; Voedisch, B.; Rulker, T.; Thie, H.; El-Ghezal, A.; Kirsch, M. I.; Schutte, M.; Helmsing, S.; Meier, D.; Schirrmann, T.; Dubel, S. A human scFv antibody generation pipeline for proteome research. *J Biotechnol,* **2011**, *152*, 159-170.

[7]   Roque, A. C. A.; Lowe, C. R. Advances and applications of *de novo* designed affinity ligands in proteomics. *Biotechnol Adv,* **2006**, *24*, 17-26.

[8]   Borrebaeck, C. A. K. Antibodies in diagnostics - from immunoassays to protein chips. *Immunol Today,* **2000**, *21*, 379-382.

[9]   Hust, M.; Dubel, S. Mating antibody phage display with proteomics. *Trends Biotechnol,* **2004**, *22*, 8-14.

[10]   Kohler, G.; Milstein, C. Continuous Cultures of Fused Cells Secreting Antibody of Predefined Specificity. *Nature,* **1975**, *256*, 495-497.

[11]   Reichert, J. M. Monoclonal antibodies in the clinic - Despite initial teething problems, the number of clinically effective monoclonal antibodies is growing. *Nat Biotechnol,* **2001**, *19*, 819-822.

[12]   Reichert, J. M.; Rosensweig, C. J.; Faden, L. B.; Dewitz, M. C. Monoclonal antibody successes in the clinic. *Nat Biotechnol,* **2005**, *23*, 1073-1078.

[13]   Brekke, O. H.; Sandlie, I. Therapeutic antibodies for human diseases at the dawn of the twenty-first century. *Nat Rev Drug Discov,* **2003**, *2*, 52-62.

[14]   Maynard, J.; Georgiou, G. Antibody engineering. *Annu Rev Biomed Eng,* **2000**, *2*, 339-376.

[15]   Chadd, H. E.; Chamow, S. M. Therapeutic antibody expression technology. *Curr Opin Biotech,* **2001**, *12*, 188-194.

[16]   van Dijk, M. A.; van de Winkel, J. G. J. Human antibodies as next generation therapeutics. *Curr Opin Chem Biol,* **2001**, *5*, 368-374.

[17]   Roque, A. C. A.; Lowe, C. R.; Taipa, M. A. Antibodies and genetically engineered related molecules: Production and purification. *Biotechnol Progr,* **2004**, *20*, 639-654.

[18]   Fassina, G.; Ruvo, M.; Palombo, G.; Verdoliva, A.; Marino, M. Novel ligands for the affinity-chromatographic purification of antibodies. *J Biochem Bioph Meth,* **2001**, *49*, 481-490.

[19]   Roque, A. C. A.; Silva, C. S. O.; Taipa, M. A. Affinity-based methodologies and ligands for antibody purification: Advances and perspectives. *J Chromatogr A,* **2007**, *1160*, 44-55.

[20]   Nord, K.; Nord, O.; Uhlen, M.; Kelley, B.; Ljungqvist, C.; Nygren, P. A. Recombinant human factor VIII-specific affinity ligands selected from phage-displayed combinatorial libraries of protein A. *Eur J Biochem,* **2001**, *268*, 4269-4277.

[21]   Fernandez, L. A. Prokaryotic expression of antibodies and affibodies. *Curr Opin Biotech,* **2004**, *15*, 364-373.

[22]   Vaughan, C. K.; Sollazzo, M. Of minibody, camel and bacteriophage. *Comb Chem High T Scr,* **2001**, *4*, 417-430.

[23]   Holt, L. J.; Herring, C.; Jespers, L. S.; Woolven, B. P.; Tomlinson, I. M. Domain antibodies: proteins for therapy. *Trends Biotechnol,* **2003**, *21*, 484-490.

[24]   Lipovsek, D.; Lippow, S. M.; Hackel, B. J.; Gregson, M. W.; Cheng, P.; Kapila, A.; Wittrup, K. D. Evolution of an interloop disulfide bond in high-affinity antibody mimics based on fibronectin type III domain and selected by yeast surface display: Molecular convergence with single-domain camelid and shark antibodies. *J Mol Biol,* **2007**, *368*, 1024-1041.

[25]   Liu, F. F.; Wang, T.; Dong, X. Y.; Sun, Y. Rational design of affinity peptide ligand by flexible docking simulation. *J Chromatogr A,* **2007**, *1146*, 41-50.

[26]   Lowe, C. R.; Lowe, A. R.; Gupta, G. New developments in affinity chromatography with potential application in the production of biopharmaceuticals. *J Biochem Bioph Meth,* **2001**, *49*, 561-574.

[27]   Atassi, M. Z.; Dolimbek, B. Z.; Deitiker, P.; Jankovic, J.; Aoki, K. R. A peptide-based immunoassay for antibodies against botulinum neurotoxin A. *J Mol Recognit,* **2007**, *20*, 15-21.

[28]   Tu, J.; Yu, Z. G.; Chu, Y. H. Combinatorial search for diagnostic agents: Lyme antibody H9724 as an example. *Clin Chem,* **1998**, *44*, 232-238.

[29]   Pinilla, C.; Appel, J. R.; Campbell, G. D.; Buencamino, J.; Benkirane, N.; Muller, S.; Greenspan, N. S. All-D peptides recognized by an anti-carbohydrate antibody identified from a positional scanning library. *J Mol Biol,* **1998**, *283*, 1013-1025.

[30]   Verdoliva, A.; Marasco, D.; De Capua, A.; Saporito, A.; Bellofiore, P.; Manfredi, V.; Fattorusso, R.; Pedone, C.; Ruvo, M. A new ligand for immunoglobulin G subdomains by screening of a synthetic peptide library. *Chembiochem,* **2005**, *6*, 1242-1253.

[31]   Friedman, M.; Nordberg, E.; Hoiden-Guthenberg, I.; Brisimar, H.; Adams, G. P.; Nilsson, F. Y.; Carlss, J.; Stahl, S. Phage display selection of Affibody molecules with specific binding to the extracellular domain of the epidermal growth factor receptor. *Protein Eng Des Sel,* **2007**, *20*, 189-199.

[32]   Scognamiglio, P. L.; Doti, N.; Grieco, P.; Pedone, C.; Ruvo, M.; Marasco, D. Discovery of Small Peptide Antagonists of PED/PEA15-D4α Interaction from Simplified Combinatorial Libraries. *Chem Biol Drug Des,* **2011**, *77*, 319-327.

[33]   Hallborn, J.; Carlsson, R. Automated screening procedure for high throughput generation of antibody fragments. *Biotechniques,* **2002**, 30-+.

[34]   Wingren, C.; Steinhauer, C.; Ingvarsson, J.; Persson, E.; Larsson, K.; Borrebaeck, C. A. K. Microarrays based on affinity-tagged single-chain Fv antibodies: Sensitive detection of analyte in complex proteomes. *Proteomics,* **2005**, *5*, 1281-1291.

[35]   Roitt, I.; Brostoff, J.; Male, D. *Immunology*, Vol. sixth ed.; Mosby: London, **2001**.

[36]   Jefferis, R. Glycosylation of recombinant antibody therapeutics. *Biotechnol Progr,* **2005**, *21*, 11-16.

[37]   Winter, G.; Milstein, C. Man-Made Antibodies. *Nature,* **1991**, *349*, 293-299.

[38]   Skerra, A. Engineered protein scaffolds for molecular recognition. *J Mol Recognit,* **2000**, *13*, 167-187.

[39]   Silverman, J.; Lu, Q.; Bakker, A.; To, W.; Duguay, A.; Alba, B. M.; Smith, R.; Rivas, A.; Li, P.; Le, H.; Whitehorn, E.; Moore, K. W.; Swimmer, C.; Perlroth, V.; Vogt, M.; Kolkman, J.; Stemmer, W. P. C. Multivalent avimer proteins evolved by exon shuffling of a family of human receptor domains. *Nat Biotechnol,* **2005**, *23*, 1556-1561.

[40] Hey, T.; Fiedler, E.; Rudolph, R.; Fiedler, M. Artificial, non-antibody binding proteins for pharmaceutical and industrial applications. *Trends Biotechnol,* **2005**, *23*, 514-522.

[41] Nygren, P. A.; Skerra, A. Binding proteins from alternative scaffolds. *J Immunol Methods,* **2004**, *290*, 3-28.

[42] Binz, H. K.; Pluckthun, A. Engineered proteins as specific binding reagents. *Curr Opin Biotech,* **2005**, *16*, 459-469.

[43] Hackel, B. J.; Kapila, A.; Wittrup, K. D. Picomolar affinity fibronectin domains engineered utilizing loop length diversity, recursive mutagenesis, and loop shuffling. *J Mol Biol,* **2008**, *381*, 1238-1252.

[44] Cwirla, S. E.; Peters, E. A.; Barrett, R. W.; Dower, W. J. Peptides on Phage - a Vast Library of Peptides for Identifying Ligands. *P Natl Acad Sci USA,* **1990**, *87*, 6378-6382.

[45] Scott, J. K.; Smith, G. P. Searching for Peptide Ligands with an Epitope Library. *Science,* **1990**, *249*, 386-390.

[46] Bellofiore, P.; Petronzelli, F.; De Martino, T.; Minenkova, O.; Bombardi, V.; Anastasi, A. M.; Lindstedt, R.; Felici, F.; De Santis, R.; Verdoliva, A. Identification and refinement of a peptide affinity ligand with unique specificity for a monoclonal anti-tenascin-C antibody by screening of a phage display library. *J Chromatogr A,* **2006**, *1107*, 182-191.

[47] Ehrlich, G. K.; Bailon, P. Identification of model peptides as affinity ligands for the purification of humanized monoclonal antibodies by means of phage display. *J Biochem Bioph Meth,* **2001**, *49*, 443-454.

[48] Gaskin, D. J. H.; Starck, K.; Turner, N. A.; Vulfson, E. N. Phage display combinatorial libraries of short peptides: ligand selection for protein purification. *Enzyme Microb Tech,* **2001**, *28*, 766-772.

[49] Hoogenboom, H. R.; de Bruine, A. P.; Hufton, S. E.; Hoet, R. M.; Arends, J. W.; Roovers, R. C. Antibody phage display technology and its applications. *Immunotechnology,* **1998**, *4*, 1-20.

[50] Ellmark, P.; Esteban, O.; Furebring, C.; Hager, A. C. M.; Ohlin, M. *In vitro* molecular evolution of antibody genes mimicking receptor revision. *Mol Immunol,* **2002**, *39*, 349-356.

[51] van Wyngaardt, W.; Malatji, T.; Mashau, C.; Fehrsen, J.; Jordaan, F.; Miltiadou, D.; du Plessis, D. H. A large semi-synthetic single-chain Fv phage display library based on chicken immunoglobulin genes. *Bmc Biotechnol,* **2004**, *4*.

[52] Hoogenboom, H. R. Selecting and screening recombinant antibody libraries. *Nat Biotechnol,* **2005**, *23*, 1105-1116.

[53] Bradbury, A. R. M.; Sidhu, S.; Dubel, S.; McCafferty, J. Beyond natural antibodies: the power of *in vitro* display technologies. *Nat Biotechnol,* **2011**, *29*, 245-254.

[54] Francisco, J. A.; Campbell, R.; Iverson, B. L.; Georgiou, G. Production and Fluorescence-Activated Cell Sorting of Escherichia-Coli Expressing a Functional Antibody Fragment on the External Surface. *P Natl Acad Sci USA,* **1993**, *90*, 10444-10448.

[55] Gunneriusson, E.; Samuelson, P.; Uhlen, M.; Nygren, P. A.; Stahl, S. Surface display of a functional single-chain Fv antibody on staphylococci. *J Bacteriol,* **1996**, *178*, 1341-1346.

[56] Jostock, T.; Dubel, S. Screening of molecular repertoires by microbial surface display. *Comb Chem High T Scr,* **2005**, *8*, 127-133.

[57] Boder, E. T.; Wittrup, K. D. Yeast surface display for screening combinatorial polypeptide libraries. *Nat Biotechnol,* **1997**, *15*, 553-557.

[58]   Boder, E. T.; Midelfort, K. S.; Wittrup, K. D. Directed evolution of antibody fragments with monovalent femtomolar antigen-binding affinity. *P Natl Acad Sci USA,* **2000**, *97*, 10701-10705.

[59]   Rajpal, A.; Beyaz, N.; Haber, L.; Cappuccilli, G.; Yee, H.; Bhatt, R. R.; Takeuchi, T.; Lerner, R. A.; Crea, R. A general method for greatly improving the affinity of antibodies by using combinatorial libraries. *P Natl Acad Sci USA,* **2005**, *102*, 8466-8471.

[60]   Gai, S. A.; Wittrup, K. D. Yeast surface display for protein engineering and characterization. *Curr Opin Struc Biol,* **2007**, *17*, 467-473.

[61]   Hanes, J.; Pluckthun, A. *In vitro* selection and evolution of functional proteins by using ribosome display. *P Natl Acad Sci USA,* **1997**, *94*, 4937-4942.

[62]   Hanes, J.; Schaffitzel, C.; Knappik, A.; Pluckthun, A. Picomolar affinity antibodies from a fully synthetic naive library selected and evolved by ribosome display. *Nat Biotechnol,* **2000**, *18*, 1287-1292.

[63]   Irving, R. A.; Coia, G.; Roberts, A.; Nuttall, S. D.; Hudson, P. J. Ribosome display and affinity maturation: from antibodies to single V-domains and steps towards cancer therapeutics. *J Immunol Methods,* **2001**, *248*, 31-45.

[64]   Roberts, R. W.; Szostak, J. W. RNA-peptide fusions for the *in vitro* selection of peptides and proteins. *P Natl Acad Sci USA,* **1997**, *94*, 12297-12302.

[65]   Soderlind, E.; Carlsson, R.; Borrebaeck, C. A. K.; Ohlin, M. The immune diversity in a test tube - Non-immunised antibody libraries and functional variability in defined protein scaffolds. *Comb Chem High Throughput Screen,* **2001**, *4*, 409-416.

[66]   Hust, M.; Toleikis, L.; Dubel, S. Antibody Phage Display. *Mod Asp Immunobiol,* **2005**, *15*, 47-49.

[67]   Ayriss, J.; Woods, T.; Bradbury, A.; Pavlik, P. High throughput screening of single-chain antibodies using multiplexed flow cytometry. *J Proteome Res,* **2007**, *6*, 1072-1082.

[68]   Clackson, T.; Wells, J. A. *In Vitro* Selection from Protein and Peptide Libraries. *Trends Biotechnol,* **1994**, *12*, 173-184.

[69]   Stewart, A.; Liu, Y. Y.; Lai, J. R. A strategy for phage display selection of functional domain-exchanged immunoglobulin scaffolds with high affinity for glycan targets. *J Immunol Methods,* **2012**, *376*, 150-155.

[70]   Lou, J. L.; Marzari, R.; Verzillo, V.; Ferrero, F.; Pak, D.; Sheng, M.; Yang, C. L.; Sblattero, D.; Bradbury, A. Antibodies in haystacks: how selection strategy influences the outcome of selection from molecular diversity libraries. *J Immunol Methods,* **2001**, *253*, 233-242.

[71]   Baek, H.; Suk, K. H.; Kim, Y. H.; Cha, S. An improved helper phage system for efficient isolation of specific antibody molecules in phage display. *Nucleic Acids Res,* **2002**, *30*.

[72]   Beaber, J. W.; Tam, E. M.; Lao, L. S.; Rondon, I. J. A new helper phage for improved monovalent display of Fab molecules. *J Immunol Methods,* **2012**, *376*, 46-54.

[73]   Chasteen, L.; Ayriss, J.; Pavlik, P.; Bradbury, A. R. M. Eliminating helper phage from phage display. *Nucleic Acids Res,* **2006**, *34*.

[74]   Huls, G.; Gestel, D.; van der Linden, J.; Moret, E.; Logtenberg, T. Tumor cell killing by *in vitro* affinity-matured recombinant human monoclonal antibodies. *Cancer Immunol Immun,* **2001**, *50*, 163-171.

[75]   Shusta, E. V.; Kieke, M. C.; Parke, E.; Kranz, D. M.; Wittrup, K. D. Yeast polypeptide fusion surface display levels predict thermal stability and soluble secretion efficiency. *J Mol Biol,* **1999**, *292*, 949-956.

[76] Feldhaus, M. J.; Siegel, R. W.; Opresko, L. K.; Coleman, J. R.; Feldhaus, J. M. W.; Yeung, Y. A.; Cochran, J. R.; Heinzelman, P.; Colby, D.; Swers, J.; Graff, C.; Wiley, H. S.; Wittrup, K. D. Flow-cytometric isolation of human antibodies from a nonimmune Saccharomyces cerevisiae surface display library. *Nat Biotechnol,* **2003**, *21,* 163-170.

[77] van den Beucken, T.; Pieters, H.; Steukers, M.; van der Vaart, M.; Ladner, R. C.; Hoogenboom, H. R.; Hufton, S. E. Affinity maturation of Fab antibody fragments by fluorescent-activated cell sorting of yeast-displayed libraries. *Febs Lett,* **2003**, *546,* 288-294.

[78] Weaver-Feldhaus, J. M.; Lou, J. L.; Coleman, J. R.; Siegel, R. W.; Marks, J. D.; Feldhaus, M. J. Yeast mating for combinatorial Fab library generation and surface display. *Febs Lett,* **2004**, *564,* 24-34.

[79] Bowley, D. R.; Labrijn, A. F.; Zwick, M. B.; Burton, D. R. Antigen selection from an HIV-1 immune antibody library displayed on yeast yields many novel antibodies compared to selection from the same library displayed on phage. *Protein Eng Des Sel,* **2007**, *20,* 81-90.

[80] Obrecht, D.; Villalgordo, J. M. *Solid-Supported Combinatorial and Paralle Synthesis of Small-Molecular-Weight Compound Libraries*, Vol. ed.; Pergamon: Oxford, **1998**.

[81] Vicennati, P.; Bensel, N.; Wagner, A.; Creminon, C.; Taran, F. Sandwich immunoassay as a high throughput screening method for cross-coupling reactions. *Angew Chem Int Edit,* **2005**, *44,* 6863-6866.

[82] Martinez-Teipel, B.; Teixido, J.; Pascual, R.; Mora, M.; Pujola, J.; Fujimoto, T.; Borrell, J. I.; Michelotti, E. L. 2-methoxy-6-oxo-1,4,5,6-tetrahydropyridine-3-carbonitriles: Versatile starting materials for the synthesis of libraries with diverse heterocyclic scaffolds. *J Comb Chem,* **2005**, *7,* 436-448.

[83] Bettinetti, L.; Lober, S.; Hubner, H.; Gmeiner, P. Parallel synthesis and biological screening of dopamine receptor ligands taking advantage of a click chemistry based BAL linker. *J Comb Chem,* **2005**, *7,* 309-316.

[84] Joshi, M.; Vargas, C.; Boisguerin, P.; Diehl, A.; Krause, G.; Schmieder, P.; Moeling, K.; Hagen, V.; Schade, M.; Oschkinat, H. Discovery of low-molecular-weight ligands for the AF6 PDZ domain. *Angew Chem Int Edit,* **2006**, *45,* 3790-3795.

[85] Cai, D.; Lee, A. Y.; Chiang, C. M.; Kodadek, T. Peptoid ligands that bind selectively to phosphoproteins. *Bioorg Med Chem Lett,* **2011**, *21,* 4960-4964.

[86] Burns, M. R.; Jenkins, S. A.; Wood, S. J.; Miller, K.; David, S. A. Structure-activity relationships in lipopolysaccharide neutralizers: Design, synthesis, and biological evaluation of a 540-membered amphipathic bisamide library. *J Comb Chem,* **2006**, *8,* 32-43.

[87] Liu, J. F.; Kaselj, M.; Isome, Y.; Ye, P.; Sargent, K.; Sprague, K.; Cherrak, D.; Wilson, C. J.; Si, Y.; Yohannes, D.; Ng, S. C. Design and synthesis of a quinazolinone natural product-templated library with cytotoxic activity. *J Comb Chem,* **2006**, *8,* 7-10.

[88] He, X. Z. G.; Gerona-Navarro, G.; Jaffrey, S. R. Ligand discovery using small molecule microarrays. *J Pharmacol Exp Ther,* **2005**, *313,* 1-7.

[89] Haigh, J. M.; Hussain, A.; Mimmack, M. L.; Lowe, C. R. Affinity ligands for immunoglobulins based on the multicomponent Ugi reaction. *Journal of Chromatography B-Analytical Technologies in the Biomedical and Life Sciences,* **2009**, *877,* 1440-1452.

[90] Houghten, R. A.; Pinilla, C.; Blondelle, S. E.; Appel, J. R.; Dooley, C. T.; Cuervo, J. H. Generation and Use of Synthetic Peptide Combinatorial Libraries for Basic Research and Drug Discovery. *Nature,* **1991**, *354,* 84-86.

[91]   Bradbury, A. Molecular library technologies at the millennium. *Trends Biotechnol,* **2000**, *18*, 131-133.

[92]   Haab, B. B.; Paulovich, A. G.; Anderson, N. L.; Clark, A. M.; Downing, G. J.; Hermjakob, H.; LaBaer, J.; Uhlen, M. A reagent resource to identify proteins and peptides of interest for the cancer community - A workshop report. *Mol Cell Proteomics,* **2006**, *5*, 1996-2007.

[93]   Denton, G.; Sekowski, M.; Price, M. R. Induction of Antibody-Responses to Breast-Carcinoma Associated Mucins Using Synthetic Peptide Constructs as Immunogens. *Cancer Lett,* **1993**, *70*, 143-150.

[94]   Pinilla, C.; Martin, R.; Gran, B.; Appel, J. R.; Boggiano, C.; Wilson, D. B.; Houghten, R. A. Exploring immunological specificity using synthetic peptide combinatorial libraries. *Curr Opin Immunol,* **1999**, *11*, 193-202.

[95]   Cauwenberghs, N.; Vanhoorelbeke, K.; Vauterin, S.; Westra, D. F.; Rome, G.; Huizinga, E. G.; Lopez, J. A.; Berndt, M. C.; Harsfalvi, J.; Deckmyn, H. Epitope mapping of inhibitory antibodies against platelet glycoprotein Ib alpha reveals interaction between the leucine-rich repeat N-terminal and C-terminal flanking domains of glycoprotein Ib alpha. *Blood,* **2001**, *98*, 652-660.

[96]   Andresen, H.; Zarse, K.; Grotzinger, C.; Hollidt, J. M.; Ehrentreich-Forster, E.; Bier, F. F.; Kreuzer, O. J. Development of peptide microarrays for epitope mapping of antibodies against the human TSH receptor. *J Immunol Methods,* **2006**, *315*, 11-18.

[97]   Ying, L. Q.; Liu, R. W.; Zhang, J. H.; Lam, K.; Lebrilla, C. B.; Gervay-Hague, J. A topologically segregated one-bead-one-compound combinatorial glycopeptide library for identification of lectin ligands. *J Comb Chem,* **2005**, *7*, 372-384.

[98]   Palmieri, G.; Cassani, G.; Fassina, G. Peptide Immobilization on Calcium Alginate Beads - Applications to Antibody Purification and Assay. *J Chromatogr B,* **1995**, *664*, 127-135.

[99]   Lund, L. N.; Gustavsson, P. E.; Michael, R.; Lindgren, J.; Norskov-Lauritsen, L.; Lund, M.; Houen, G.; Staby, A.; St Hilaire, P. M. Novel peptide ligand with high binding capacity for antibody purification. *J Chromatogr A,* **2012**, *1225*, 158-167.

[100]   Fassina, G.; Verdoliva, A.; Odierna, M. R.; Ruvo, M.; Cassini, G. Protein a mimetic peptide ligand for affinity purification of antibodies. *J Mol Recognit,* **1996**, *9*, 564-569.

[101]   Lowe, C. R. Combinatorial approaches to affinity chromatography. *Curr Opin Chem Biol,* **2001**, *5*, 248-256.

[102]   Sproule, K.; Morrill, P.; Pearson, J. C.; Burton, S. J.; Hejnaes, K. R.; Valore, H.; Ludvigsen, S.; Lowe, C. R. New strategy for the design of ligands for the purification of pharmaceutical proteins by affinity chromatography. *Journal of Chromatography B,* **2000**, *740*, 17-33.

[103]   Li, R. X.; Dowd, V.; Stewart, D. J.; Burton, S. J.; Lowe, C. R. Design, synthesis, and application of a Protein A mimetic. *Nat Biotechnol,* **1998**, *16*, 190-195.

[104]   Teng, S. F.; Sproule, K.; Husain, A.; Lowe, C. R. Affinity chromatography on immobilized "biomimetic" ligands synthesis, immobilization and chromatographic assessment of an immunoglobulin G-binding ligand. *Journal of Chromatography B,* **2000**, *740*, 1-15.

[105]   Palanisamy, U. D.; Winzor, D. J.; Lowe, C. R. Synthesis and evaluation of affinity adsorbents for glycoproteins: an artificial lectin. *Journal of Chromatography B,* **2000**, *746*, 265-281.

[106]   Renou, E. N. S.; Gupta, G.; Young, D. S.; Dear, D. V.; Lowe, C. R. The design, synthesis and evaluation of affinity ligands for prion proteins. *J Mol Recognit,* **2004**, *17*, 248-261.

[107] Roque, A. C. A.; Taipa, M. A.; Lowe, C. R. Synthesis and screening of a rationally designed combinatorial library of affinity ligands mimicking protein L from Peptostreptococcus magnus. *J Mol Recognit,* **2005**, *18*, 213-224.

[108] Lee, W. C.; Lee, K. H. Applications of affinity chromatography in proteomics. *Anal Biochem,* **2004**, *324*, 1-10.

[109] Roque, A. C. A.; Taipa, M. A.; Lowe, C. R. An artificial protein L for the purification of immunoglobulins and Fab fragments by affinity chromatography. *J Chromatogr A,* **2005**, *1064*, 157-167.

[110] Roque, A. C. A. Ph.D. Thesis. Design, Synthesis and Evaluation of Immunoglobulin-Binding Ligands: An Artificial Protein L, Instituto Superior Técnico, Universidade Técnica de Lisboa, 2004.

[111] Buckler, D. R.; Park, A.; Viswanathan, M.; Hoet, R. M.; Ladner, R. C. Screening isolates from antibody phage-display libraries. *Drug Discov Today,* **2008**, *13*, 318-324.

[112] Miyamoto, S.; Liu, R. W.; Hung, S. S.; Wang, X. B.; Lam, K. S. Screening of a one bead-one compound combinatorial library for beta-actin identifies molecules active toward Ramos B-lymphoma cells. *Anal Biochem,* **2008**, *374*, 112-120.

[113] Nielsen, A. L.; Jorgensen, F. S.; Olsen, L.; Christensen, S. F.; Benie, A. J.; Bjornholm, T.; Hilaire, P. M. S. A Diversity Optimized Combinatorial Library for the Identification of Fc-Fragment Binding Ligands. *Biopolymers,* **2010**, *94*, 192-205.

[114] Linthicum, D. S.; Tetin, S. Y.; Anchin, J. M.; Ioerger, T. R. Antibody-ligand interactions: Computational modeling and correlation with biophysical measurements. *Comb Chem High T Scr,* **2001**, *4*, 439-449.

[115] Saraswat, L. D.; Zhang, H. Y.; Hardy, L. W.; Jones, S. S.; Bhikhabhai, R.; Brink, C.; Bergenstrahle, A.; Haglund, R.; Gallion, S. L. Affinity ligand selection from a library of small molecules: Assay development, screening, and application. *Biotechnol Progr,* **2005**, *21*, 300-308.

[116] Bradbury, A.; Velappan, N.; Verzillo, V.; Ovecka, M.; Chasteen, L.; Sblattero, D.; Marzari, O.; Lou, J. L.; Siegel, R.; Pavlik, P. Antibodies in proteomics II: screening, high throughput characterization and downstream applications. *Trends Biotechnol,* **2003**, *21*, 312-317.

[117] Catty, D. *Antibodies: A Practical Approach*, Vol. II, ed.; IRL Press: Oxford, **1989**.

[118] Roque, A. C. A.; Taipa, A. M.; Lowe, C. R. A new method for the screening of solid-phase combinatorial libraries for affinity chromatography. *J Mol Recognit,* **2004**, *17*, 262-267.

[119] Wassaf, D.; Kuang, G. N.; Kopacz, K.; Wu, Q. L.; Nguyen, Q.; Toews, M.; Cosic, J.; Jacques, J.; Wiltshire, S.; Lambert, J.; Pazmany, C. C.; Hogan, S.; Ladner, R. C.; Nixon, A. E.; Sexton, D. J. High throughput affinity ranking of antibodies using surface plasmon resonance microarrays. *Anal Biochem,* **2006**, *351*, 241-253.

[120] Edmundson, A. B.; Tribbick, G.; Plompen, S.; Geysen, H. M.; Yuriev, E.; Ramsland, P. A. Binding of synthetic peptides by a human monoclonal IgM with an unusual combining site structure. *J Mol Recognit,* **2001**, *14*, 229-238.

CHAPTER 6

Screening and Mechanism-Based Evaluation of Estrogenic Botanical Extracts

Cassia R. Overk[1] **and Judy L. Bolton**[*,2]

[1]*Department of Neurosciences, School of Medicine, University of California at San Diego, La Jolla, California, CA 92093, USA;* [2]*UIC/NIH Center for Botanical Dietary Supplements Research and Department of Medicinal Chemistry and Pharmacognosy, College of Pharmacy, University of Illinois at Chicago, 833 S. Wood St., M/C 781, Chicago, IL 60612, USA*

Abstract: Symptoms associated with menopause can greatly affect the quality of life for women. Botanical dietary supplements have been viewed by the public as safe and effective despite a lack of evidence indicating an urgent necessity to standardize these supplements chemically and biologically. Seventeen plants were evaluated for estrogenic biological activity using standard assays: competitive estrogen receptor (ER) binding assay for both alpha and beta subtypes, transient transfection of the estrogen response element (ERE) luciferase plasmid into MCF-7 cells expressing either ER alpha or ER beta, and the Ishikawa alkaline phosphatase induction assay for both estrogenic and antiestrogenic activities. Based on the combination of data pooled from these assays, the following was determined: a) a high rate of false positive activity for the competitive binding assays, b) some extracts had estrogenic activity despite a lack of ability to bind the ER, c) one extract exhibited selective estrogen receptor modulator (SERM) activity, and d) several extracts show additive/synergistic activity. Taken together, these data indicate a need to reprioritize the order in which the bioassays are performed for maximal efficiency of programs using bioassay-guided fractionation. In addition, possible explanations for the conflicts in the literature over the estrogenicity of *Cimicifuga racemosa* (black cohosh) are suggested.

Keywords: Botanicals, estrogen, dietary supplements, menopause, selective estrogen receptor modulators.

INTRODUCTION

Symptoms associated with menopause such as insomnia, loss of libido, vaginal atrophy, depression, and hot flashes can greatly affect the quality of life for

*Address correspondence to Judy L. Bolton: UIC/NIH Center for Botanical Dietary Supplements Research and Department of Medicinal Chemistry and Pharmacognosy, College of Pharmacy, University of Illinois at Chicago, 833 S. Wood St., M/C 781, Chicago, IL 60612, USA; Tel: (312) 996-5280; Fax: (312) 996-7107; E-mail: judy.bolton@uic.edu

women. Many women have used hormone replacement therapy (HRT) to alleviate menopausal symptoms; however, with the publication of the Women's Health Initiative in 2002, the number of women using HRT has dramatically decreased [1, 2]. Even before the publication of several large studies of HRT, many women have been turning to herbal remedies for the relief of menopausal symptoms [3-5], perhaps because they are viewed as safe [6]. Among the herbal remedies currently being used, many contain phytochemicals that mimic the effects of estrogens, commonly referred to as phytoestrogens. Unfortunately, few botanicals are chemically and biologically standardized to relevant active compounds using an appropriate mechanism of action [7-9].

There are a number of different targets within the signaling process that can be segmented into *in vitro* assays (Scheme **1**). In the classical estrogen signaling pathway, the first step is the binding of a ligand to the ER, which can be replicated using the ERα and ERβ [^{3}H]-estradiol competitive binding assay [10] (Scheme **1A**). Once the ligand binds the estrogen receptor (ER), the dimerized complex will move from the cytosol of the cell through the nuclear pore and into the nucleus. After entering the nucleus, the ER-dimer will bind to the estrogen response element (ERE) which is located upstream of estrogen-controlled genes. This process is modeled *in vitro* using ERE-luciferase (ERE-luc) induction [11] in MCF-7 ERα positive and ERβ positive cell lines (Scheme **1B**). Once bound to the ERE, coactivators, corepressors, and transcription factors will bind, and transcription will be initiated. During translation, the mRNA will be converted into protein, and posttranslational modifications will then complete the signaling pathway to produce a functional protein. This last step can be evaluated using the Ishikawa alkaline phosphatase assay (Scheme **1C**) [12].

The purpose of this research was to evaluate seventeen plants using a variety of established cell-free and cell-based estrogenic assays to biologically characterize the botanicals [13]. Since both estrogens and selective serotonin reuptake inhibitors (SSRIs) have been known to alleviate symptoms associated with menopause [14], plants were selected based on a literature search of the Natural Products Alert (NAPRALERT) database for the most widely used remedies for menopausal symptoms, menstrual disorders, or reported estrogenic, serotonergic,

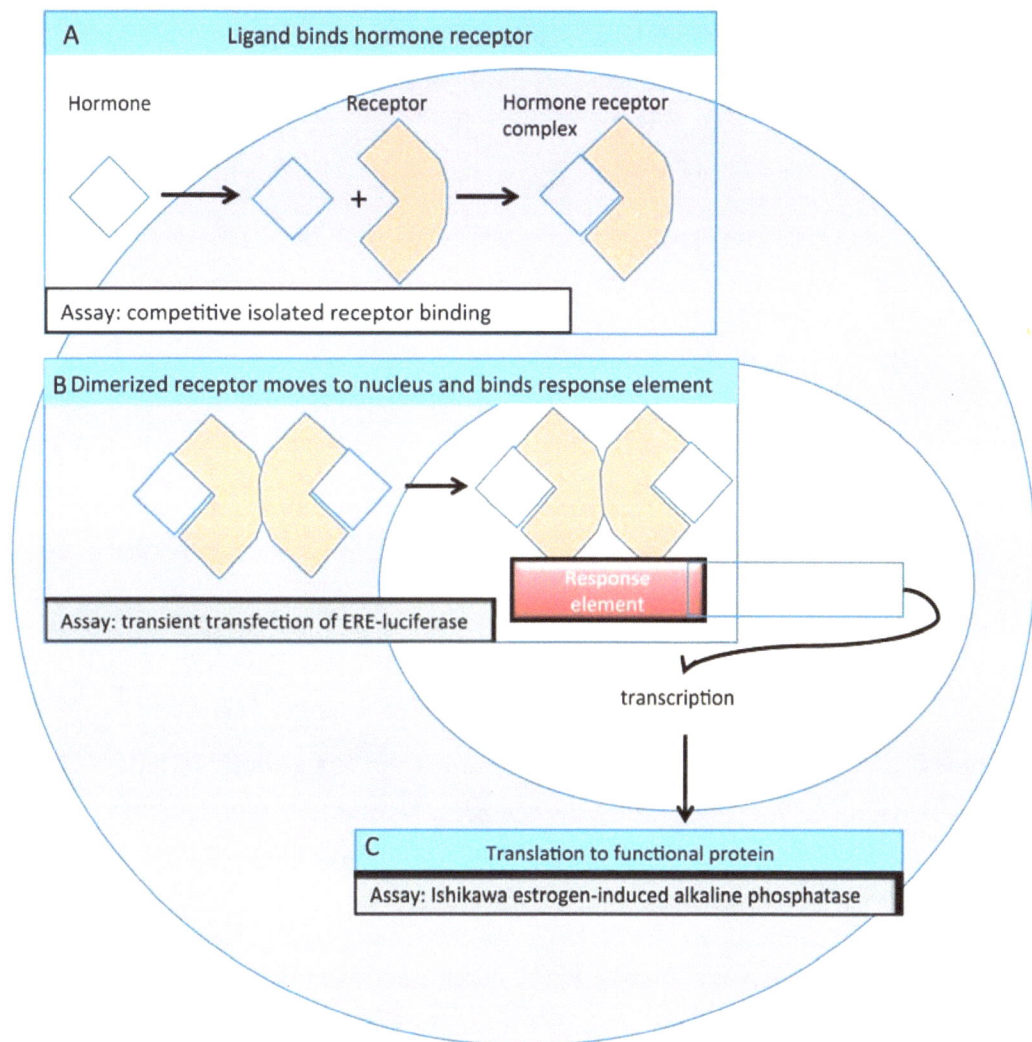

Scheme 1. Simplified depiction of the classical estrogen receptor-signaling pathway from ligand binding to functional protein and corresponding *in vitro* assays. (**A**) The hormone enters the cell and binds with a receptor to form a hormone-receptor complex, which can be modeled using a competitive binding assay. (**B**) Once the hormone-receptor complex dimerizes, it will translocate from the cytoplasm to the nucleus where it binds to the response element. This step was evaluated using a transient transfection of the ERE-luciferase plasmid in to cells that were ERα or ERβ positive. (**C**) Then the target gene is transcribed and translated into a protein. The estrogen-inducible protein, alkaline phosphatase was measured in the Ishikawa cell line.

anti-steroidogenic, and anti-fertility activity. These plants were obtained, extracted, and evaluated for estrogenic activity *in vitro* using the ERα and ERβ

[³H]-estradiol competitive binding assay, ERE-luc induction in MCF-7 ERα positive and ERβ positive cell lines, and the Ishikawa alkaline phosphatase assay. The results of these studies were originally published in reference [13] and are updated in this ebook chapter.

MATERIALS AND METHODS

All chemicals were purchased from Sigma-Aldrich (St. Louis, MO) or Fisher Scientific (Itasca, IL) unless stated otherwise.

Plant Material

As indicated in Table **1**, *Angelica sinensis* (roots) was purchased at Yin Wall City, Chicago, IL (2001). *Asclepias tuberosa* was bought from www.blessedherbs.com. *Cimicifuga americana* (aerial parts) was collected in Swain County, NC. *Cimicifuga racemosa* (aerial parts) was collected in Sevier County, TN. *Cimicifuga rubifolia* (aerial parts) was collected in Hancock County, TN. *Pueraria lobata* was collected in Evanston, IL. *Alisma plantago-aquatica* (rhizomes and roots), *Cimicifuga racemosa* (rhizomes and roots), *Cornus officinalis* (fruits), *Paeonia moutan* (bark), *Polygonum multiflorum* (roots), *Pueraria mirifica* (bark), *Valeriana officinalis* (roots), *Viburnum opulus* (bark), *Viburnum prunifolium* (bark), and *Vitex agnus-castus* (berries), were provided by PureWorld Botanicals, now NATUREX (South Hackensack, NJ). *Beta vulgaris* (roots) and *Daucus carota* (roots) were purchased from a local grocery store. Voucher specimens have been deposited at the Pharmacognosy Field Station, Department of Medicinal Chemistry and Pharmacognosy, College of Pharmacy, University of Illinois at Chicago, and verified by botanist Dr. D. D. Soejarto.

Extraction

Using the industry-standard operating procedures for botanical extractions, sequential percolation was used to extract the botanicals as follows: Plant material (200 g each) except *A. sinesis* was minced and macerated in petroleum ether (PE, 600 mL) overnight, and percolated exhaustively with the same solvent (total 8 L). The marc was macerated in dichloromethane (CH₂Cl₂, 600 mL) overnight, and

Table 1.   Plants Selected for Estrogenic Screening

Botanical Name	Plant Part	Common Name	Source	Rational
Alisma plantago-aquatica L.	rhizomes	Water Plantain	[1]PureWorld	menopausal symptoms [41]
Angelica sinensis (Oliv.) Diels	roots	Dang-Gui	[2]Yin Wall City, Inc	menopausal symptoms; [30] serotonergic [42]
Asclepias tuberosa L.	roots	Butterfly Weed	[2]www.blessedherbs	contains steroidal compounds [43]
Beta vulgaris L.	roots	Beets	[2]local grocery store	menopausal symptoms [44]
Cimicifuga americana Michaux	roots	Yellow Cohosh	[3]Swain County, NC	chemotaxonomic relationship [45]
Cimicifuga racemosa (L.) Nutt.	aerial	Black Cohosh	[3]Sevier County, TN	related plant part
Cimicifuga racemosa (L.) Nutt.	roots	Black Cohosh	[3]Sevier County, TN	menopausal symptoms [46-57] serotonergic [18, 58]
Cimicifuga rubifolia (Kearney) Kartesz	aerial	Appalachian Bugbane	[3]Hancock County; TN	chemotaxonomic relationship [45]
Cornus officinalis Sieb. & Zucc.	fruits	Dogwood	[1]PureWorld	menopausal symptoms [41]
Daucus carota L.	roots	Queen Anne's lace; Carrot	[2]Local grocery store	anti-steroidogenic activity [59]; estrogenic activity [60]; antifertility activity [61]
Paeonia moutan Sims	bark	Peony	[1]Yakima Chief, Inc.,	menopausal symptoms [62]
Pueraria lobata (Willd.) Ohwi.	aerial	Kudzu	[1]PureWorld	estrogenic activity [27, 33]; menopausal symptoms [63]; serotonergic activity [64]
Pueraria mirifica Airy, Shaw & Suvatabandhu	bark	Kwao Keur	[3]Evanston, Illinois	estrogenic activity [65-68]
Valeriana officinalis L.	roots	Valerian	[1]PureWorld	serotonergic activity [69]
Viburnum opulus L.	bark	Cramp Bark	[1]PureWorld	smooth muscle antispasmodic [70]
Viburnum prunifolium L.	bark	Black Haw	[1]PureWorld	menopausal symptoms [71]
Vitex agnus-castus L.	fruits	Chasteberry	[1]PureWorld	menstrual disorders [72-77]; menopausal symptoms [78]

[1] Provided.
[2] Purchased.
[3] Collected.

percolated exhaustively with CH_2Cl_2 (8 L). Finally, the marc was macerated in 75% ethanol (EtOH, 600 mL) overnight, and percolated exhaustively with of 75% EtOH (8 L). The PE, CH_2Cl_2, and 75% EtOH percolates were combined separately, and the solvents were removed *in vacuo* to yield respective extracts of different polarity.

The minced sample of *A. sinensis* (2 kg) was macerated in CH_3OH (2 L, 24 h) and percolated exhaustively with CH_3OH (6 L). The percolates were combined and the solvent evaporated *in vacuo* to yield the CH_3OH extract. The CH_3OH extract was dissolved in 15% aqueous CH_3OH and partitioned with petroleum ether (3 × 1 L). The aqueous-methanol partition was evaporated *in vacuo* to remove CH_3OH, and the remaining aqueous partition was successively partitioned with $CHCl_3$ (3 × 1 L) and n-butanol (BuOH, 3 × 1 L). Removal of the solvent yielded the petroleum ether, $CHCl_3$, BuOH, and H_2O soluble partition.

ERα and ERβ Competitive Binding Assays

The competitive ERα and ERβ binding assays were used with tritiated estradiol based on the method of Obourn *et al.* [10] with minor modifications [15], to determine *in vitro* binding affinities of the substrates with the receptors. The reaction mixture consisted of sample in DMSO (5 µL), pure human recombinant diluted ERα and ERβ (0.5 pmol, 5 µL) in ER binding buffer, "Hot Mix" [400 nM, 5 µL prepared fresh using 95 Ci/mmol [^{3}H] estradiol, diluted in 1:1 ethanol:ER binding buffer; obtained from NEN Life Science Products (Boston, MA)], and ER binding buffer (85 µL). The incubation was carried out at room temperature for 2 h before a hydroxyapatite slurry (HAPs, 50%, 100 µL) was added. The tubes were incubated on ice for 15 min with vortexing every 5 min. The appropriate ER wash buffer was added (1 mL), and the tubes were vortexed before centrifuging at 10,000 × g for 1 min. The supernatant was discarded, and this wash step was repeated three times. The HAPs pellet containing the ligand-receptor complex was resuspended in ethanol (200 µL) and transferred to scintillation vials. An additional volume of ethanol (200 µL) was used to rinse the centrifuge tube. Cytoscint [4 mL/vial; ICN (Costa Mesa, CA)] was added, and the radioactivity was counted using a Beckman LS 5801 liquid scintillation counter (Schaumburg, IL). The percent inhibition of [^{3}H] estradiol binding to each ER was determined using Equation 1. The binding capability (percent) of the sample was calculated in comparison with that of estradiol (50 nM, 90%).

$$[1 - (dpm_{sample} - dpm_{blank})/(dpm_{DMSO} - dpm_{blank})] \times 100 = \% \text{ sample binding} \quad \textbf{(1)}$$

Cell Culture Conditions

The Ishikawa cell line was provided by R. B. Hochberg (Yale University, New Haven, CT) and was maintained in Dulbecco's Modified Eagle medium (DMEM/F12) containing sodium pyruvate (1%), non-essential amino acids (NEAA, 1%), glutamax-1 (1%), insulin (0.05%), and heat-inactivated fetal bovine serum (FBS, 10%). MCF-7 WS8 cells were provided by V. C. Jordan (Fox Chase Cancer Center) and were grown in RPMI 1640 media containing glutamax-1 (1%), NEAA (1%), insulin (0.05%), and heat-inactivated FBS (5%). The MCF-7 C4-12-5 ERβ positive stable cell line ([16] referred to as MCF-7 ERβ) was provided by D. B. Lubahn (University of Missouri) and was grown in MEM (catalogue number 3024) supplemented with stripped calf bovine serum (CBS, 5%), Pen/Strep (2%), insulin (6 ng/mL), sodium carbonate (2.2 g/L), HEPES (1.25 M, 8 mL), glutamax (1%), and G418S (50 mg/mL stock, 6 mL). Stripped serum was prepared by incubating the serum with acetone-washed activated charcoal (100 mg/mL) at 4 °C for 30 min, and centrifuged at 4,000 RPM for 15 min at 4 °C. This step was repeated in triplicate. DMSO concentrations for all cell culture assays were below 0.1%.

Induction of Alkaline Phosphatase in Cultured Ishikawa Cells

The procedure of Pisha *et al.* [12] with minor modifications [15] was used. Ishikawa cells (1.5×10^4 cells/190 μL/well) were preincubated in 96-well plates overnight in estrogen-free medium. Test samples (20 μg/mL final concentration in DMSO) were added to the cells in a total volume (200 μL media/well) were incubated at 37 °C for 4 days. For the determination of antiestrogenic activity, media used to dilute the test samples was supplemented with estradiol (2 nM). The induction plates were processed by washing the plates with PBS and adding Triton x 100 (0.01%, 50 μL) in Tris buffer (pH 9.8, 0.1 M). Plates were subjected to a freeze thaw (-80 °C for at least 24 h before warming to 37 °C). An aliquot (150 μL) of *p*-nitrophenylphosphate (phosphatase substrate, 1 mg/mL) in Tris buffer (pH. 9.8, 0.1 M) was added to each well. The enzyme activity was measured by reading the release of *p*-nitrophenol at 405 nm every 15 s with a 10 s shake between readings for 16 readings using a Power Wave 200 microplate scanning spectrophotometer (Bio-Tek Instruments, Winooski, VT). The maximal slopes of the lines generated by the kinetic readings were calculated. Estrogenic induction was calculated using Equation 2, and for antiestrogenic determination, the percent induction as compared with the background induction control was calculated using Equation 3.

$$[(slope_{sample} - slope_{cells})/(slope_{estrogen} - slope_{cells})] \times 100 = \% \text{ estrogenic induction} \quad \textbf{(2)}$$

$$[1-((slope_{sample} - slope_{cells})/(slope_{estrogen} - slope_{cells}))] \times 100 = \% \text{ antiestrogenic induction} \quad \textbf{(3)}$$

Cytotoxicity Assay

Cytotoxicity can be a false negative in the agonist assay and a false positive in the antagonism assay. To identify false responses, a 96-hour assay is run in parallel with the induction assay. Cytotoxicity was determined for the Ishikawa alkaline phosphatase induction assays using ~1,000 cells per well in a 96-well plate (5 x 106 cells/mL). Cells were treated with the same samples used in the inductions assays for the respective cell lines. The plates were harvested after 96 h. A day zero plate was prepared by plating at least half of a 96-well plate and allowing the cells to settle overnight. To process the plates 50 μL of cold 50% trichloroacetic acid (TCA) was added to the media (final concentration 20%) and stored at 4 °C for 30 min. The plates were washed with tap water and dried overnight. The following day the plates were stained with 100 μL of sulphohodamine B (SRB) at room temperature and washed with 1% acetic acid and dried in the dark overnight. The dye was suspended in 200 μL Tris buffer (0.1 mM) and mixed using a plate shaker until the dye was completely solubilized. The plates were read using the endpoint mode at 515 nm. Calculation of the percent cytotoxicity was determined using Equation 4. Generally cytotoxicity greater that 20% will interfere with cell-based alkaline phosphatase assays.

$$[1-((OD_{sample} - OD_{day0})/(OD_{DMSO} - OD_{day0}))] \times 100 = \% \text{ cytotoxic.} \quad \textbf{(4)}$$

Measure of ERE Activation

The Dual-Luciferase Reporter Assay System from Promega (Madison, WI) was used to evaluate the functional formation of the ER-ERE complex and luciferase protein expression. Both MCF-7 WS8 and MCF-7 ERβ cell lines were cultured in estrogen-free media 96 h before transfection. The cells were transfected with the pERE-luciferase plasmid (2 µg), which contains three copies of the *Xenopus laevis* vitellogenin A2 ERE upstream of fire fly luciferase (a gift from Dr. V.C. Jordan). To normalize transfection efficiency, pRL-TK plasmid (1 µg, Promega) was co-transfected. Cells (5×10^6) in serum-free media were transfected by electroporation in a 0.4 cm cuvette (Bio-Rad Laboratories) at a voltage (0.320 kV) and a high capacitance (950 µF) in a GenePulser X-cell (Bio-Rad Laboratories). The cells were resuspended in estrogen-free media, transferred to 6-well plates immediately after electroporation, and incubated overnight. The cells were treated with the extracts for 24 h. The luciferase activities in the cell lysates were measured using the Dual-Luciferase Reporter Assay System from Promega (Madison, WI) with a FLUOstar OPTIMA (BMG LABTECH, Durham, NC).

RESULTS

ER Alpha and ER Beta Competitive Assay

The plant extracts were screened in the ERα and ERβ competitive assay at 200 µg/mL (n = 5, two independent measurements). Extracts where the mean percent inhibitory activity was within one standard deviation of 50% or greater were considered active (Table **2**). Six plants were active in the ERα assay (Fig. **1**): the petroleum ether extracts of *A. sinensis* (81%), *A. tuberosa* (69%), *C. racemosa* (aerial parts; 55%), *V. officinalis* (43%), the petroleum ether and the dichloromethane extracts of *C. rubifolia* (56% and 51%, respectively), and the dichloromethane extract of *P. lobata* (63%). There were thirteen plant extracts that bound the ERβ (Fig. **1**): the 75% ethanol and the dichloromethane extracts of *P. lobata* (63 and 45%, respectively), the petroleum ether extract and chloroform partition of *A. sinensis* (99 and 46%, respectively), the petroleum ether extracts of *A. plantago-aquatica* (51%), *A. tuberosa* (85%), *B. vulgaris* (44%), *C. americana* (47%), *C. officinalis* (48%), *C. racemosa* (aerial parts; 64%), *C. rubifolia* (66%),

P. moutan (60%), *P. multiflorum* (BC 268; 50%), *V. officinalis* (81%), and *V. agnus-castus* (68%).

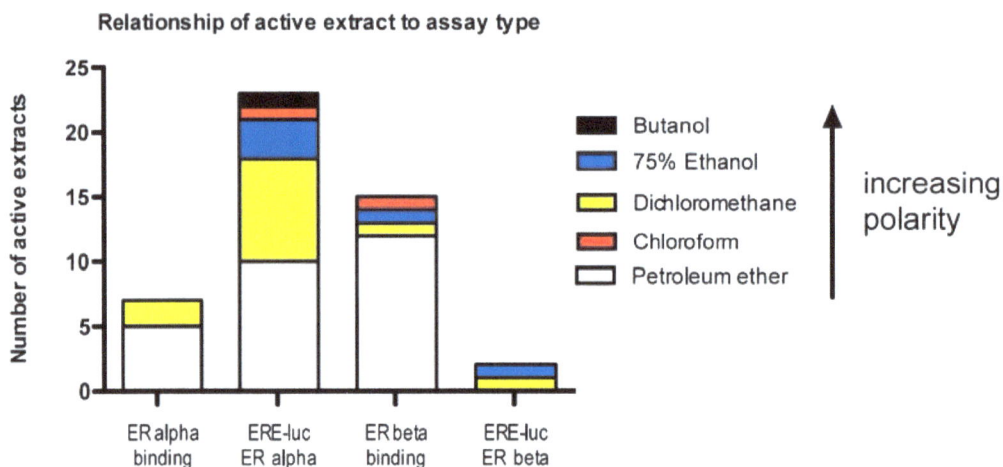

Fig. (1). Bar chart representation of the frequency of extracts with activity in either ER alpha binding or ER beta binding assays and ERE-luc ER alpha or ERE-luc ER beta positive cell lines.

Estrogenic Alkaline Phosphatase Induction in the Ishikawa Cell Line

The ALP assay was used to distinguish agonist activity from antagonist activity [12]. At the tested concentration of 20 µg/mL (n = 3 independent times in triplicate), the petroleum ether extract of *A. sinensis* (49%), and the dichloromethane extracts of *B. vulgaris* (44%), *C. americana* (79%) had antiestrogenic activity that did not appear to be caused by cytotoxicity (Fig. **2**). Cytotoxicity can have a negative impact on the results of the alkaline phosphatase assays when it is above 20%, as well as on the ERE-luc assay. In the estrogenic ALP assay, cytotoxicity was apparent at 20 µg/mL for the dichloromethane extracts of *A. tuberosa* (106%), *C. officinalis* (31%), *P. multiflorum* (BC 286; 31%), *P. mirifica* (115%), *V. prunifolium* (42%), and *V. agnus-castus* (112%), the petroleum ether extracts of *V. agnus-castus* (39%), *P. mirifica* (100%), and *C. officinalis* (31%), the 75% ethanol extracts of *A. tuberosa* (99%), *P. mirifica* (60%), and the chloroform partition of *A. sinensis* (31%). In the ERE-luc assay ERα positive cell line, cytotoxicity was apparent in all of the *A. tuberosa* treated samples as indicated by the high standard deviation and low value for the transfection control vector, pRL-TK.

Table 2.   Screening Results in ER Binding, ERE-luciferase, and Alkaline Phosphatase Assays

Plant or Sample Name	Extract Type	[4]Binding	[5,6]ERE-luc	[6,7]ALP	[6,7]ALP	[6,7]SRB	[4]Binding	[5,6]ERE-luc
		ERα	ERα	Estrogenic	Antiestrogenic	Cytotoxicity	ERβ	ERβ
17 β estradiol	Control	**95 ± 1**	**21 ± 7**	**100 ± 10**	-10 ± 8	4 ± 2	**90 ± 1**	**7 ± 3**
4-hydroxy-tamoxifen	Control	**96 ± 1**	1 ± 0.1	0 ± 5	**94 ± 1**	2 ± 1	**94 ± 1**	1 ± 0.2
DMSO	Control	0	1 ± 0.2	1 ± 2	0 ± 5	1 ± 2	0	1 ± 0.2
Alisma plantago-aquatica	Petroleum ether	31 ± 5	0.9 ± 0.4	-7.3 ± 4.2	-1.5 ± 13.3	10.7 ± 16.2	**51 ± 7**	2.3 ± 1.9
Angelica sinensis	Petroleum ether	[8]**81 ± 10**	**9.7 ± 3.5**	-4.7 ± 6.2	**48.7 ± 9.5**	16.4 ± 13.6	**99 ± 9**	1.1 ± 0.4
	Chloroform	19 ± 9	**5.9 ± 3**	12.3 ± 4.7	19.4 ± 9.8	30.6 ± 8	**46 ± 6**	0.8 ± 0.5
	Butanol	4 ± 3	**9.5 ± 2.2**	12.4 ± 8.2	-16.3 ± 4.5	3.3 ± 15.2	0 ± 2	1.2 ± 0.2
Asclepias tuberosa	Petroleum ether	**69 ± 7**	1.1 ± 0.4	-17.9 ± 15.5	-9.1 ± 21	-19.9 ± 9.8	**85 ± 5**	1.1 ± 0.8
	Dichloro-methane	16 ± 9	4.1 ± 3.1	-26.5 ± 4.5	114.6 ± 6.7	**106.1 ± 5.3**	32 ± 10	**10.7 ± 0.8**
	75% ethanol	0 ± 2	3.9 ± 4.3	-10.3 ± 8.9	110.1 ± 8.5	**99.1 ± 3.6**	5 ± 8	1.3 ± 1.1
Beta vulgaris	Petroleum ether	35 ± 10	**4.7 ± 1.9**	-2.4 ± 17.4	4.4 ± 14.8	-15.9 ± 29.4	**44 ± 7**	1.1 ± 0.3
	Dichloro-methane	33 ± 11	3.2 ± 2	-2.3 ± 5.5	**43.7 ± 8.6**	-20.2 ± 30.9	30 ± 5	1.1 ± 0.1
Cimicifuga americana	Petroleum ether	37 ± 8	1.6 ± 0.7	-5.8 ± 14.4	27.1 ± 27.5	-3.7 ± 37	**47 ± 10**	1.4 ± 0.3
	Dichloro-methane	31 ± 5	2.8 ± 1.1	-16.2 ± 14.9	**79.2 ± 19.5**	5.8 ± 32.1	21 ± 9	1.4 ± 0.6
Cimicifuga racemosa	DCM extract	34 ± 7	**6.6 ± 3.2**	-2.5 ± 1.3	3.5 ± 8.7	8.3 ± 5.2	45 ± 3	1.1 ± 0.4
	75% EtOH	21 ± 5	**5.8 ± 1.8**	4.3 ± 5.3	**-48.4 ± 4.9**	0.1 ± 5.7	23 ± 3	1 ± 0.2
	PE extract	**55 ± 8**	3.2 ± 1.5	-4.5 ± 4.8	21.3 ± 7.2	4.8 ± 10.9	**64 ± 8**	1 ± 0.5
Cimicifuga rubifolia	Petroleum ether	**56 ± 2**	2.5 ± 2	-13.1 ± 12.4	33.2 ± 17	-23.2 ± 25	**66 ± 6**	1.1 ± 0.5

---

[4] Percent binding at 200 µg/mL.
[5] Fold induction where DMSO is 1.
[6] Tested at 20 µg/mL.
[7] Percent induction (estrogenic), inhibition (antiestrogenic), or cytotoxic.
[8] Bold face-type indicates extract with assay activity.

(Table 2) *contd*.....

Plant or Sample Name	Extract Type	[9]Binding	[10,11]ERE-luc	[6,12]ALP	[6,7]ALP	[6,7]SRB	[4]Binding	[5,6]ERE-luc
		ERα	ERα	Estrogenic	Antiestrogenic	Cytotoxicity	ERβ	ERβ
	Dichloro-methane	**51 ± 5**	**7.4 ± 1.1**	0.3 ± 15.9	21 ± 14.2	4.4 ± 19.8	40 ± 8	0.8 ± 0.3
Cornus officinalis	Petroleum ether	26 ± 16	**24.3 ± 6.6**	2.6 ± 5.9	41.8 ± 4.1	11.5 ± 3.4	**48 ± 4**	1.5 ± 0.5
	Dichloro-methane	0 ± 4	2.5 ± 0.7	-2.2 ± 4.8	24.9 ± 5.2	**30.6 ± 19.1**	0 ± 2	1.3 ± 0.3
Daucus carota	Petroleum ether	21 ± 6	**12.3 ± 5.1**	-6.8 ± 11.2	-26.6 ± 14.6	-18.7 ± 19.1	36 ± 6	0.9 ± 0.5
	Dichloro-methane	22 ± 5	**5.4 ± 1.8**	-5.7 ± 8	-11.5 ± 16.1	13.7 ± 14.1	26 ± 6	0.7 ± 0.2
Paeonia moutan	Petroleum ether	21 ± 5	2.8 ± 1	-9.7 ± 11.4	8.7 ± 16.7	7.7 ± 19.7	**60 ± 6**	2 ± 1.7
Polygonum multiflorum	Petroleum ether	15 ± 8	**21.7 ± 8.5**	6.5 ± 1.8	1.2 ± 3.3	7.6 ± 14.3	**50 ± 9**	1.4 ± 0.7
	Dichloro-methane	30 ± 7	**25.8±11.2**	12.3 ± 3.4	0.4 ± 12.3	12.7 ± 17.8	36 ± 13	3 ± 2.5
Pueraria lobata	Petroleum ether	32 ± 2	**10.3 ± 2.8**	-10.8 ± 12.6	**46.9 ± 18.9**	12.7 ± 30	36 ± 2	1.5 ± 1.4
	Dichloro-methane	**63 ± 6**	**24.5 ± 9.9**	-0.3 ± 25.6	36 ± 32.6	-3.3 ± 5.5	**45 ± 19**	1.2 ± 0.5
	75% ethanol	17 ± 16	**26.9± 12.7**	-1.3 ± 13.8	28.9 ± 40.7	7.6 ± 50.1	**63 ± 3**	**3.5 ± 1.3**
Pueraria mirifica	Petroleum ether	12 ± 8	[6]N.T.	-2.7 ± 2.6	58.5 ± 37	**100.6 ± 38.8**	18 ± 12	N.T.
	Dichloro-methane	12 ± 5	**25.7 ± 9.5**	-4.2 ± 1.1	101.1 ± 8.6	**115 ± 2.2**	19 ± 11	3.2 ± 1.6
	75% ethanol	18 ± 7	**28 ± 8.6**	-1.4 ± 4.8	52.2 ± 61	**60.4 ± 62.2**	31 ± 8	2.4 ± 1.3
Valeriana officinalis	Petroleum ether	**43 ± 11**	**7.2 ± 3.1**	-7.8 ± 7.6	33.7 ± 16.8	12.5 ± 19.3	**81 ± 4**	1.3 ± 0.8
	Dichloro-methane	13 ± 12	**5.9 ± 1.6**	-8.5 ± 7.7	19.4 ± 8.2	-7.6 ± 21	13 ± 6	0.7 ± 0.1
Viburnum prunifolium	Dichloro-methane	3 ± 6	0.7 ± 0.5	-3.5 ± 3	68.5 ± 6.3	**42 ± 4.4**	3 ± 4	1 ± 0
Vitex agnus-castus	Petroleum ether	37 ± 6	**9.5 ± 2.5**	-1.1 ± 1.6	43.5 ± 3.3	**39.1 ± 8.3**	68 ± 9	1.1 ± 0.7
	Dichloro-methane	13 ± 3	**9.6 ± 6.6**	-2.3 ± 1.5	94.5 ± 7.2	**112.5 ± 8.8**	20 ± 19	0.8 ± 0.3

---

[9] Percent binding at 200 μg/mL.
[10] Fold induction where DMSO is 1.
[11] Tested at 20 μg/mL.
[12] Percent induction (estrogenic), inhibition (antiestrogenic), or cytotoxic.
[6] Not tested due to the limited quantities available.

Ishikawa Alkaline Phosphatase

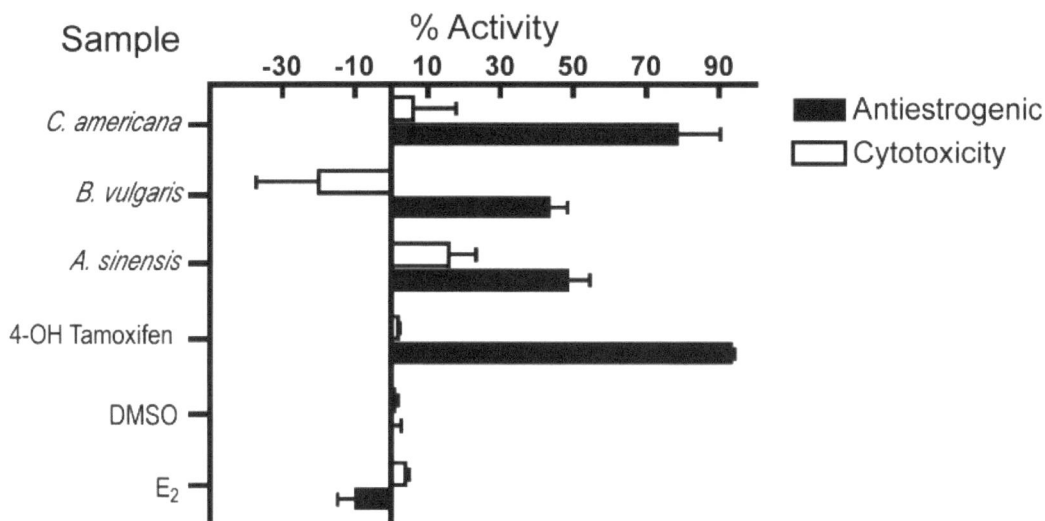

Fig. (2). Antiestrogenic and cytotoxic graph of active samples and controls. The petroleum ether extract of *A. sinensis*, and the dichloromethane extracts of *B. vulgaris* and *C. americana* had antiestrogenic activity without apparent cytotoxicity in the Ishikawa alkaline phosphatase inhibition assay when tested at 20 μg/mL. Samples were tested at least three independent times in triplicate. Samples were considered active if they were within one standard deviation or greater of 50% for the antiestrogenic assay. Samples with greater than 20% cytotoxicity are know to interfere with the accuracy of the antiestrogenic assay by causing a false positive result.

Interestingly, some extracts when combined with estradiol appear to have a greater estrogenic response than the extract alone (Fig. **3**). The 75% ethanol extract of *C. racemosa*, aerial parts had a -48% antiestrogenic response indicating that more alkaline phosphatase enzyme was induced when compared to the extract without 17β estradiol (4%). Also, the 75% ethanol extract of *V. prunifolium* also had a similar effect when combined with 17β estradiol (-33%) compared to the extract alone (0%). This may also be true for the 75% ethanol extract of *C. rubifolia* (-31%) and the petroleum ether extract of *D. carota* (-27%), but the standard deviations for both samples were large (22 and 15, respectively).

ERE-Luciferase Induction in ER Alpha and ER Beta Positive Cell Lines

Following the ERα and ERβ competitive assay, the transient transfection of MCF-7 WS8 or MCF-7 C4-12-5 ERβ+ cell lines with the ERE-luciferase plasmid was used to help confirm functional activity of the extracts that had activity in the isolated ER competitive assay. The cells were treated with extracts (20 µg/mL) for 24 h, and the activity was first normalized for transfection efficiency using the pRL-TK vector, and then normalizing to the DMSO control (DMSO = 1). Extracts with > two–fold induction were considered active (Table **1**; n ≥ 3).

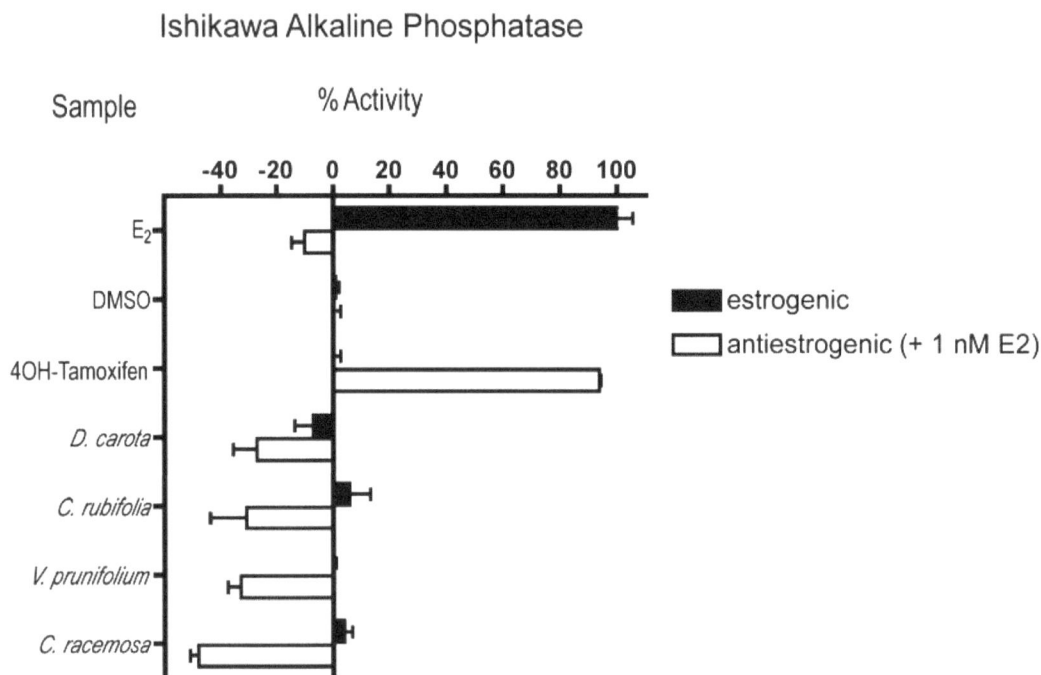

Fig. (3). Samples with possible synergistic activity when combined with 17β estradiol in the Ishikawa alkaline phosphatase assay. Samples were treated alone to determine the inherent estrogenic activity of the sample, or in combination with 1 nM E_2 to determine if there was an antiestrogenic effect of the samples. 17β Estradiol shows the stereotypical estrogenic response alone and is therefore estrogenic, while 4-hydroxytamoxifen shows a stereotypical antiestrogenic response when combined with 1 nM E_2, but it is not estrogenic alone. The 75% ethanol extract of *C. racemosa* (aerial parts) was not estrogenic alone, but when combined with 1 nM E_2 was actually more estrogenic than when E_2 is tested in the presence of 1 nM E_2. The 75% ethanol extract of *V. prunifolium* and *C. rubifolia*, and the petroleum ether extract of *D. carota* are also displayed. Samples were tested in triplicate at least three independent times, and are represented as averages ± standard deviation.

In the MCF-7 WS8 (ERα positive) cell line, the petroleum ether, chloroform, and butanol partitions of *A. sinensis* (9.7, 5.9, and 9.5, respectively), the petroleum ether, dichloromethane, and 75% ethanol extracts of *P. lobata* (10.3, 24.5, and

26.9, respectively), the dichloromethane and 75% ethanol extracts of *C. racemosa* (aerial parts; 6.6 and 5.8, respectively) and *P. mirifica* (25.7 and 28.0, respectively), the petroleum ether and dichloromethane extracts of *D. carota* (12.3 and 5.4, respectively), *P. multiflorum* (BC 268: 21.7 and 25.8; BC 286: 15.1 and 16, respectively), *V. officinalis* (7.2 and 5.9), and *V. agnus-castus* (9.5 and 9.6, respectively), the petroleum ether extracts of *B. vulgaris* (4.7), *C. rubifolia* (7.4), *C. officinalis* (24.3) were active (Fig. **1**). All extracts of *A. tuberosa* were cytotoxic at 20 µg/mL, which caused some false positive activity. In the MCF-7 ERβ cell line, none of the petroleum ether extracts were active. The 75% ethanol extract of *P. lobata* (3.5) and the dichloromethane extracts of *A. tuberosa* (10.7) and *P. mirifica* (3.2) had activity (Fig. **1**).

DISCUSSION

The UIC/NIH Center for Botanical Dietary Supplements Research (the Center) was established in the fall of 1999 to address issues of standardization, quality, safety, and efficacy of botanical dietary supplements. Using a multidisciplinary strategy to achieve its basic and clinical research objectives, the Center has focused on botanicals with potential benefits for women's health. Based on these data, a paradigm shift occurred within the Center for how the assays are utilized to provide the most meaningful data. For example, during the initial years when testing botanicals in the Center, the competitive estrogen receptor binding assays were primarily performed, and then the induction of alkaline phosphatase in Ishikawa cells was used to determine the extent to which the botanicals were acting as antagonists or agonists. The logical thought was that botanicals that were not active in the competitive binding assay would not be active in the alkaline phosphatase assay. This presumption was first challenged by the identification of botanicals that do not bind to the isolated receptor, but exhibited activity in the cell-based assays. *C. racemosa* (roots) is another example of a botanical that questioned this paradigm since it was not estrogenic either *in vitro* or *in vivo* [17, 18]. However, it is one of the most popular botanicals used by menopausal women despite conflicting results from clinical trials [19-21]. Alternative mechanisms such as the serotonergic [18, 22] and opioid [23, 24] pathways have since been implicated.

While the selected assays are not new, and the throughput for these assays has not been improved, the summation of data from these assays has improved the total information content and enabled possible new combinations of activity for plant extracts. The ER binding assays provide information on binding to the receptor in a cell-free environment. While this information is important, it does not indicate to what extent the activity is functional, or what type of activity it has; *i.e.*, agonist or antagonist activity [10]. The Ishikawa ALP assay is used to distinguish agonist activity from antagonist activity, but it is only ERα positive, and is affected by cytotoxicity [12] whereas the cell-free ER-binding assays are not affected by cytotoxicity. Finally, the ERE-luc assay provides functional agonist activity in ERα or ERβ cell lines and is capable of some metabolism [25]. While not incorporated into the panel of assays to identify antagonist activity, it is fully capable of distinguishing agonist from antagonists and compounds without functional activity [25]. It is also affected by cytotoxicity, and it is significantly slower to process agonist and antagonist assays in parallel to the same extent as the ALP assays. Therefore each assay provides some information about mechanism of action individually. However, due to the inherent limitations of each assay, the characterization of a sample is significantly improved by incorporating these assays into a panel.

In order to identify plants that might have a beneficial effect on symptoms associated with menopause, a NAPRALERT search was conducted to select plants with previously reported hormonal or neurotransmitter activity. Among the plants selected, nine plants had previously been reported to be useful for the relief of menopause symptoms, and four plants were specified as having estrogenic activity. In general, to cover the whole range of polarity for phytoconstituents three extracts of each plant, petroleum ether, dichloromethane, and 75% ethanol, were prepared and tested in the isolated ERα and ERβ competitive assay, ERE-luc fold-induction in both ERα and ERβ positive breast cancer cell lines, and the induction of alkaline phosphatase in the Ishikawa endometrial cell line. The extracts were also tested in a secondary assay for cytotoxicity. Due to the nature of *in vitro* assays the terms "false positive" and "false negative" are defined as follows. False positive activity is attributed to a mechanism assumed to be independent of the tested mechanism such as non-specific binding of lipophilic material to an isolated receptor which blocks the tritiated-estradiol from specifically binding to the binding pocket. A false negative result has been defined to mean a lack of activity, which is attributed to a mechanism assumed to be independent of the tested mechanism such as a sample that is cytotoxic rather than specifically inhibits a specific enzymatic reaction in the cell.

The initial biological characterization of the extracts indicated that many of the petroleum ether extracts had activity in the ER binding assays, particularly in the ERβ assay, where twelve of the fifteen active extracts were petroleum ether extracts (Fig. **1**). Petroleum ether extracts are known to contain fatty acids, which have been reported in the literature to non-specifically bind isolated receptors and/or may only have weak estrogenic activity [26]. In addition to fatty acid, there appeared to be a pattern of activity where the first solvent, petroleum ether, used in the sequential extraction has more activity (Fig. **1**). This may be partially related to the sequential extraction method, which is industry-standard. Since compounds can dissolve in more than one solvent to varying degrees, it is possible that as the first extracting solvent, petroleum ether extracted more of the active constituents compared to the following solvents. Based on in-house evaluation of specific plants subject to individual extraction and sequential extraction, the order, parallel or sequential did not show as cause for concern. In general, the relative proportions of activity would be maintained if the plant material had been extracted in parallel. At best, a change in extraction method may change the degree to which an extract is active, but it would not cause an inactive extract to become active or vice-versa.

This hypothesis was supported by the observation that the 75% ethanol extract of *P. lobata* had activity in the ERβ competitive assay and in the ERE-luc assay in the ERβ positive cell line (Fig. **1**). Recently, *P. lobata* was reported to contain compounds known to preferentially act through the ERβ including, formononetin, biochanin A, genistein, daidzein, and puerarin [27-29]. Similarly, in the ERα assays, there was a correlation between ERα competitive binding activity and ERE-luc assay for the petroleum ether extract of *A. sinensis* (81% binding affinity and 9.7 fold induction) and the dichloromethane extracts of *C. rubifolia*, and *P. lobata*. For the latter two extracts, there was relatively weak activity (51% and 63%, respectively) in the ERα competitive binding assay, but relatively strong fold-induction (7.4 and 25, respectively) at one-tenth the concentration in the ERE-luc assay.

One of the more surprising observations from the current study was the number of extracts that did not bind the receptor, but did have activity in the ERE-luc assay (Fig. 1). In the ERβ assays, *A. tuberosa* and *P. lobata* were the only plants that did not have activity in the competitive ERβ binding assay despite having activity in the ERE-luc assay in ERβ cells. In the ERα assays, this was also the case for the tested chloroform and butanol partitions of *A. sinensis* and certain extracts of *C. racemosa*, aerial parts, *C. officinalis*, *D. carota*, *P. multiflorum*, *P. lobata*, *P. mirifica*, *V. officinalis*, and *V. agnus-castus*. One important example is *A. sinensis,* which did not have activity in the isolated ERα receptor competitive binding for any of the partitions except for petroleum ether, but showed potent activity in the ERE-luc assay for both the chloroform and butanol partitions. While the petroleum ether extracts generally did not have functional activity, *A. sinensis* did have functional activity in the ERE-luc assay and antiestrogenic activity in the Ishikawa alkaline phosphatase assay. *Angelica sinensis* is one of the most commonly used herbs in China for relief of PMS and menopause [30]; however, the mechanism of action has not clearly been identified. This challenge has been addressed very recently with the use of quantitative NMR to identify the chemically unstable compound, ligustilide, as one of the biologically active constituents in *A. sinensis* [9].

Classically, extracts that do not bind the estrogen receptor would not be thought to be active in the ERE-luc assay, which is generally used to identify ligand-ER complexes that bind the ERE causing gene transcription. The disconnection between the binding to the ER and the activation of the ERE has raised many hypotheses. One is that these plant extracts do not bind to the ER, but instead use a non-classical estrogen pathway that phosphorylates the ER [31]. The non-liganded, phosphorylated ER has been demonstrated to bind the ERE and recruit transcription factors. This could explain why it has been challenging to explain the mechanism of action for plants that have been traditionally used by women, but when the extracts were scientifically tested, they do not contain ER ligands. Another hypothesis is that the extract contains ligands that do not competitively bind to the ER, but instead bind to a second binding site [32]. The full implication of this situation with different structures has yet to be explored, but does raise some interesting possibilities. These hypotheses raise the concern that women taking extracts that are considered to be "non-estrogenic" due to health concerns of hormone-dependent cancers might still be at risk. A third explanation might be metabolism of botanical compounds in the cells to estrogenic products. It has previously been shown that *Trifolium pratense* [15] and *P. lobata* [27, 33] contain some compounds that are not inherently estrogenic, but can be metabolically converted to estrogenic compounds such as, genistein and daidzein. Similarly isoxanthohumol, a compound found in *Humulus lupulus*, can also be metabolized to the potent phytoestrogen, 8-prenylnaringenin [15, 34], although not in rats [35].

In the Ishikawa assay, none of the extracts had agonist activity, but some had antiestrogenic activity that appeared to be unrelated to cytotoxicity or direct inhibition of alkaline phosphatase. At 20 μg/mL the petroleum ether partition of *A. sinensis* was the only sample that had ERα activity in all three assays. In fact, *A. sinensis* might be considered a true SERM since it was an agonist in the ERE-luc assay, but had antagonistic activity in the Ishikawa assay. *Angelica sinensis* may also have selective tissue effects since the antiestrogenic Ishikawa assay was in an endometrial cell line, while the ERE-luc was in a breast cancer cell line. More recently, the antiestrogenic observation was confirmed using a lipohilic fraction of a methanolic extract that is rich in ligustilide [9]. Other extracts that appear to have antiestrogenic activity in the ALP assay were the dichloromethane extracts of *B. vulgaris* and *C. americana*. It is noteworthy that the ethnobotanical use of *B. vulgaris* has generally been attributed to its vitamin content rather than the presence of antiestrogenic compound(s). *Cimicifuga americana* has not been reported in the literature as being used to alleviate any hormone-related symptoms; however, it is taxonomically closely related to *C. racemosa*.

There were some unexpected results in the antagonist alkaline phosphatase assay for some extracts, which had large negative values. This initially was overlooked, but after carefully review of the data, and multiple independent repetitions of the assay, the values were determined to be reproducible. The antagonist assay was designed to identify compounds that blocked the effect of estrogen. When this occurs, the ALP enzyme is not produced, and substrate is not converted to product. Therefore, a score of 100% in this assay would indicate complete blockage of the estrogenic activity, and 0% would indicate that the estrogenic activity led to the ALP enzyme converting substrate to product. When the data go into the negative range, it might indicate a synergistic effect with estrogen where even more ALP is produced causing faster conversion of substrate to product (Fig. **3**).

Four plants had negative values in the antagonist ALP assay indicating a possible synergistic activity: the 75% ethanol extracts of *C. racemosa* (aerial parts), *C. rubifolia*, and *V. prunifolium*, and the petroleum ether extracts of *D. carota*. The activity was most prevalent in the 75% ethanol extract of *C. racemosa* (aerial parts). The extract alone at 20 μg/mL was not estrogenic in the Ishikawa assay (4.3%); however, when combined with estrogen it had a -48% activity in the antiestrogenic ALP assay. These data might be interpreted as the extract combined with estrogen induced enough ALP enzyme to convert substrate to product twice as fast as estrogen alone. This may have occurred by the extract working through a non-estrogen pathway that upregulates the ER, such as the progesterone pathway. A striking observation was that this extract also had activity in the ERE-luc assay (5.8 fold-induction) in the ERα positive cell line, but did not bind to the estrogen receptor (24%). Investigation of the contribution of aerial parts to the biological profile of adulterated *C. racemosa* preparations may also contribute to the resolution of the conflict in the literature revolving around the estrogenic activity of *C. racemosa* [36, 37]. While completely unexplored in the literature, the biological activities of the aerial parts of *C. racemosa* clearly deserve further investigation.

Cytotoxicity can play a large role in the false positive and false negative interpretation of cell based assays. In the ALP assays, cytotoxicity could cause a false negative in the agonist assay, and similarly, cytotoxicity can cause a false positive in the antagonist ALP assay where a lack of enzyme is really caused by a lack of cell viability. At 20 μg/mL, the dichloromethane and the 75% ethanol extracts of *A. tuberosa* were completely cytotoxic in the Ishikawa cell line, and was the likely cause of the large standard deviations in the ERE-luc assay. The vector control for *A. tuberosa* in the ERE-luc assay was low for a few of the replicates compared with other samples tested in parallel. This indicates that the cells were not able to utilize the control vector, because the cells were not viable rather than due to poor transfection efficiency, which is what the control vector is supposed to indicate. This resulted in dividing by a small denominator, which resulted in a large product with a large standard deviation. The dichloromethane and the 75% ethanol extracts had activity in the ERE-luc in ERα positive cells, and the dichloromethane extract had activity in the ERβ positive cells. When the values for the transfection control vector were evaluated, data were consistent with other plant extracts that did not have cytotoxicity. Another case where cytotoxicity was a factor was for *P. mirifica*, which was completely cytotoxic for the petroleum ether and dichloromethane extracts, and 60% cytotoxic for the 75% ethanol extract when tested in the Ishikawa cell line. It is also possible that the lack of estrogenic activity may be a result of the cytotoxicity, extraction procedure, or a combination of both. Although *P. mirifica* [38] and its substituents, puerarin [39], and miroestrol [40] have been reported to be estrogenic in the literature, the discrepancy between the activity of the extracts is most likely due to a different and more gentle extraction procedure was used in the present study. Of interest was that, while the extracts were toxic in the four-day assays, they did not appear to have a significant effect on the 24-hour assays.

In summary, based on these findings, the Center has moved toward cell-based assays for evaluating the hormonal activities of botanicals. Employing a diverse panel of bioassays, several extracts have been identified which do not conform to the classical estrogen receptor signaling pathway as they do not bind the estrogen receptor, but appear to have cell-based activity in the ERE-luc assay. Extracts that are not inherently estrogenic, but appear to increase the estrogenic activity in the presence of estrogen have also been identified. Finally, the synergistic activity opens avenues that need to be explored concerning the overall biological activity of the plant extracts and the value of multiple bioassays in characterizing the plant activities.

ABBREVIATIONS

CBS   =   Calf bovine serum

DMEM/F12   =   Dulbecco's Modified Eagle medium

ER	=	Estrogen receptor
ERE	=	Estrogen response element
ERE-luc	=	ERE-luciferase
FBS	=	Fetal bovine serum
HRT	=	Hormone replacement therapy
HAPs	=	Hydroxyapatite slurry
NAPRALERT	=	Natural Products Alert
NEAAs	=	Non-essential amino acids
PE	=	Petroleum ether
SERMs	=	Selective estrogen receptor modulators
SSRIs	=	Selective serotonin reuptake inhibitors

ACKNOWLEDGEMENTS

This work was supported, in part, by grant P50 AT00155 provided jointly by the National Center for Complementary and Alternative Medicine (NCCAM), the Office of Dietary Supplements (ODS), the Office for Research on Women's Health (ORWH), and the National Institute of General Medicine NIGMS) of the National Institutes of Health (NIH). C.R.O. is grateful for a Ruther L. Kirschstein NCCAM Predoctoral fellowship F31 AT 24232. The contents of this paper are solely the responsibility of the authors and do not necessarily represent the official views of NIH.

CONFLICT OF INTEREST

The authors state that there is no conflict of interest.

DISCLOSURE

This chapter has been updated from its original publication: Overk, C. R., Yao, P., Chen, S., Deng, S., Imai, A., Main, M., Schinkovitz, A., Farnsworth, N. R., Pauli, G. F., and Bolton, J. L. (2008) High-content screening and mechanism-based evaluation of estrogenic botanical extracts. *Comb. Chem. High Throughput Screen.* *11*, 283-293.

REFERENCES

[1]   Hersh, A.L.; Stefanick, M.L. Stafford, R.S. National use of postmenopausal hormone therapy: annual trends and response to recent evidence. *JAMA*, **2004**, *291*(1), 47-53.

[2]   Haas, J.S.; Kaplan, C.P.; Gerstenberger, E.P. Kerlikowske, K. Changes in the use of postmenopausal hormone therapy after the publication of clinical trial results. *Ann. Intern. Med.*, **2004**, *140*(3), 184-188.

[3]   Murkies, A.L.; Wilcox, G. Davis, S.R. Clinical review 92: Phytoestrogens. *Clin. Endocrinol. Metab.*, **1998**, *83*(2), 297-303.

[4]   Setchell, K.D. Phytoestrogens: the biochemistry, physiology, and implications for human health of soy isoflavones. *Am. J. Clin. Nutr.*, **1998**, *68*(6 Suppl), 1333S-1346S.

[5]   Setchell, K.D. Cassidy, A. Dietary isoflavones: biological effects and relevance to human health. *J. Nutr.*, **1999**, *129*(3), 758S-767S.

[6]   Anonymous What are bioidentical hormones. *Harvard Women's Health Watch*, **2006**, *13*(12), 1-3.

[7]   Calixto, J.B. Efficacy, safety, quality control, marketing and regulatory guidelines for herbal medicines (phytotherapeutic agents). *Braz. J. Med. Biol. Res.*, **2000**, *33*(2), 179-189.

[8]   van Breemen, R.B.; Fong, H.H.S. Farnsworth, N.R. The Role of Quality Assurance and Standardization in the Safety of Botanical Dietary Supplements. *Chem. Res. Toxicol.*, **2007**, *20*(4), 577-582.

[9]   Godecke, T.; Yao, P.; Napolitano, J.G.; Nikolic, D.; Dietz, B.M.; Bolton, J.L.; van Breemen, R.B.; Farnsworth, N.R.; Chen, S.N.; Lankin, D.C. Pauli, G.F. Integrated standardization concept for Angelica botanicals using quantitative NMR. *Fitoterapia*, **2012**, *83*(1), 18-32.

[10]   Obourn, J.D.; Koszewski, N.J. Notides, A.C. Hormone- and DNA-binding mechanisms of the recombinant human estrogen receptor. *Biochemistry*, **1993**, *32*(24), 6229-6236.

[11]   Catherino, W.H. Jordan, V.C. Increasing the number of tandem estrogen response elements increases the estrogenic activity of a tamoxifen analogue. *Cancer Lett.*, **1995**, *92*(1), 39-47.

[12]   Pisha, E.P., J.M. Cell-based assay for the determination of estrogenic and anti-estrogenic activities. *Methods Cell Sci.*, **1997**, *19*(37-43.

[13]   Overk, C.R.; Yao, P.; Chen, S.; Deng, S.; Imai, A.; Main, M.; Schinkovitz, A.; Farnsworth, N.R.; Pauli, G.F. Bolton, J.L. High-content screening and mechanism-based evaluation of estrogenic botanical extracts. *Comb. Chem. High Throughput Screen.*, **2008**, *11*(4), 283-293.

[14]   Nelson, H.D.; Vesco, K.K.; Haney, E.; Fu, R.; Nedrow, A.; Miller, J.; Nicolaidis, C.; Walker, M. Humphrey, L. Nonhormonal therapies for menopausal hot flashes: systematic review and meta-analysis. *JAMA*, **2006**, *295*(17), 2057-2071.

[15]   Overk, C.R.; Yao, P.; Chadwick, L.R.; Nikolic, D.; Sun, Y.; Cuendet, M.A.; Deng, Y.; Hedayat, A.S.; Pauli, G.F.; Farnsworth, N.R.; van Breemen, R.B. Bolton, J.L. Comparison of the *in vitro* estrogenic activities of compounds from hops (Humulus lupulus) and red clover (Trifolium pratense). *J. Agric. Food. Chem.*, **2005**, *53*(16), 6246-6253.

[16]   Oesterreich, S.; Zhang, P.; Guler, R.L.; Sun, X.; Curran, E.M.; Welshons, W.V.; Osborne, C.K. Lee, A.V. Re-expression of Estrogen Receptor alpha in Estrogen Receptor alpha-negative MCF-7 Cells Restores both Estrogen and Insulin-like Growth Factor-mediated Signaling and Growth. *Cancer Res.*, **2001**, *61*(5771-5777.

[17]   Liu, J.; Burdette, J.E.; Xu, H.; Gu, C.; van Breemen, R.B.; Bhat, K.P.; Booth, N.; Constantinou, A.I.; Pezzuto, J.M.; Fong, H.H.; Farnsworth, N.R. Bolton, J.L. Evaluation of estrogenic activity of plant extracts for the potential treatment of menopausal symptoms. *J Agric. Food Chem.*, **2001**, *49*(5), 2472-2479.

[18]   Burdette, J.E.; Liu, J.; Chen, S.N.; Fabricant, D.S.; Piersen, C.E.; Barker, E.L.; Pezzuto, J.M.; Mesecar, A.; Van Breemen, R.B.; Farnsworth, N.R. Bolton, J.L. Black cohosh acts as a mixed competitive ligand and partial agonist of the serotonin receptor. *J. Agric. Food Chem.*, **2003**, *51*(19), 5661-5670.

[19]   Geller, S.E.; Shulman, L.P.; van Breemen, R.B.; Banuvar, S.; Zhou, Y.; Epstein, G.; Hedayat, S.; Nikolic, D.; Krause, E.C.; Piersen, C.E.; Bolton, J.L.; Pauli, G.F. Farnsworth, N.R. Safety and efficacy of black cohosh and red clover for the management of vasomotor symptoms: a randomized controlled trial. *Menopause*, **2009**, *16*(6), 1156-1166.

[20]   Wuttke, W.; Seidlova-Wuttke, D. Gorkow. C. The Cimicifuga preparation BNO 1055 *vs* conjugated estrogens in a double-blind placebo-controlled study: effects on menopause symptoms and bone markers. *Maturitas*, **2003**, *44 Suppl 1*(S67-77.

[21]   Liske, E.; Hanggi, W.; Henneicke-von Zepelin, H.H.; Boblitz, N.; Wustenberg, P. Rahlfs, V.W. Physiological investigation of a unique extract of black cohosh (Cimicifugae racemosae rhizoma): a 6-month clinical study demonstrates no systemic estrogenic effect. *J. Womens Health Gend. Based Med.*, **2002**, *11*(2), 163-174.

[22]   Powell, S.L.; Godecke, T.; Nikolic, D.; Chen, S.N.; Ahn, S.; Dietz, B.; Farnsworth, N.R.; van Breemen, R.B.; Lankin, D.C.; Pauli, G.F. Bolton, J.L. *In vitro* serotonergic activity of black cohosh and identification of N(omega)-methylserotonin as a potential active constituent. *J. Agric. Food Chem.*, **2008**, *56*(24), 11718-11726.

[23]   Rhyu, M.R.; Lu, J.; Webster, D.E.; Fabricant, D.S.; Farnsworth, N.R. Wang, Z.J. Black cohosh (Actaea racemosa, Cimicifuga racemosa) behaves as a mixed competitive ligand and partial agonist at the human mu opiate receptor. *J. Agric. Food Chem.*, **2006**, *54*(26), 9852-9857.

[24]   Reame, N.E.; Lukacs, J.L.; Padmanabhan, V.; Eyvazzadeh, A.D.; Smith, Y.R. Zubieta, J.K. Black cohosh has central opioid activity in postmenopausal women: evidence from naloxone blockade and positron emission tomography neuroimaging. *Menopause*, **2008**, *15*(5), 832-840.

[25]   Overk, C.R.; Peng, K.W.; Asghodom, R.T.; Kastrati, I.; Lantvit, D.D.; Qin, Z.; Frasor, J.; Bolton, J.L. Thatcher, G.R. Structure-activity relationships for a family of benzothiophene selective estrogen receptor modulators including raloxifene and arzoxifene. *ChemMedChem*, **2007**, *2*(10), 1520-1526.

[26]	Liu, J.; Burdette, J.E.; Sun, Y.; Deng, S.; Schlecht, S.M.; Zheng, W.; Nikolic, D.; Mahady, G.; van Breemen, R.B.; Fong, H.H.; Pezzuto, J.M.; Bolton, J.L. Farnsworth, N.R. Isolation of linoleic acid as an estrogenic compound from the fruits of Vitex agnus-castus L. (chasteberry). *Phytomedicine*, **2004**, *11*(1), 18-23.

[27]	Zhang, D.; Ren, Y.; Dai, S.; Liu, W. Li, G. [Isoflavones from vines of Pueraria lobata]. *Zhongguo Zhong Yao Za Zhi*, **2009**, *34*(24), 3217-3220.

[28]	Zhang, C.Z.; Wang, S.X.; Zhang, Y.; Chen, J.P. Liang, X.M. *In vitro* estrogenic activities of Chinese medicinal plants traditionally used for the management of menopausal symptoms. *J. Ethnopharmacol.*, **2005**, *98*(3), 295-300.

[29]	Boue, S.M.; Wiese, T.E.; Nehls, S.; Burow, M.E.; Elliott, S.; Carter-Wientjes, C.H.; Shih, B.Y.; McLachlan, J.A. Cleveland, T.E. Evaluation of the estrogenic effects of legume extracts containing phytoestrogens. *J. Agric. Food Chem.*, **2003**, *51*(8), 2193-2199.

[30]	Taylor, M. Complementary and alternative medicine preparations used to treat symptoms of menopause. *Menopausal Medicine*, **2012**, *20*(1), S1-S8.

[31]	Pettersson, K. Gustafsson, J.A. Role of estrogen receptor beta in estrogen action. *Annu. Rev. Physiol.*, **2001**, *63*(165-192.

[32]	Wang, Y.; Chirgadze, N.Y.; Briggs, S.L.; Khan, S.; Jensen, E.V. Burris, T.P. A second binding site for hydroxytamoxifen within the coactivator-binding groove of estrogen receptor beta. *Proc. Natl. Acad. Sci. U S A*, **2006**, *103*(26), 9908-9911.

[33]	Park, E.K.; Shin, J.; Bae, E.A.; Lee, Y.C. Kim, D.H. Intestinal bacteria activate estrogenic effect of main constituents puerarin and daidzin of Pueraria thunbergiana. *Biol Pharm Bull*, **2006**, *29*(12), 2432-2435.

[34]	Nikolic, D.; Li, Y.; Chadwick, L.R.; Pauli, G.F. van Breemen, R.B. Metabolism of xanthohumol and isoxanthohumol, prenylated flavonoids from hops (Humulus lupulus L.), by human liver microsomes. *J. Mass Spectrom.*, **2005**, *40*(3), 289-299.

[35]	Overk, C.R.; Guo, J.; Chadwick, L.R.; Lantvit, D.D.; Minassi, A.; Appendino, G.; Chen, S.N.; Lankin, D.C.; Farnsworth, N.R.; Pauli, G.F.; van Breemen, R.B. Bolton, J.L. *In vivo* estrogenic comparisons of Trifolium pratense (red clover) Humulus lupulus (hops), and the pure compounds isoxanthohumol and 8-prenylnaringenin. *Chem. Biol. Interact.*, **2008**, *176*(1), 30-39.

[36]	Piersen, C.E. Phytoestrogens in botanical dietary supplements: implications for cancer. *Integr. Cancer Ther.*, **2003**, *2*(2), 120-138.

[37]	Kronenberg, F. Fugh-Berman, A. Complementary and alternative medicine for menopausal symptoms: a review of randomized, controlled trials. *Ann. Intern. Med.*, **2002**, *137*(10), 805-813.

[38]	Cherdshewasart, W.; Subtang, S. Dahlan, W. Major isoflavonoid contents of the phytoestrogen rich-herb Pueraria mirifica in comparison with Pueraria lobata. *J. Pharm. Biomed. Anal.*, **2007**, *43*(2), 428-434.

[39]	Cherdshewasart, W.; Traisup, V. Picha, P. Determination of the estrogenic activity of wild phytoestrogen-rich Pueraria mirifica by MCF-7 proliferation assay. *J Reprod Dev*, **2008**, *54*(1), 63-67.

[40]	Matsumura, A.; Ghosh, A.; Pope, G.S. Darbre, P.D. Comparative study of oestrogenic properties of eight phytoestrogens in MCF7 human breast cancer cells. *J Steroid Biochem Mol Biol*, **2005**, *94*(5), 431-443.

[41]   Liang, R.; Chen, M.R. Xu, X. [Effect of dandi tablet on blood lipids and sex hormones in women of postmenopausal stage]. *Zhongguo Zhong Xi Yi Jie He Za Zhi*, **2003**, *23*(8), 593-595.

[42]   Deng, S.; Chen, S.N.; Yao, P.; Nikolic, D.; van Breemen, R.B.; Bolton, J.L.; Fong, H.H.; Farnsworth, N.R. Pauli, G.F. Serotonergic activity-guided phytochemical investigation of the roots of Angelica sinensis. *J. Nat. Prod.*, **2006**, *69*(4), 536-541.

[43]   Abe, F. Yamauchi, T. An androstane bioside and 3'-thiazolidinone derivatives of doubly-linked cardenolide glycosides from the roots of Asclepias tuberosa. *Chem. Pharm. Bull.*, **2000**, *48*(7), 991-993.

[44]   Rosso, R.; Brema, F.; Porcile, G.F. Santi, L. [Antitumoral activity of calusterone in advanced mammary carcinoma (author's transl)]. *Tumori*, **1976**, *62*(1), 79-84.

[45]   He, K.; Pauli, G.F.; Zheng, B.; Wang, H.; Bai, N.; Peng, T.; Roller, M. Zheng, Q. *Cimicifuga* species identification by high performance liquid chromatography-photodiode array/mass spectrometric/evaporative light scattering detection for quality control of black cohosh products. *J. Chromatogr.*, **2006**, *1112*(1-2), 241-254.

[46]   Garita-Hernandez, M.; Calzado, M.A.; Caballero, F.J.; Macho, A.; Muänoz, E.; Meier, B.; Brattstrèom, A.; Fiebich, B.L. Appel, K. The growth inhibitory activity of the *Cimicifuga racemosa* extract Ze 450 is mediated through estrogen and progesterone receptors-independent pathways. *Planta Med.*, **2006**, *72*(4), 317-323.

[47]   Nappi, R.E.; Malavasi, B.; Brundu, B. Facchinetti, F. Efficacy of *Cimicifuga racemosa* on climacteric complaints: a randomized study *versus* low-dose transdermal estradiol. *Gynecol. Endocrinol.*, **2005**, *20*(1), 30-35.

[48]   Mahady, G.B. Black cohosh (*Actaea/Cimicifuga racemosa*): review of the clinical data for safety and efficacy in menopausal symptoms. *Treat. Endocrinol.*, **2005**, *4*(3), 177-184.

[49]   Huntley, A. The safety of black cohosh (*Actaea racemosa, Cimicifuga racemosa*). *Expert Opin. Drug Saf.*, **2004**, *3*(6), 615-623.

[50]   Dog, T.L.; Powell, K.L. Weisman, S.M. Critical evaluation of the safety of Cimicifuga racemosa in menopause symptom relief. *Menopause*, **2003**, *10*(4), 299-313.

[51]   Borrelli, F.; Izzo, A.A. Ernst, E. Pharmacological effects of Cimicifuga racemosa. *Life Sci.*, **2003**, *73*(10), 1215-1229.

[52]   Anonymous Cimicifuga racemosa. Monograph. *Altern. Med. Rev.*, **2003**, *8*(2), 186-189.

[53]   Popp, M.; Schenk, R. Abel, G. Cultivation of *Cimicifuga racemosa* (L.) nuttal and quality of CR extract BNO 1055. *Maturitas*, **2003**, *44*(S1-7.

[54]   Borrelli, F. Ernst, E. *Cimicifuga racemosa*: a systematic review of its clinical efficacy. *Eur. J. Clin. Pharmacol.*, **2002**, *58*(4), 235-241.

[55]   Pepping, J. Black cohosh: Cimicifuga racemosa. *Am. J. Health. Syst. Pharm.*, **1999**, *56*(14), 1400-1402.

[56]   Lieberman, S. A review of the effectiveness of *Cimicifuga racemosa* (black cohosh) for the symptoms of menopause. *J. Womens Health*, **1998**, *7*(5), 525-529.

[57]   Dèuker, E.M.; Kopanski, L.; Jarry, H. Wuttke, W. Effects of extracts from *Cimicifuga racemosa* on gonadotropin release in menopausal women and ovariectomized rats. *Planta Med.*, **1991**, *57*(5), 420-424.

[58]   Fabricant, D.S.; Nikolic, D.; Lankin, D.C.; Chen, S.N.; Jaki, B.U.; Krunic, A.; van Breemen, R.B.; Fong, H.H.; Farnsworth, N.R. Pauli, G.F. Cimipronidine, a cyclic guanidine alkaloid from Cimicifuga racemosa. *J Nat Prod*, **2005**, *68*(8), 1266-1270.

[59]   Majumder, P.K.; Dasgupta, S.; Mukhopadhaya, R.K.; Mazumdar, U.K. Gupta, M. Anti-steroidogenic activity of the petroleum ether extract and fraction 5 (fatty acids) of carrot (*Daucus carota* L.) seeds in mouse ovary. *J. Ethnopharmacol.*, **1997**, *57*(3), 209-212.

[60]   Sharma, M.M.; Lal, G. Jacob, D. Estrogenic and pregnancy interceptory effects of carrot *daucus carota* seeds. *Indian J. Exp. Biol.*, **1976**, *14*(4), 506-508.

[61]   Kapoor, M.; Garg, S.K. Mathur, V.S. Anthiovulatory activity of five indigenous plants in rabbits. *Indian J. Med. Res.*, **1974**, *62*(8), 1225-1227.

[62]   Miller-Martini, D.M.; Chan, R.Y.; Ip, N.Y.; Sheu, S.J. Wong, Y.H. A reporter gene assay for the detection of phytoestrogens in traditional Chinese medicine. *Phytother. Res.*, **2001**, *15*(6), 487-492.

[63]   Woo, J.; Lau, E.; Ho, S.C.; Cheng, F.; Chan, C.; Chan, A.S.; Haines, C.J.; Chan, T.Y.; Li, M. Sham, A. Comparison of *Pueraria lobata* with hormone replacement therapy in treating the adverse health consequences of menopause. *Menopause*, **2003**, *10*(4), 352-361.

[64]   Chueh, F.S.; Chang, C.P.; Chio, C.C. Lin, M.T. Puerarin acts through brain serotonergic mechanisms to induce thermal effects. *J. Pharmacol. Sci.*, **2004**, *96*(4), 420-427.

[65]   Chansakaow, S.; Ishikawa, T.; Seki, H.; Sekine, K.; Okada, M. Chaichantipyuth, C. Identification of deoxymiroestrol as the actual rejuvenating principle of "Kwao Keur", *Pueraria mirifica*. The known miroestrol may be an artifact. *J. Nat. Prod.*, **2000**, *63*(2), 173-175.

[66]   Malaivijitnond, S.; Kiatthaipipat, P.; Cherdshewasart, W.; Watanabe, G. Taya, K. Different effects of *Pueraria mirifica,* a herb containing phytoestrogens, on LH and FSH secretion in gonadectomized female and male rats. *J. Pharmacol. Sci.*, **2004**, *96*(4), 428-435.

[67]   Trisomboon, H.; Malaivijitnond, S.; Watanabe, G. Taya, K. Estrogenic effects of Pueraria mirifica on the menstrual cycle and hormone-related ovarian functions in cyclic female cynomolgus monkeys. *J. Pharmacol. Sci.*, **2004**, *94*(1), 51-59.

[68]   Cherdshewasart, W.; Cheewasopit, W. Picha, P. The differential anti-proliferation effect of white (*Pueraria mirifica*), red (*Butea superba*), and black (*Mucuna collettii*) Kwao Krua plants on the growth of MCF-7 cells. *J. Ethnopharmacol.*, **2004**, *93*(2-3), 255-260.

[69]   Dietz, B.M.; Mahady, G.B.; Pauli, G.F. Farnsworth, N.R. Valerian extract and valerenic acid are partial agonists of the 5-HT5a receptor *in vitro*. *Brain Res. Mol. Brain Res.*, **2005**, *138*(2), 191-197.

[70]   Nicholson, J.A.; Darby, T.D. Jarboe, C.H. Viopudial, a hypotensive and smooth muscle antispasmodic from Viburnum opulus. *Proc. Soc. Exp. Biol. Med.*, **1972**, *140*(2), 457-461.

[71]   Xu, H.; Fabricant, D.S.; Piersen, C.E.; Bolton, J.L.; Pezzuto, J.M.; Fong, H.; Totura, S.; Farnsworth, N.R. Constantinou, A.I. A preliminary RAPD-PCR analysis of *Cimicifuga* species and other botanicals used for women's health. *Phytomedicine*, **2002**, *9*(8), 757-762.

[72]   Daniele, C.; Thompson Coon, J.; Pittler, M.H. Ernst, E. *Vitex agnus castus*: a systematic review of adverse events. *Drug Saf.*, **2005**, *28*(4), 319-332.

[73]   Wuttke, W.; Jarry, H.; Christoffel, V.; Spengler, B. Seidlovâa-Wuttke, D. Chaste tree (*Vitex agnus-castus*)--pharmacology and clinical indications. *Phytomedicine*, **2003**, *10*(4), 348-357.

[74]   Atmaca, M.; Kumru, S. Tezcan, E. Fluoxetine *versus Vitex agnus castus* extract in the treatment of premenstrual dysphoric disorder. *Hum. Psychopharmacol.*, **2003**, *18*(3), 191-195.

[75]   Schellenberg, R. Treatment for the premenstrual syndrome with agnus castus fruit extract: prospective, randomised, placebo controlled study. *Brit. Med. J.*, **2001**, *322*(7279), 134-137.

[76]   Berger, D.; Schaffner, W.; Schrader, E.; Meier, B. Brattstrèom, A. Efficacy of *Vitex agnus castus* L. extract Ze 440 in patients with pre-menstrual syndrome (PMS). *Arch. Gynecol. Obstet.*, **2000**, *264*(3), 150-153.

[77]   Loch, E.G.; Selle, H. Boblitz, N. Treatment of premenstrual syndrome with a phytopharmaceutical formulation containing Vitex agnus castus. *J. Womens Health Gend. Based Med.*, **2000**, *9*(3), 315-320.

[78]   Chopin Lucks, B. *Vitex agnus castus* essential oil and menopausal balance: a research update (Complementary Therapies in Nursing and Midwifery 8 (2003) 148-154). *Complement. Ther. Nurs. Midwifery*, **2003**, *9*(3), 157-160.

Send Orders for Reprints to reprints@benthamscience.net

CHAPTER 7

A Homogeneous Platform to Discover Inhibitors of the GoLoco Motif/G-alpha Interaction

Adam J. Kimple[1], Adam Yasgar[2], Mark Hughes[3], Ajit Jadhav[2], Francis S. Willard[1,§], Robin E. Muller[1], Christopher P. Austin[2], James Inglese[2], Gordon C. Ibeanu[3], David P. Siderovski[1] and Anton Simeonov[*,2]

[1]*Department of Pharmacology, Lineberger Comprehensive Cancer Center, and UNC Neuroscience Center, The University of North Carolina at Chapel Hill, Chapel Hill, NC 27599-7365, USA;* [2]*NIH Chemical Genomics Center, National Center for Advancing Translational Sciences, National Institutes of Health, Bethesda, MD 20892-3370, USA;* [3]*BRITE Institute, North Carolina Central University, Durham, NC 27707, USA*

Abstract: The GoLoco motif is a short $G\alpha$-binding polypeptide sequence. It is often found in proteins that regulate cell-surface receptor signaling, such as RGS12, as well as in proteins that regulate mitotic spindle orientation and force generation during cell division, such as GPSM2/LGN. Here, we describe a high-throughput fluorescence polarization (FP) assay using fluorophore-labeled GoLoco motif peptides for identifying inhibitors of the GoLoco motif interaction with the G-protein alpha subunit $G\alpha_{i1}$. The assay exhibits considerable stability over time and is tolerant to DMSO up to 5%. The Z´-factors for robustness of the GPSM2 and RGS12 GoLoco motif assays in a 96-well plate format were determined to be 0.81 and 0.84, respectively; the latter assay was run in a 384-well plate format and produced a Z´-factor of 0.80. To determine the screening factor window (Z-factor) of the RGS12 GoLoco motif screen using a small molecule library, the NCI Diversity Set was screened. The Z-factor was determined to be 0.66, suggesting that this FP assay would perform well when developed for 1,536-well format and scaled up to larger libraries. We then miniaturized to a 4 µL final volume a pair of FP assays utilizing fluorescein- (green) and rhodamine- (red) labeled RGS12 GoLoco motif peptides. In a fully-automated run, the Sigma-Aldrich LOPAC[1280] collection was screened three times with every library compound being tested over a range of concentrations following the quantitative high-throughput screening (qHTS) paradigm; excellent assay performance was noted with average Z-factors of 0.84 and 0.66 for the green- and red-label assays, respectively.

Keywords: Fluorescence anisotropy, fluorescence polarization, GoLoco motif, heterotrimeric G-proteins, high-throughput screening.

---

*Address correspondence to Anton Simeonov: NIH Chemical Genomics Center, National Center for Advancing Translational Sciences, 9800 Medical Center Drive, MSC 3370, Bethesda, MD 20892-3370 USA. Tel: 1-301-217-5721; Email: asimeono@mail.nih.gov

§Present address: Lilly Research Laboratories, Eli Lilly and Company, Indianapolis, IN, USA

INTRODUCTION

Many extracellular signals, including hormones, neurotransmitters, growth factors, and sensory stimuli relay information intracellularly by activation of plasma membrane-bound receptors. The largest class of such receptors is the superfamily of seven transmembrane-domain G protein-coupled receptors (GPCRs), so named because these cell-surface proteins were originally found to couple extracellular stimuli into intracellular changes *via* activation of G-protein heterotrimers (Gαβγ) [1]. GPCRs represent a major therapeutic target giving rise to the largest single fraction of the prescription drug market with annual sales of several billion dollars [2]; however, opportunities to develop therapeutics that target the intracellular regulatory machinery controlling the kinetics and duration of GPCR signal transduction have been relatively ignored by comparison.

A diverse family of Gα-interacting proteins has been shown to share a common GoLoco ("Gα$_{i/o}$-Loco" interaction) motif (Fig. **1**) (reviewed in [3, 4]). GoLoco motif-containing proteins generally bind to GDP-bound Gα subunits of the Gi (adenylyl-cyclase inhibitory) class and act as GDP dissociation inhibitors (GDIs), slowing the spontaneous exchange of GDP for GTP and preventing re-association with Gβγ subunits [5-13]. Determination of the crystallographic structure of Gα$_{i1}$·GDP in complex with the GoLoco motif of RGS14 [7] revealed critical determinants of Gα subunit specificity and GDI activity. The N-terminal alpha-helix of the GoLoco motif peptide binds between switch II and the α3 helix of the Gα$_{i1}$ Ras-like domain (Fig. **1B**), grossly deforming the normal site of Gβγ interaction [7]. The aspartate-glutamine-arginine triad, which defines the final residues of the highly-conserved 19 amino-acid GoLoco motif signature (Fig. **1A**), orients the arginine residue into the guanine nucleotide-binding pocket of Gα, allowing contacts to be made between its basic δ-guanido group and the α- and β-phosphates of GDP [7]. Mutation of this single arginine residue within the Asp-Gln-Arg triad causes a loss of GDI activity [7, 8, 11].

A well-characterized physiological function of GoLoco motif proteins is in the regulation of asymmetric cell division in worm, fruit fly, and mammalian development (reviewed in [4, 14]). For example, GPSM2 (a quadruple GoLoco motif-containing protein, previously known as LGN) binds to nuclear mitotic apparatus protein (NuMA) and regulates mitotic spindle assembly; altering endogenous cellular levels of GPSM2, either *via* overexpression or RNA interference-mediated knockdown, leads to aberrant chromosomal segregation during mitosis [15]. Similar functions have also been ascribed to Drosophila and *C. elegans* homologs of GPSM2 (Pins and GPR-1/-2, respectively; refs. [16-20]).

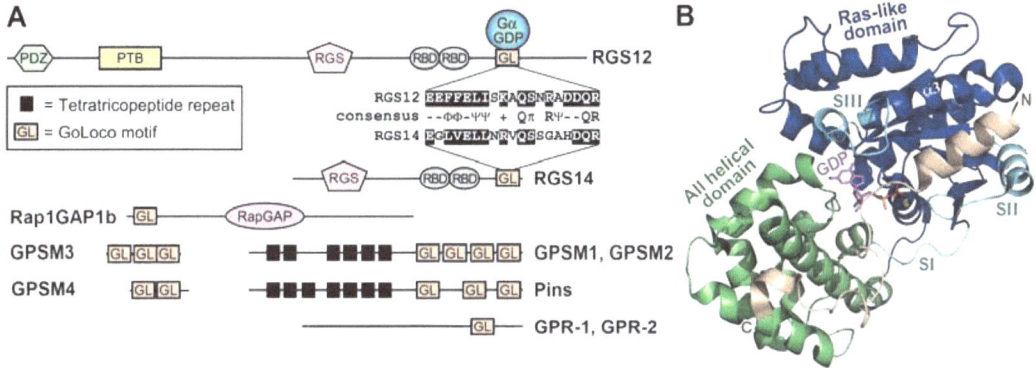

Fig. (1). The GoLoco motif is a Gα$_i$·GDP -interacting polypeptide found singly or in arrays in various proteins. (**A**) Domain architecture of representative GoLoco motif proteins and a sequence alignment of the conserved core of the RGS12 and RGS14 GoLoco motifs. Domain abbreviations: GPSM, G-protein signaling modulator; PDZ, PSD-95/Discs large/ZO-1 homology; PTB, phosphotyrosine-binding domain; RGS, regulator of G-protein signaling box; RBD, Ras-binding domain; RapGAP, Rap-specific GTPase-activating protein domain. (**B**) The crystal structure of Gα$_{i1}$ (Ras-like domain in *blue*, all α-helical domain in *green*, switch regions in *cyan*) bound to the GoLoco motif of RGS14 (PDB ID 2OM2). The GoLoco motif peptide (*tan*) binds across the Ras-like and all-helical domains of Gα$_{i1}$, trapping GDP (*magenta*, with α- and β-phosphates in *orange*) within its binding site. The bound magnesium ion is illustrated in *lime green*.

Evidence is also emerging that GoLoco motif-containing proteins act as critical components of cell-surface receptor-mediated signal transduction pathways. GPSM2 over-expression has been found to affect both basal and GPCR-activated potassium currents from GIRK channels [21], the latter effect similar to what we previously observed *via* cellular microinjection of GoLoco motif peptides [22]. We have recently shown RGS12 to be a receptor-selective scaffold for components of the mitogen-activated protein kinase (MAPK) cascade [23]. RNA interference-mediated knockdown of RGS12 protein levels in primary mouse dorsal root ganglion neurons blunts nerve growth factor-stimulated axonogenesis [23]. Mutating the arginine residue within the Asp-Gln-Arg triad of the RGS12

GoLoco motif leads to a mislocalization of RGS12 to the nucleus, away from its normally punctate endosomal pattern of expression [24]. This latter finding suggests that small molecule inhibition of the GoLoco motif/Gα_i interaction could serve to abrogate the normal signaling regulatory properties of GoLoco motif proteins, not only for RGS12 in the context of inhibiting sustained MAPK signal output, but also for GPSM2 and homologs in the context of dysregulating cell division processes in cancerous states of unchecked cellular proliferation [25, 26].

In this article, we describe the development of high-throughput screening (HTS) assays based on fluorescence polarization (FP) for the identification of small molecule inhibitors of the GoLoco motif/Gα protein interaction (Fig. **2**). FP is often used to detect the binding of fluorescently-labeled small ligands to larger binding partners (*e.g.*, refs. [27-32]). FP is based on the physical principle that fluorescein and other fluorophores are only excited by incident light that is polarized parallel to their axis. If this fluorophore is stationary or only slowly rotating, subsequent emission remains polarized along the same axis. Conversely, if polarized light excites a fluorophore rapidly tumbling in solution (Fig. **2A**), the resulting emission is depolarized by the rapid rotational diffusion that occurs during the lifetime of the excited state (~4 ns for fluorescein, ref. [33]). This depolarization is quantified as fluorescence anisotropy (FA) or fluorescence polarization (FP) by measuring the intensity of the emission perpendicular ($I_\perp$) and parallel ($I_\parallel$) to the plane of excitation (Equation (1); for a more comprehensive explanation of FA and FP, we refer the reader to ref. [33]). FA and FP are not equal but can be interconverted using equation (2). FP is a unitless ratio; however, it is often expressed as "milliP" (mP).

$$FA = \frac{I_\parallel - I_\perp}{I_\parallel + 2I_\perp} \qquad FP = \frac{I_\parallel - I_\perp}{I_\parallel + I_\perp} \qquad\qquad (1)$$

$$FA = \frac{2(FP)}{3 - FP} \qquad\qquad (2)$$

Because the depolarization of fluorophore-emitted light is directly related to the rotational motion of the fluorophore-labeled molecule, and thus inversely related

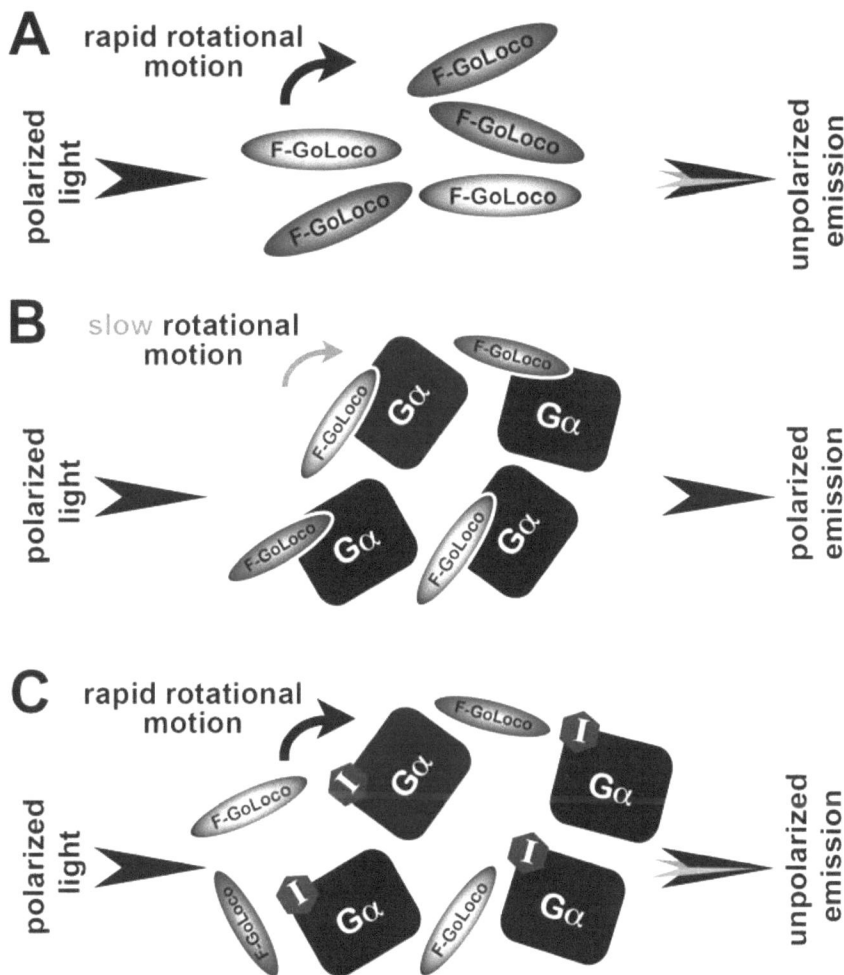

Fig. (2). Schematic of a fluorescence polarization assay for detection of FITC-GoLoco motif probe binding to its Gα$_{i1}$ subunit target. (**A**) When excited by plane-polarized light, the rapid rotational motion of the unbound FITC-GoLoco motif probe decorrelates the light. (**B**) The rotational diffusion of the FITC-GoLoco motif probe dramatically decreases as its effective molecular weight changes upon binding to Gα$_{i1}$. Consequently, polarized excitation results in polarized emission. (**C**) A small molecule inhibitor ("I") that binds to Gα$_{i1}$ in competition with the FITC-GoLoco motif probe increases the concentration of unbound (and rapidly rotating) probe, resulting in a decreased polarization signal.

to its total molecular weight (MW), FA and FP are theoretically limited to measuring the binding of a lower MW ligand to a higher MW substrate (*e.g.*, Fig. **2B**). While in theory this technique can be used to measure binding of any fluorescently-labeled ligand to a substrate as long as the MW$_{ligand}$ is much less than $MW_{substrate}$, in practice these assays are limited to ligands that have a MW of less than 5,000 Da. This limitation arises because of the short half-life (~4 ns) of the excited state of fluorescein isothiocyanate (FITC) [33], the most readily used dye in fluorescence polarization assays. (Rhodamine-based dyes have an even shorter half-life in the excited state; ref. [33]). While other dyes with longer lifetimes can be used to measure binding between larger molecules [34-36], their use has not been widespread.

Fluorescence polarization assays have been developed to detect various biological events such as phosphorylation, proteolytic cleavage, single nucleotide polymorphism detection, cAMP production, protein-protein interactions, and protein-DNA interactions [28, 29, 31, 32, 37-41]. This article focuses on our development and validation of a ligand displacement assay to screen for inhibitors of RGS12/$G\alpha_{i1}$ and GPSM2/$G\alpha_{i1}$ interactions (Fig. **2C**).

MATERIALS AND METHODS

Chemicals and Assay Material

Unless otherwise noted, all chemicals used were the highest grade available from Sigma Aldrich (St. Louis, MO) or Fisher Scientific (Pittsburgh, PA). Tris-HCl used in the 1,536-well plate format assay was procured from Invitrogen. 96-well black bottom plates were obtained from Costar (Corning, NY). The LOPAC[1280] library of known bioactives (1280 compounds from Sigma-Aldrich; arrayed for screening as 8 concentrations at 5 μL each in 1,536-well Greiner polypropylene compound plates) was received as DMSO solutions at initial concentration of 10 mM. Plate-to-plate (vertical) dilutions in 384-well format and 384-to-1,536 compressions were performed on an Evolution P[3] dispense system equipped with 384-tip pipetting head and two RapidStak units (Perkin-Elmer; Wellesley, MA). Additional details on the preparation of the compound library for quantitative high-throughput screening (qHTS) are provided elsewhere [42, 43].

Protein Expression and Purification

Expression and purification of human His_6-$G\alpha_{i1}$ from the expression plasmid pProEXHTb-h$G\alpha_{i1}$ was performed essentially as previously described [7].

Briefly, BL21 (DE3) *E. coli* (Novagen; San Diego, CA) were grown to an $OD_{600\,nm}$ of 0.6-0.8 at 37°C before induction with 0.5 mM isopropyl-β-D-thiogalactopyranoside. After culture for 14-16 hours at 20°C, cells were pelleted by centrifugation and frozen at -80°C. Prior to purification, bacterial cell pellets were resuspended in N1 buffer (50 mM Tris pH 8.0, 300 mM NaCl, 10 mM $MgCl_2$, 10 mM NaF, 30 μM $AlCl_3$, 50 μM GDP, 30 mM imidazole, 5% (w/v) glycerol). Bacteria were lysed at 10 MPa using an Emulsiflex pressure homogenizer (Avestin; Ottawa, Canada). Cellular lysates were centrifuged at 100,000 x g for 30 minutes at 4°C. The supernatant was applied to a nickel-nitrilotriacetic acid resin FPLC column (FF HisTrap; GE Healthcare), washed with 7 column volumes of N1 buffer then 3 column volumes of N1 buffer containing an additional 30 mM of imidazole before eluting with N1 buffer containing an additional 300 mM of imidazole. Eluted protein was incubated with tobacco etch virus (TEV) protease and dialyzed into low imidazole buffer (N1 buffer with 5 mM DTT) overnight at 4°C (to cleave the N-terminal hexahistidine tag) before being passed over a second HisTrap column to separate the untagged $G\alpha_{i1}$ from contaminants and cleavage products. The column flow-through was pooled and resolved using a calibrated 150 ml size exclusion column (Sephacryl S200, GE Healthcare) with S200 buffer (50 mM Tris pH 7.5, 150 mM NaCl, 10 μM GDP, 5% (w/v) glycerol). Protein was then concentrated to approximately 1 mM, as determined by $A_{280\,nm}$ measurements upon denaturation in guanidine hydrochloride. Concentration was calculated based on the predicted extinction coefficient obtained using the ProtParam webtool [44]. His_6-$G\alpha_{oA}$ was purified using similar chromatographic methods as previously described [45].

Peptide Synthesis

Unless otherwise denoted, peptides were synthesized by Fmoc-group protection, purified *via* HPLC, and confirmed using mass spectrometry by the Tufts University Core Facility (Medford, MA). Peptide sequences were as follows:

FITC-RGS12:

FITC-β-alanine-DEAEEFFELISKAQSNRADDQRGLLRKEDLVLPEFLR-amide;

FITC-GPSM2(GL2):

FITC-β-alanine-NTDEFLDLLASSQSRRLDDQRASFSNLPGLRLTQNSQS-amide;

GPSM1 GoLoco consensus:

TMGEEDFFDLLAKSQSKRMDDQRVDLAG-amide;

GPR-1(GoLoco wildtype)

EPVDMMDLIFSMSSRMDDQRTELPAARFIPPRPVSSASK-amide;

GPR-1(GoLoco R>F):

EPVDMMDLIFSMSSRMDDQFTELPAARFIPPRPVSSASK-amide.

The 5-carboxytetramethylrhodamine (TAMRA)-labeled peptide (TAMRA-DEAE EFFELISKAQSNRADDQRGLLRKEDLVLPEFLR-amide) was synthesized and HPLC-purified by Invitrogen (Carlsbad, CA).

Fluorescence Polarization Measurements in 96-Well and 384-Well Plate Formats

Polarization measurements during assay pilot trials were conducted using a PHERAstar microplate reader (BMG Labtech; Offenburg, Germany) with the fluorescence polarization module. Excitation wavelength was 485 ± 6 nm and emission was detected at 520 ± 15 nm. For each independent experiment, the gain of the parallel and perpendicular channel was calibrated so that 5 nM of FITC-RGS12 peptide had a polarization value of ~35 mP. The final volume of each 96-well plate well was brought to 180 µl with PheraBuffer (10 mM Tris pH 7.5, 150 mM NaCl, 10 mM $MgCl_2$, 100 µM GDP, and 0.05% (v/v) NP40); the final volume per well in the 384-well plate format was 50 µL. For nucleotide selectivity studies, PheraBuffer was alternatively supplemented with aluminum tetrafluoride (*i.e.*, addition of 10 mM NaF and 30 µM $AlCl_3$). Data analysis for these assay pilot trials was conducted using PHERAstar software V1.60 (BMG LABTECH, Germany), as well as Excel version X for Macintosh (Microsoft,

Seattle, Washington) and GraphPad Prism v4.0 (San Diego, CA). All dissociation constant (K_D) values were determined with non-linear regression and fitting to Equation 3, in which *FP* is the fluorescence polarization (measured in mP), [$G\alpha$] is the concentration of $G\alpha_{i1}$, B_{max} is the maximum polarization, and FP_{zero} is a correction factor to account for the polarization of unbound peptide (~35 mP).

$$FP = \frac{B_{\max}[G\alpha]}{K_D + [G\alpha]} + FP_{zero} \tag{3}$$

Surface Plasmon Resonance (SPR) Binding Assay

As a secondary screen for compounds that demonstrated at least partial concentration-dependent responses in the primary FP screen, optical detection of surface plasmon resonance (SPR) was performed using a Biacore 3000 (GE Healthcare; Piscataway, NJ). Surfaces of carboxymethylated dextran (CM5) biosensors (GE Healthcare) were covalently derivatized with anti-GST antibody as previously described [45, 46]. A GST fusion protein containing the minimal GoLoco motif of RGS12 [24] and GST protein alone (the latter as a negative control) were separately loaded onto anti-GST antibody surfaces to levels of ~900 resonance units (RUs) before 200 μL of 40 nM $G\alpha_{i1}$·GDP protein (preincubated in either test compound or DMSO vehicle only) was injected over all flow cells using the KINJECT command at a flow-rate of 40 μL/minute with a dissociation phase of 2000 seconds. The biosensor surface was then stripped with a 40 μL injection of 10 mM Glycine pH 2.2 before being reloaded with GST-RGS12 fusion protein or GST alone for subsequent $G\alpha_{i1}$·GDP injections. Non-specific binding to the GST alone surface was subtracted from each sensorgram curve using BIAevaluation software v.3.0 (Biacore). Percent inhibition of binding was calculated as the maximal RUs of specific binding observed (just before the dissociation phase) from a compound-treated $G\alpha_{i1}$·GDP injection compared to a paired DMSO control-treated $G\alpha_{i1}$·GDP injection.

qHTS Validation in 1,536-Well Plate Format

Control plate set-up. Titration of the unlabeled control peptide was delivered *via* pin transfer [47] of 23 nL of solution per well from a separate source plate into column 2 of each assay plate. The starting concentration of the control peptide was 10 mM and 20 mM for the FITC (green) and TAMRA (red) assay, respectively, followed by two-fold dilution points in duplicate, for a total of sixteen concentrations, resulting in final assay concentration range from 57.2 μM to 1.74 nM, and 114 μM to 3.49 nM, for the green and red assay, respectively.

Pre-screen assay miniaturization and optimization. Titration samples containing a constant amount of fluorophore-labeled peptide and variable concentrations of $G\alpha_{i1}$ protein were prepared in 384-well plates and transferred into 1,536-well black solid bottom plates by the use of CyBiWell 384-tip pipeting system (CyBio Boston, MA). For the subsequent 1,536-well-based experiments, a Flying Reagent Dispenser (FRD, Aurora Discovery, presently Beckman-Coulter) [48] was used to dispense reagents into the assay plates.

qHTS protocol. Four μL of reagents (10 nM FITC- or 15 nM TAMRA-labeled peptide in columns 3 and 4 as negative control; a mixture of 10 nM FITC- or 15 nM TAMRA-labeled peptide with $G\alpha_{i1}$ [50 nM in the green assay and 25 nM in the red assay, respectively] in columns 1, 2, 5 – 48) were dispensed into 1,536-well Greiner black assay plates. Compounds and control peptide (23 nL) were transferred *via* Kalypsys pintool equipped with a 1,536-pin array (10 nL slotted pins, V&P Scientific, San Diego, CA) [47]. The plate was incubated for 10 min at room temperature, and then read on a ViewLux high-throughput CCD imager (Perkin-Elmer, Wellesley, MA) using FITC polarization filter sets (excitation 480 nm, emission 540 nm) for the green assay and BODIPY sets (excitation 525 nm, emission 598 nm) for the red assay, respectively. During reagent dispensing, reagent bottles were kept submerged in a 4 °C recirculating chiller bath and all liquid lines were covered with aluminum foil to minimize probe and protein degradation. All screening operations were performed on a fully integrated robotic system (Kalypsys, San Diego, CA) containing one RX-130 and two RX-90 anthropomorphic robotic arms (Staubli, Duncan, SC). Library plates were screened starting from the lowest and proceeding to the highest concentration. Vehicle-only plates, with DMSO being pin-transferred to the entire column 5 – 48 compound area, were included at the beginning, middle, and the end of the validation run in order to record any systematic shifts in assay signal.

Analysis of qHTS data. Screening data were corrected and normalized, and concentration-effect relationships derived by using NCGC in-house developed algorithms. Percent activity was computed after normalization using the median values of the uninhibited, or neutral, control (32 wells located in column 1) and the free-probe, or 100% inhibited, control (64 wells, entire columns 3 and 4), respectively. An in-house database was used to track sample concentrations across plates, while ActivityBase (ID Business Solutions Ltd, Guildford, UK) was used for compound and plate registrations. A four-parameter Hill equation [49] was fitted to the concentration-response data by minimizing the residual error between the modeled and observed responses.

RESULTS

Detection of Gα/GoLoco Motif Interactions Using Fluorescence Polarization

We previously described the use of a fluorescein isothiocyanate-labeled RGS12 GoLoco motif peptide (FITC-RGS12) as a probe to measure Gα/GoLoco motif interactions using FP [13]. Our aim in this present study was to validate this FP assay, and develop a corresponding $G\alpha_{i1}$/GPSM2 interaction assay, as robust techniques for high-throughput screening for small molecule inhibitors of the Gα/GoLoco motif interaction. To establish an assay for $G\alpha_{i1}$ binding to a GoLoco motif from GPSM2 (Fig. **3**), we first incubated increasing concentrations of $G\alpha_{i1}$ protein (up to 10 μM) with constant amounts (either 0.1, 1.0, or 10 nM) of FITC-GPSM2(GL2) peptide encoding the second GoLoco motif of GPSM2. We observed robust interaction of $G\alpha_{i1}$ with FITC-GPSM2(GL2), whereby addition of saturating amounts of $G\alpha_{i1}$ caused an increase in FP from ~35 mP to ~160 mP (Fig. **3A**). Saturation binding isotherms illustrated that signal strength was optimal at FITC-GPSM2(GL2) probe concentrations of 1 nM and above (Fig. **3A**). Non-linear regression was used to fit the binding isotherms from experiments using two different probe concentrations to Equation 3, yielding dissociation constants (K_D) of 34 nM (using 1.0 nM probe) and 38 nM (using 10 nM FITC-GPSM2(GL2) probe). Saturation binding isotherms of FITC-RGS12 binding to $G\alpha_{i1}$ were also generated (Fig. **4**). FITC-RGS12 levels were held constant at 0.01, 0.1, 1, or 10 nM while the concentration of $G\alpha_{i1}$ was increased up to 3 μM. While binding was observable with sub-nanomolar concentrations of FITC-RGS12 probe, maximal polarization (~200 mP) was observed at FITC-RGS12 concentrations ≥ 1 nM (Fig. **4A**); however, at probe concentrations below 5 nM, increased noise was observed upon the addition of DMSO (data not shown). The binding affinity for the FITC-RGS12 to $G\alpha_{i1}$ was 3.8 nM (using 1 nM FITC-RGS12 probe).

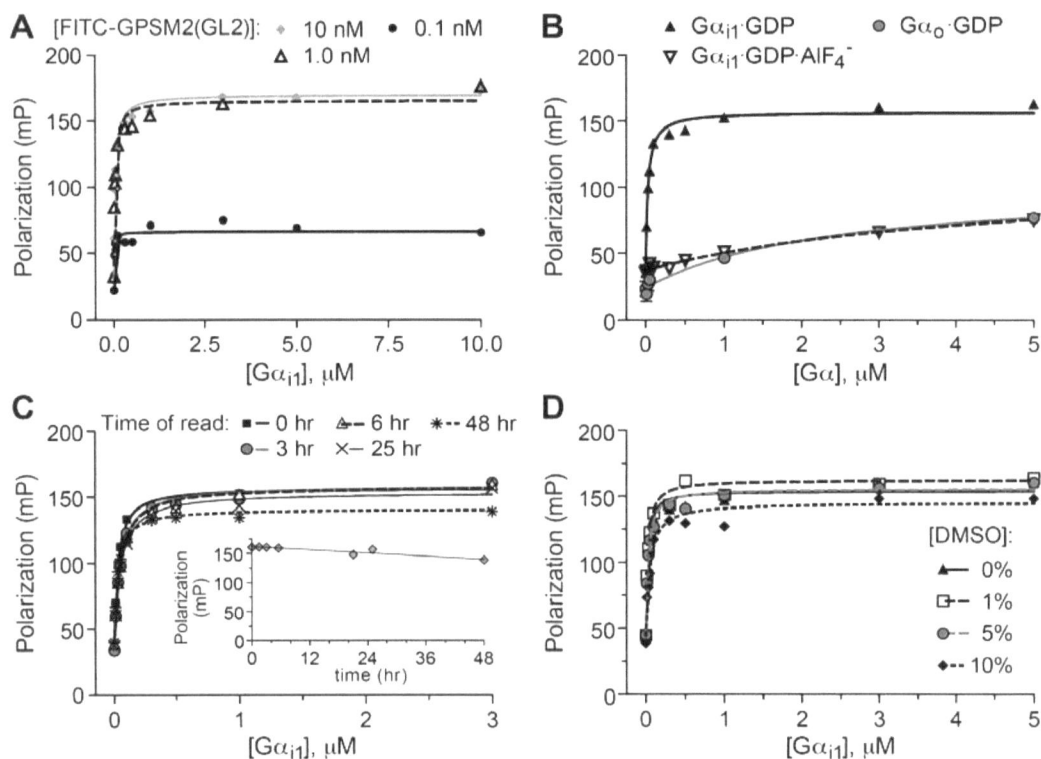

Fig. (3). **96-well microtiter plate-formatted fluorescence polarization assay for FITC-GPSM2(GL2) probe binding to $G\alpha_{i1}$.** (**A**) Concentration dependence and saturability of binding. Indicated concentrations of FITC-GPSM2(GL2) probe were incubated with indicated concentrations of $G\alpha_{i1} \cdot$GDP prior to measuring fluorescence polarization at equilibrium. (**B**) Nucleotide and $G\alpha$ subunit dependence of polarization signal. 1 nM of FITC-GPSM2(GL2) probe was incubated with indicated concentrations of $G\alpha_{i1} \cdot$GDP (ground-state), $G\alpha_{i1} \cdot$GDP$\cdot$AlF$_4^-$ (transition-state-mimetic form), or $G\alpha_o \cdot$GDP prior to measuring fluorescence polarization at equilibrium. (**C**) Time-stability studies. 1 nM of FITC-GPSM2(GL2) probe was incubated with indicated concentrations of $G\alpha_{i1} \cdot$GDP in 96-well microtiter plate wells for indicated times prior to measuring fluorescence polarization. *Inset*, Time-dependence of polarization signal from 1 nM of FITC-GPSM2(GL2) probe incubated with 3 μM $G\alpha_{i1} \cdot$GDP. (**D**) DMSO tolerance. 1 nM of FITC-GPSM2(GL2) probe was incubated with indicated concentrations of $G\alpha_{i1} \cdot$GDP and indicated final concentrations (v/v) of DMSO prior to measuring fluorescence polarization.

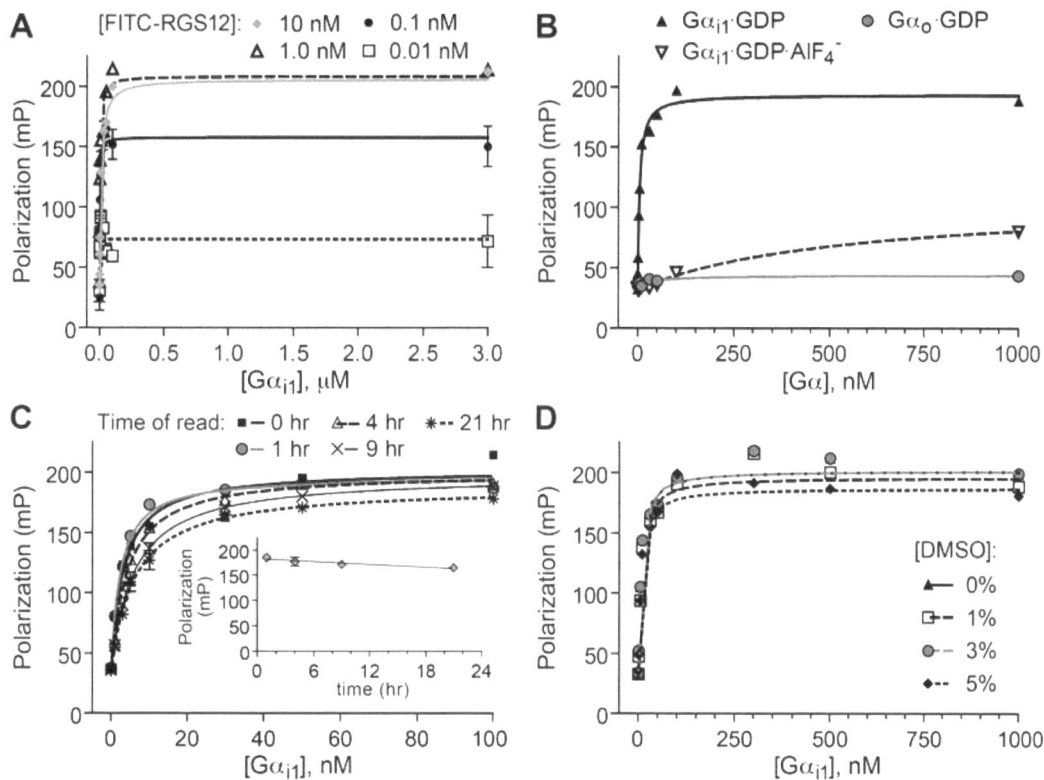

Fig. (4). 96-well microtiter plate-formatted fluorescence anisotropy assay for FITC-RGS12 GoLoco motif probe binding to $G\alpha_{i1}$. (**A**) Concentration dependence and saturability of binding. Indicated concentrations of FITC-RGS12 probe were incubated with indicated concentrations of $G\alpha_{i1}$·GDP prior to measuring fluorescence polarization at equilibrium. (**B**) Nucleotide and $G\alpha$ subunit dependence of polarization signal. 5 nM of FITC-RGS12 peptide was incubated with indicated concentrations of $G\alpha_{i1}$·GDP (ground-state), $G\alpha_{i1}$·GDP·AlF_4^- (transition-state-mimetic form), or $G\alpha_o$·GDP prior to measuring fluorescence polarization at equilibrium. (**C**) Time-stability studies. 5 nM of FITC-RGS12 probe was incubated with indicated concentrations of $G\alpha_{i1}$·GDP in 96-well microtiter plate wells for indicated times prior to measuring fluorescence polarization. *Inset*, Time-dependence of polarization signal from 5 nM of FITC-RGS12 probe incubated with 30 nM $G\alpha_{i1}$·GDP. (**D**) DMSO tolerance. 5 nM of FITC-RGS12 peptide was incubated with indicated concentrations of $G\alpha_{i1}$·GDP and indicated final concentrations (v/v) of DMSO prior to measuring fluorescence polarization. Error bars are mean ± SEM from triplicate samples.

To verify that these FP assays truly detect binding of the FITC-GPSM2(GL2) and FITC-RGS12 probes to their intended target of $G\alpha_{i1}$·GDP (consistent with the known biochemistry of GoLoco motif/$G\alpha$ interactions [4]), we tested the nucleotide dependence of the interaction. Saturation binding isotherms were generated at a constant concentration of 1 nM of FITC-GPSM2(GL2) probe with increasing concentrations of $G\alpha_{i1}$ in either PheraBuffer or PheraBuffer with aluminum tetrafluoride (AlF_4^-, which binds $G\alpha$ to create a transition state-mimetic form). As expected, upon the addition of AlF_4^-, there was a dramatic decrease in observed binding affinity. FITC-GPSM2(GL2) probe bound three orders of magnitude more avidly to $G\alpha_{i1} \cdot GDP$ than to $G\alpha_{i1} \cdot GDP \cdot AlF_4^-$ (K_D of 35 nM *versus* 15 μM, respectively; Fig. **3B**). Binding of FITC-RGS12 probe demonstrated a similar preference for $G\alpha_{i1} \cdot GDP$ (K_D of 4.3 nM *versus* 1.6 μM for $G\alpha_{i1} \cdot GDP \cdot AlF_4^-$; Fig. **4B**).

While GoLoco motifs were originally described as $G\alpha_{i/o}$-binding peptides [3], subsequent biochemical characterization has demonstrated preferential binding to the $G\alpha_i$ subfamily ($G\alpha_{i1}$, $G\alpha_{i2}$, $G\alpha_{i3}$) and not to $G\alpha_o$ (*e.g.*, refs. [6, 9]). To assess $G\alpha$ subunit specificity, the binding of FITC-GPSM2(GL2) and FITC-RGS12 probes to $G\alpha_{i1} \cdot GDP$ and $G\alpha_o \cdot GDP$ proteins was compared. Both GoLoco motif probes exhibited significantly higher binding affinities for $G\alpha_{i1} \cdot GDP$ than for $G\alpha_o \cdot GDP$. The K_D for the FITC-GPSM2(GL2)/$G\alpha_o \cdot GDP$ interaction was determined to be 3 μM, nearly two orders of magnitude higher than the affinity of the FITC-GPSM2(GL2)/$G\alpha_{i1} \cdot GDP$ interaction (Fig. **3B**). Similarly, the binding affinity of FITC-RGS12 for $G\alpha_o \cdot GDP$ was observed to be 70 μM *versus* 4.3 nM for $G\alpha_{i1} \cdot GDP$ (Fig. **4B**).

To further validate this GoLoco motif/$G\alpha_{i1}$ interaction assay for use in HTS, we characterized time dependence and dimethylsulfoxide (DMSO) tolerance of the assay. To assess the stability of the assay over extended periods of time, we measured saturation binding isotherms using 96-well plates containing FITC-GPSM2(GL2) or FITC-RGS12 probes (and increasing concentration of $G\alpha_{i1}$) that were rescanned at several hour time intervals. The K_D of the FITC-GPSM2(GL2)/$G\alpha_{i1}$ interaction was consistent over the first 25 hours of repeated measurements and increased marginally only at 48 hours (Fig. **3C**). Similar long-term stability was also observed for FITC-RGS12/$G\alpha_{i1}$ interaction (Fig. **4C**). Additional tests were made to establish the sensitivity of the assay to the standard HTS compound solvent DMSO; the FP assay using either FITC-GPSM2(GL2) or FITC-RGS12 probe demonstrated remarkable tolerance up to at least 5% (v/v) DMSO (Figs. **3D, 4D**).

Competitive Binding Studies

To confirm that the FITC-GPSM2(GL2) and FITC-RGS12 probes bound in a reversible manner, unlabeled GoLoco motif peptides were used as "cold competitors" (Fig. **5**). For these competition binding assays, the concentrations of the FITC-labeled probe and $G\alpha_{i1}$ were chosen so that the polarization signal was at ~80% of the maximal response [50]. Addition of unlabeled competitor peptide, derived from GPSM1 GoLoco motifs [5], to a mixture of 5 nM FITC-RGS12 probe and 30 nM $G\alpha_{i1}$ resulted in a dose-dependent decrease in polarization (Fig. **5A**) with an IC_{50} of 1 μM. In separate tests using 1 nM FITC-GPSM2(GL2) probe and 600 nM $G\alpha_{i1}$, the IC_{50} for the unlabeled GoLoco motif competitor (based on GPR-1; [20]) was determined to be 177 nM (Fig. **5B**). To rule out the possibility of the apparent competition being an artifact of high peptide concentrations, we also titrated in a GoLoco motif peptide with the critical arginine of the Asp-Gln-Arg triad mutated to phenylalanine ("GoLoco R>F"; Fig. **5B**). As expected, this mutant peptide had no inhibitory effect at the same or higher concentrations.

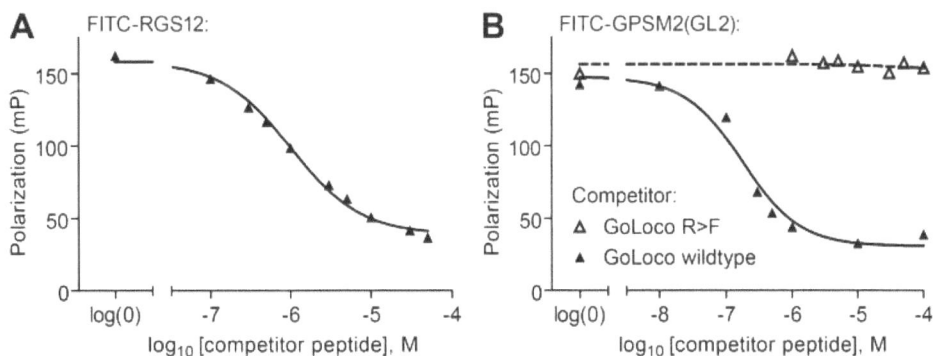

Fig. (5). Competitive inhibition of fluorescence polarization signal by unlabeled GoLoco motif peptides. (**A**) Indicated concentrations of the unlabeled GPSM1 GoLoco motif consensus peptide was added to 5 nM FITC-RGS12 probe and 30 nM $G\alpha_{i1}$. (**B**) Indicated concentrations of the unlabeled GPR-1 GoLoco motif peptide ("GoLoco wildtype) or the same peptide with the critical arginine mutated to phenylalanine ("GoLoco R>F"). Peptides were incubated with 1 nM of FITC-GPSM2(GL2) probe and 600 nM of $G\alpha_{i1}$·GDP protein prior to measuring fluorescence polarization at equilibrium.

Estimation of Screening Window

The next step in validation of this FP-based HTS assay was determining its screening window. An initial screening window can be crudely estimated by measuring many samples that only contain positive or negative controls for inhibition of the probe/Gα interaction [51, 52]. The mean and standard deviation of these controls were used to determine a Z′-factor for the assay using Equation (4), where σ is the standard deviation of the positive or negative control for inhibition and μ is the mean of the positive or negative control FP measurement [52]. Unlike other methods for quantifying the quality of an assay, the Z′-factor accounts for both the dynamic range (denominator) of the assay as well as the variation from well-to-well (numerator). The Z′-factor for the FP assay using 5 nM FITC-RGS12 probe, 30 nM Gα$_{i1}$, and 30 μM unlabeled GoLoco motif competitor peptide was calculated to be 0.84. This value was obtained by running one 96-well plate of positive controls and one 96-well plate of negative controls at 175 μl final volume and 1% (v/v) DMSO. Reduction in well volume below 175 μl was found to increase the standard deviation of both positive and negative controls (data not shown). Performing the same analysis with the FITC-GPSM2(GL2) probe resulted in a Z′-factor of 0.81. To assess the scalability of this FP assay to higher density plates, the same replicates of positive and negative controls for inhibition were also run with the FITC-RGS12 probe using 384 well plates. The Z′-factor was found to be 0.80 using a final volume of 50 μl.

$$Z' = 1 - \frac{3\sigma_+ + 3\sigma_-}{|\mu_- - \mu_+|} \tag{4}$$

Initial Small Molecule Screen in 96-Well Plate Format

While computing a Z′-factor is useful in assay development, screening window data from an actual compound library screen is more informative [51]. Towards this goal, we first obtained the National Cancer Institute (NCI) Developmental Therapeutics Program's Diversity Set (http://dtp.nci.nih.gov/branches/dscb/diversity_explanation.html) and screened 1976 compounds from this collection at 100 μM final concentration (Fig. **6**). Raw fluorescence polarization data was first normalized to the mean polarization signal from negative controls (1% (v/v) DMSO only; set to 100% binding signal) and from positive controls for inhibition (30 μM GPSM1 competitor peptide; set to 0% binding signal) (Fig. **6A**). A total of 286 compounds were excluded based on non-specific effects on the fluorescence polarization and total fluorescence intensity readouts. First, compounds were excluded if the obtained polarization value was 5 standard deviations higher than the negative control or 5 standard deviations lower than the positive control for inhibition ("Polarization filter", Fig. **6B**). Next, compounds were excluded if the raw fluorescence intensity value was 6 standard deviations higher than the negative control or 6 standard deviations lower than the positive control for inhibition ("Intensity filter", Fig. **6C**). The screening window Z-factor from normalized data for the remaining 1690 compounds was 0.66, with a hit-rate of 0.3% (6 out of 1976 compounds tested; 'hit' defined as >75% inhibition) (Fig. **6C**). A parallel screening of the Diversity Set at 50 µM final compound concentration gave a screening window Z-factor of 0.69 with a hit rate of 0.48 % (data not shown).

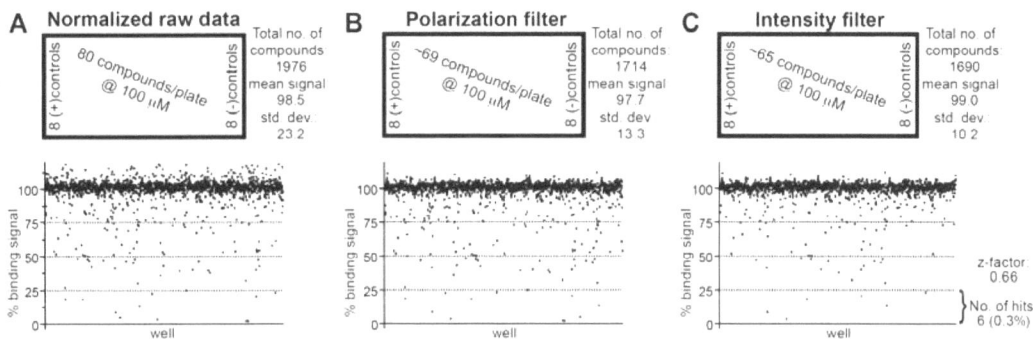

Fig. (6). Data from pilot screen of the NCI Diversity Set to establish a screening window Z-factor. (**A**) Plot of normalized fluorescence polarization data from entire 1976 compound set run in 96-well plate format with 5 nM of FITC-RGS12 probe and 30 nM of $G\alpha_{i1}$·GDP protein. Positive control wells contained 30 µM of competitor GPSM1 peptide; negative control wells contained vehicle only (1% (v/v) DMSO). (**B**) Data after exclusion of wells with polarization values 5 standard deviations outside control values (as described in text). (**C**) Data after additional exclusion of wells with raw fluorescence intensity values 6 standard deviations outside control values (as described in text).

Screening in the 384-Well Plate Format and Hit Validation by SPR

To examine the performance of the FP assay against a larger compound collection, we used the 384-well plate formatted assay in a screen of a 33,600-compound subset of the Biogen Idec 350,000 compound library. Thirty-two compounds were identified as inhibiting the assay by at least 30% (~1% hit rate); most of the hits were found in four clusters (Fig. **7A**), reflecting the grouping of compounds sharing similar chemistry on the same plates which is inherent to the design of the library subset derived from the original 350,000 compound library. Subsequent re-testing of each hit revealed 17 compounds exhibiting at least partial concentration-dependent inhibition of the primary FP assay. To validate these hits as inhibitors of the protein/peptide interaction, a secondary assay was performed based on optical detection of changes in surface plasmon resonance (*e.g.*, Fig. **7B**, **C**) upon binding $G\alpha_{i1}$ to immobilized GST-RGS12(GoLoco motif) fusion protein [24]. One of the hits from the primary FP assay was also found to inhibit the secondary SPR assay in a dose-dependent fashion (Fig. **7D**, **E**). This compound is now the subject of further analysis.

Assay Miniaturization to 1,536-Well Plates and Evaluation of Red-Shifted Peptide Probes

The FP assay was further miniaturized to a final volume of 4 µL in 1,536-well plate format by direct volume reduction. Retaining the inclusion of NP-40 in the assay buffer helped prevent peptide and protein absorption to the polystyrene wells due to the increased surface-to-volume ratio and also served to minimize the interfering effect of promiscuous inhibitors acting *via* colloidal aggregate formation [43, 47]. In a titration experiment using 10 nM FITC-RGS12 probe (hereinafter referred to as *green probe*), a robust FP signal change was observed (Fig. **8**) and a $G\alpha_{i1}$ protein concentration of 50 nM was selected for subsequent validation experiments. When the complex of 10 nM green probe and 50 nM $G\alpha_{i1}$ protein was incubated with varying concentrations of unlabeled peptide in the 1,536-well plate, a concentration-response curve was observed (Fig. **8B**) whose associated IC_{50} value matched closely that obtained from 96- and 384-well based experiments.

In parallel with the miniaturization of the original green assay, a red-shifted probe was explored. Prior experience and our recent profiling of the NIH Molecular Libraries Small Molecule Repository (MLSMR) compound library with respect to autofluorescence [53] prompted us to seek a red-shifted assay system in order to minimize the fraction of fluorescent compounds interfering with the fluorescent readout. Thus, a peptide of the same RGS12 GoLoco motif sequence was labeled

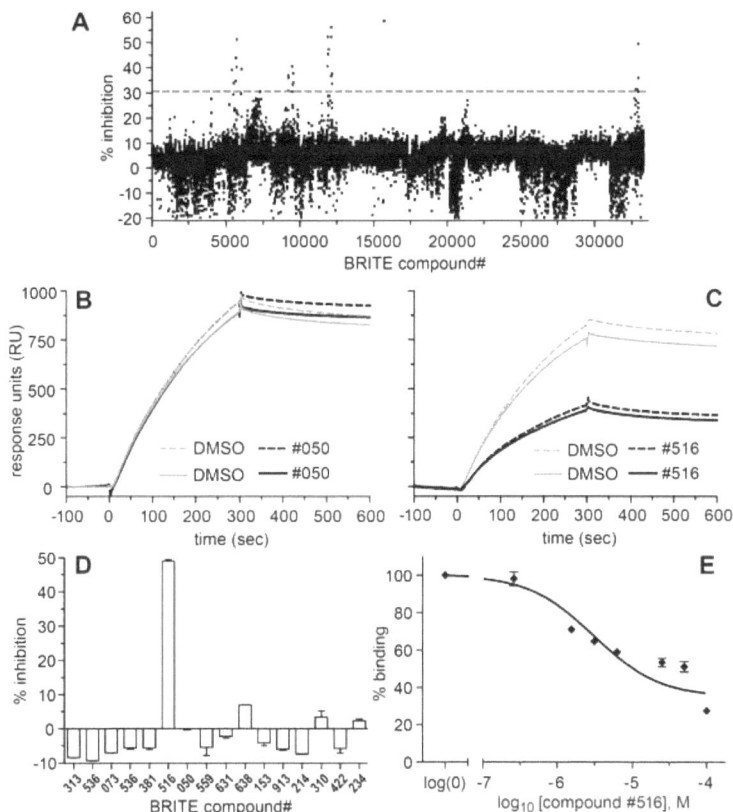

Fig. (7). Data from pilot primary screen of the BRITE Biogen Idec library subset and hit validation using an SPR-based secondary assay. (A) Plot of percent inhibition for each compound in the 33,600-element BRITE Biogen Idec library subset run at 10 μM final concentration in 384-well plate format with 5 nM of FITC-RGS12 probe and 46 nM of $G\alpha_{i1}$·GDP protein. Note the clustering of inhibitory activity reflecting plate-wise grouping of similar compound chemistry within the 33,600 compound subset of the larger 350,000 Biogen Idec library. Gray dashed line represents cut-off of greater than 30% inhibition used to select compounds for subsequent dose-response testing in the same primary FP assay. (B, C) Representative SPR data from two compounds exhibiting at least partial concentration-dependent responses in the primary FP assay. Panel **B** represents negative data from compounds (such as #050) that, after preincubation with $G\alpha_{i1}$·GDP, did not inhibit the latter binding to a GST-RGS12(GoLoco motif) biosensor surface during a 5 minute association phase (0 – 300 seconds). Panel **C** represents positive data from a confirmed inhibitor of the $G\alpha_{i1}$·GDP / GST-RGS12(GoLoco motif) interaction (compound #516). (**D**) Results of single-dose testing (13.3 μM final concentration) of 16 hits from the primary FP assay in the SPR-based secondary assay, performed as described in Materials and Methods. (**E**) Dose-response curve of the sole confirmed hit (compound #516) from the SPR-based secondary assay, performed as described in Materials and Methods, except with 50 μL injections of $G\alpha_{i1}$·GDP at a flow-rate of 20 μL/min and a subsequent 200 second dissociation time.

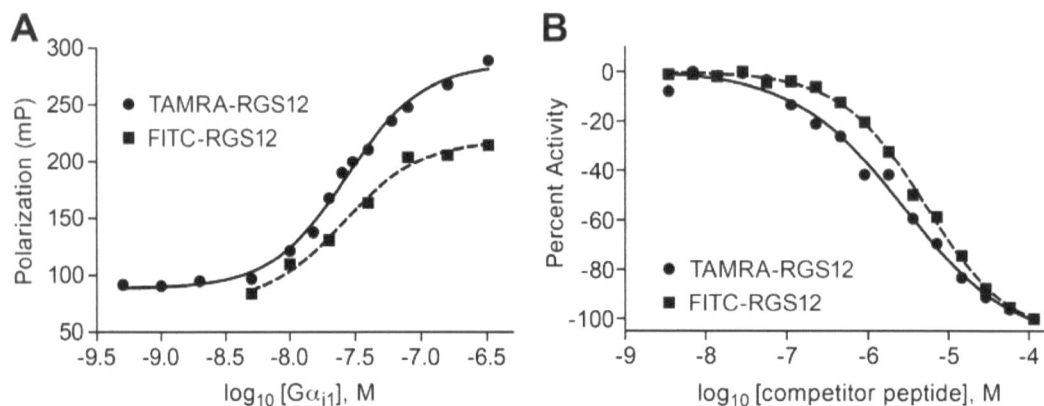

Fig. (8). Miniaturization of FP assay to 1,536-well plate format and evaluation of FITC-
versus **TAMRA-labeled RGS12 GoLoco motif peptide probe.** (A) Fluorescence polarization signal of FITC- (green) and TAMRA- (red) probes in the 1,536-well plate format. Protein-concentration dependence of the FP signal of 10 nM green probe (solid squares) and 15 nM red probe (solid circles) was measured in titrations with $G\alpha_{i1}$. Evident from the plots is the greater FP signal obtained from the red probe. (B) Probe displacement by unlabeled peptide control in the 1,536-well plate format. Green (solid squares, 10 nM FITC-RGS12 probe plus 50 nM $G\alpha_{i1}$) and red (solid circles, 15 nM TAMRA-RGS12 probe plus 25 nM $G\alpha_{i1}$) protein complexes were allowed to interact with series of concentrations of unlabeled peptide (pin-transferred from DMSO stock solutions) for 15 min at room temperature. The leftward-shift in dose-response of the red probe curve is a reflection of the slight increase in assay sensitivity afforded by the decreased protein concentration.

with 5-carboxytetramethyl rhodamine (TAMRA, hereinafter referred to as *red probe*) and subjected to the same assay optimization experiments. In order to maintain robust fluorescence intensity signal with this fluorophore, the red probe concentration was increased slightly to 15 nM. In protein titration experiments, the FP signal change observed with the red-labeled peptide was higher, in the range of 180-190 mP, as previously experienced with this fluorophore [53] (Fig. **8A**). The increased FP window per same protein concentration allowed us to decrease the $G\alpha_{i1}$ protein concentration in the red assay to half that of the green assay (25 nM *versus* 50 nM) while maintaining a sufficient signal window. Consistent with lowered probe and protein concentrations for the red system (15 nM red probe with 25 nM $G\alpha_{i1}$ protein *versus* 10 nM green probe and 50 nM $G\alpha_{i1}$ protein), the displacement of the red probe from its complex by the unlabeled competitor peptide resulted in left-shifted concentration-response curve (Fig. **8B**), thus confirming that the lowered protein load resulted in slightly improved assay sensitivity.

During the course of our red probe exploration, we evaluated two red-shifted fluorophores. An RGS12 GoLoco motif peptide of the same sequence labeled with BODIPY Texas Red failed to yield a change in fluorescence polarization when titrated with $G\alpha_{i1}$ protein (data not shown), presumably due to an adverse effect of the fluorophore on the RGS12 peptide binding affinity and/or increased self-aggregation of the probe due to the hydrophobic nature of the BODIPY moiety. Thus, not every combination of peptide probe and fluorophore should be expected to yield a readily-optimizable binding assay and, as the present limited example suggests, fluorophores of an overly-hydrophobic nature might be problematic when used with peptides (as opposed to oligonucleotide or DNA probes, for example) while those containing a number of ionizable groups such as TAMRA might offer a better chance for developing a good peptide-based FP assay [53].

qHTS Robotic Validations Using the LOPAC[1280] Library

Once the peptide probe and $G\alpha_{i1}$ protein concentrations were optimized for the green and red assays, we proceeded to run fully-automated, 1,536-well based robotic validations. For each fluorophore system, the LOPAC[1280] collection was screened three consecutive times in concentration-response mode [42]. A total of 30 plates were run per fluorophore assay: 24 compound plates (*i.e.*, three iterations of the LOPAC[1280] eight-concentrations set) and 6 control DMSO plates,. The assay signal windows, as expressed by the difference between mean FP values for the bound and unbound labeled peptide controls, were stable throughout the robotic validation (Fig. **9A**). Both assays performed robustly, yielding an average Z' factor of 0.84 for the green assay and 0.66 for the red assay, respectively (Fig. **9B**). The intra-plate peptide control titration curves remained nearly overlapping throughout the screen progression (Fig. **9C**), yielding average IC_{50} values of 7.8 µM and 0.6 µM for the green and red assays, respectively. During these qHTS experiments, each library compound was tested as an eight-point titration, with concentrations ranging from 2 nM to 57 µM, and for each well and each assay system, fluorescence polarization values, as well as parallel- and perpendicular-plane fluorescence intensity values, were collected and stored in the database.

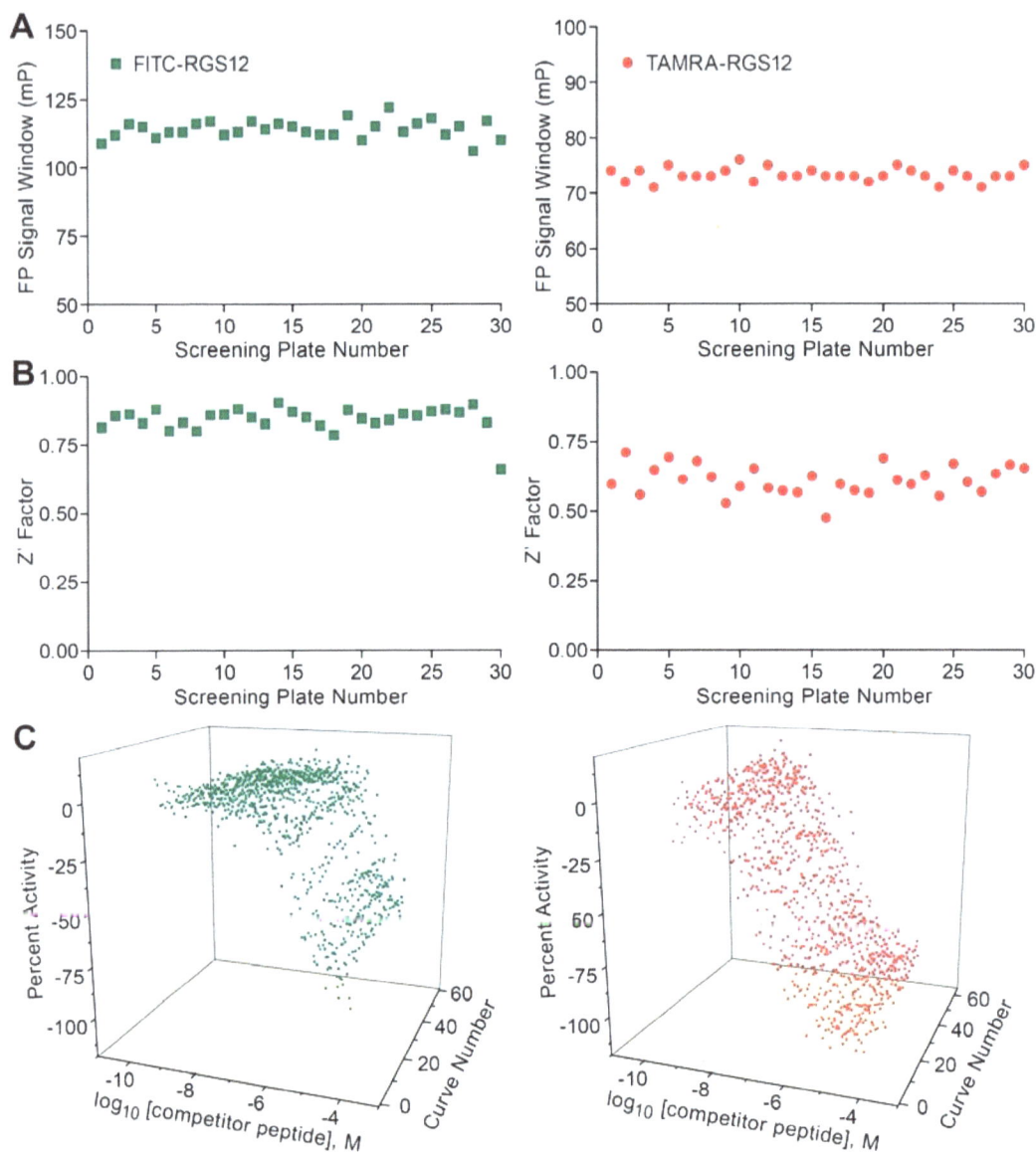

Fig. (9). qHTS Performance. Shown for both the green and red probe FP assays are the (**A**) FP signal window, (**B**) Z' factor trend, and (**C**) intra-plate control titrations (duplicate curves per plate) as a function of screening plate number.

Unlike traditional HTS, qHTS provides concentration responses for all the compounds screened and allows determination of the half-maximal activity concentrations associated with each active compound. Additionally, compound effect can be described with respect to the shape, efficacy, and goodness-of-fit of its concentration-response curve [42]. Our LOPAC[1280] library validation runs revealed 8 active compounds shared by the green and red screens, some of which were associated with complete concentration-response curves while others showed single-point inhibition at the highest concentration and, as such, the sigmoidal dose-response curves fitted through their data were of the lowest quality and reproducibility. However, for most of the active compounds identified in the LOPAC[1280] library, there was excellent reproducibility within the triplicate runs, as well as good agreement between the outcomes from the green and red assays. Four examples of triplicate green and red concentration-response curves derived from the validations are shown in Fig. (**10**).

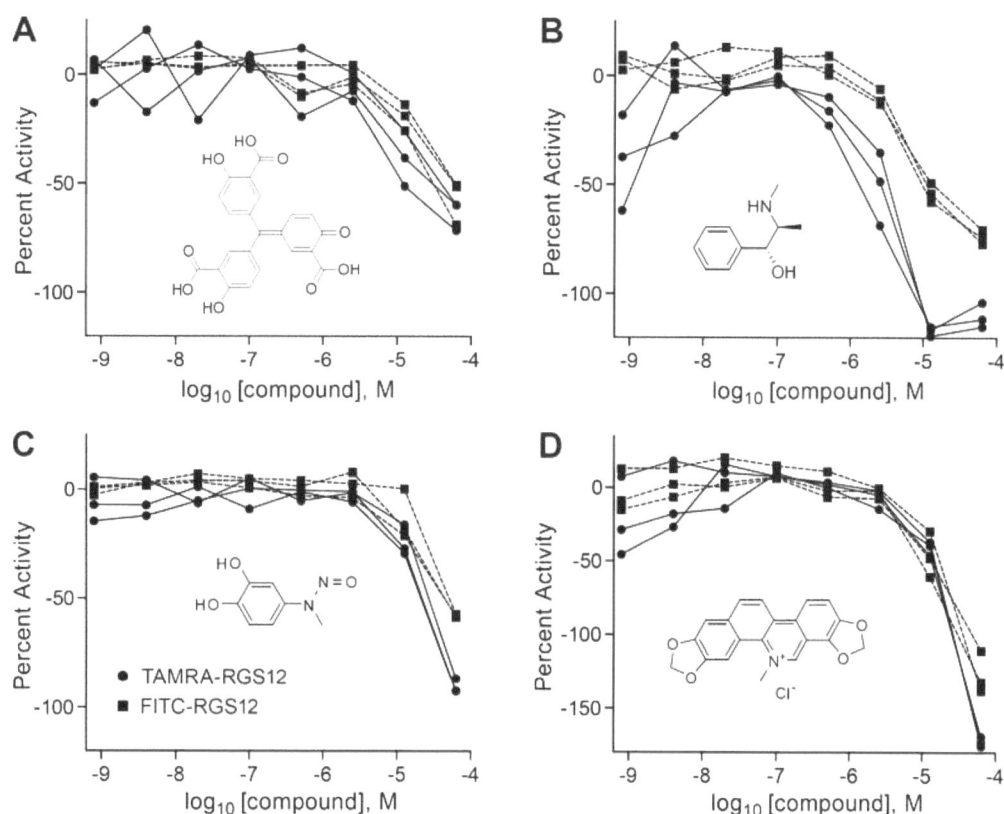

Fig. (10). Examples of validation-derived active compounds. The concentration-response curves (triplicate runs in both colors, with green probe data in solid squares and red probe data in solid circles) are shown for (**A**) NCGC00093568 (PubChem SID 11110719), (**B**) NCGC00093901 (PubChem SID 11111142), (**C**) NCGC00094195 (PubChem SID 11111500), and (**D**) NCGC00094379 (PubChem SID 11111810).

DISCUSSION

Sensitivity of Binding Detection and Screening Window Optimization

In agreement with several previous studies of $G\alpha_i$/GoLoco interactions (reviewed in [4]), the FP assay we have developed clearly demonstrates preferential binding of the GoLoco motif to the inactive, ground state of $G\alpha_i$ (*i.e.*, $G\alpha_{i1} \cdot$GDP) and selective binding of $G\alpha_{i1}$ *versus* $G\alpha_o$ (Figs. **3B**, **4B**). The equilibrium binding-based FP assay was also found to detect binding with affinities that are consistent, but higher, than previously published and unpublished results using kinetic measurements (k_{on}, k_{off}) obtained by surface plasmon resonance [7, 9]. These higher observed affinities for the RGS12 and GPSM2 interactions with $G\alpha_{i1} \cdot$GDP are most likely the result of the highly-sensitive probe detection technique being used in the FP assay, allowing use of probe concentrations that are less than the observed K_D values. Additionally, the relatively hydrophobic FITC moiety added to these GoLoco motif peptides is likely to bind to $G\alpha_{i1}$ and thereby increase the overall affinity of the labeled GoLoco motif peptide for its $G\alpha_i$ substrate.

From Equation (2), one can see that the Z′-factor is dependent on the standard deviation of the positive and negative controls. We found that the standard deviation could be decreased by increasing the amount of FITC-GoLoco motif probe in the assay as well as increasing the number of excitation flashes per well during fluorescence polarization measurements. However, these two factors must be balanced with competing considerations of increasing reagent consumption and the time to scan plates. An alternative way to increase the screening window would be to increase the $G\alpha_{i1}$ concentration to increase the difference between the minimum and maximum FP signal; however, this change would concomitantly increase the amount of unbound $G\alpha_{i1}$ and thus require more cold competitor peptide or compound to cause inhibition in the signal, resulting in a less sensitive assay.

Small-Scale Library Screens and Strategies for Minimizing Compound Interference

The fundamentally ratiometric nature of the fluorescence polarization measurement theoretically reduces the effects of interference from compounds that have overlapping spectra with the FITC-labeled probe [30]. Interference from compounds with overlapping absorbance spectra should not change an FP reading so long as the absorbance is <u>proportional</u> (P) along both axes (Eq. 5); however, the robustness of this ratiometric measurement cannot compensate for compounds that interfere by increasing or decreasing the signal in an <u>additive</u> (A) manner (Eq. 5).

$$FP_{prop} = \frac{(P)(I_{\parallel}) - (P)(I_{\perp})}{(P)(I_{\parallel}) + (P)(I_{\perp})} = \frac{I_{\parallel} - I_{\perp}}{I_{\parallel} + I_{\perp}} \text{ but } FP_{add} = \frac{(I_{\parallel} + A) - (I_{\perp} + A)}{(I_{\parallel} + A) + (I_{\perp} + A)} \neq \frac{I_{\parallel} - I_{\perp}}{I_{\parallel} + I_{\perp}} \quad (5)$$

While a high tolerance to interference from compound absorbance is an advantage of FP assays, 226 compounds (11.4% of the set) were excluded from our pilot screening data of the NCI Diversity Set, based on FP measurement interference. With the FITC-labeled (green) assay, we developed a systematic way to exclude interfering compounds as shown in Fig. (**6**). From the initial raw data, each plate was normalized so that the average of eight positive control wells for inhibition were set to 0% binding and the average of eight negative control wells were set to 100% binding. After this normalization, compounds that resulted in polarization values 5 standard deviations above 100% binding or 5 standard deviations below 0% binding were excluded. Compounds giving readings above the threshold likely interfered by causing aggregation of either probe or substrate. Compounds giving readings significantly below 0% binding were excluded because these compounds clearly interfered with probe fluorescence. Following this "polarization filter", additional compounds were excluded based on intensity values [30]. While we observed that the fluorescence intensity of the FITC-GoLoco motif probes increased upon binding to Gα, this change in intensity was consistent between wells and across plates. Compounds that resulted in a total intensity value ($2I_{\perp}+I_{\parallel}$) falling 6 standard deviations outside of the intensity window established from the controls were also excluded. As the result of these exclusions, 1690 of 1976 compounds remained within the NCI Diversity Set for consideration as Gα$_i$/GoLoco motif binding inhibitors, with 6 of these compounds demonstrating inhibition of greater than 75 percent. From this single-concentration screen of nearly two thousand compounds at 100 μM final concentration and 1% (v/v) DMSO, the Z-factor was 0.66. This FP screen was also conducted at 50 μM final compound concentration and very little improvement in Z-factor was noted (data not shown). While the Z-factor was significantly lower than the Z′-factor calculated from controls, the Z-factor derived from this pilot library screen represents the actual screening window and, at a value of 0.66, still reflects an excellent assay robustness amenable to HTS of larger compound collections.

Another aspect of our optimization of this FP screening strategy was the development and implementation of a red-shifted fluorophore assay employing a TAMRA-labeled version of the RGS12 GoLoco motif peptide. The rationale for this change was to move farther away from the autofluorescence-sensitive regions of the light spectrum. In fact, our recently-completed fluorescent spectroscopic profiling of the NIH Molecular Libraries Small Molecule Repository (MLSMR) and other compound libraries demonstrated that, in blue-shifted fluorophore regions such as the frequently-utilized UV/vis spectrum (excitations near 360 nm and emissions near 450 nm) and fluorescein spectrum (excitations around 480 nm and emissions near 520 nm), a significant proportion of library compounds are expected to interfere with the fluorescent assay readout (as high as 3% in the UV/vis region and 0.1% in the fluorescein region, respectively) [53]. In the present work, the transition to a red-shifted fluorophore resulted in an additional two-fold benefit of lowering the protein requirement for the screen and improving the sensitivity of the binding assay, both due to the fact that the rhodamine-based probe afforded greater FP signal change for the same protein concentration.

Benefits of the qHTS Approach

The robotic validation screen for inhibitors of the RGS12 GoLoco motif/$G\alpha_{i1}$ complex was performed in qHTS format, with every compound tested over a range of concentrations, spanning from tens of micromolar to low nanomolar, to generate a broad concentration-response profile. Thus, in addition to potencies and efficacies being assigned to each active compound immediately out of the primary screen, false positives and negatives due to single-point outliers are easily identified in the context of compound titration. Stated differently, after performing qHTS, the selection of active compounds is based on the premise that the biological effect of an active compound is a function of its concentration, rather than on pure statistical arguments and application of cutoffs. While the library preparation and the primary screen are "front-loaded" with an increased number of plates, the savings associated with reduced cherry-picking, re-arraying, and retesting steps tend to make up for those elevated initial costs, due to the increased robustness and higher information content of the screening data. Additionally, the higher quality of such screening datasets is expected to make them more valuable for data-mining in recently-established public databases such as PubChem.

In both robotic validations, the green and red assays performed robustly in the 1,536-well plate format, with Z′-factors remaining flat with the screen progression. The intra-plate unlabeled peptide control titration, which can be viewed as a combined internal standard for both the underlying assay biology and the reproducibility of compound transfer, yielded concentration-response curves that remained stable and reproducible throughout the screens (Fig. **9C**). Of note, miniaturization of this and other assays all the way to the 1,536-well plate format not only leads to reagent savings but also allows one to employ *additional* controls such as the intra-plate titration described here. The application of such controls that measure "the pulse" of the assay, while not necessarily required for signal normalization purposes, is made possible by the availability of so many additional wells in the 1,536-well plate. In lower plate densities, such as the 96-well plate, allocating 8 or 16 wells to low and high normalization controls is frequently barely enough to provide good statistics during large-scale screening. In contrast, in 1,536-well plates, the simple propagation of one empty 96-well plate column (equivalent to 8 wells) to the higher density plate leads to the natural creation of 128 wells (sixteen 96-well source plates feeding into one 1,536-well final plate) [43]. This 16-fold increase in the potentially-available wells makes it possible to add information content to each assay plate (by further partitioning the controls area) during large-collection miniaturized screens without placing undue burden on library preparation or otherwise compromising the outcome of the screens.

During our qHTS validations, each library compound was tested at eight concentrations and, for each well and fluorophore-type assay, three measurements were collected for a combined total of ~200,000 data points. The observed top active compounds in the green and red screens reproduced well upon repeated primary screening and across fluorophores (Fig. **10**). Our successful robotic validation screen suggests that this FP assay is robust and sensitive enough to be utilized in a large-scale, 1,536-well based screen. Most recently, the red assay was scaled up to a fully-automated robotic screen of 241,637 compounds arrayed as seven-point dilution series: the average Z' factors for the 998-plate screen was 0.61, leading to the generation of 1,465,344 data points (PubChem Assay Identifier 880). The complete analysis of this screen and the follow-up on top hits is ongoing and the results will be presented in due course.

ACKNOWLEDGEMENTS

Work performed at the NCATS was supported by the Molecular Libraries Initiative of the National Institutes of Health Roadmap for Medical Research. Work performed in the Siderovski lab was funded by NIH grants F30 MH074266 (to A.J.K.) and R03 NS053754 (to D.P.S.).

CONFLICT OF INTEREST

The authors state that there is no conflict of interest.

DATA DEPOSITION

The two 1,536-well plate formatted bioassays reported in this paper, as well as active compounds identified in the LOPAC[1280] library validation screens, have been deposited in the PubChem database, http://pubchem.ncbi.nlm.nih.gov (AID codes 879 for the green assay and 880 for the red assay).

DISCLOSURE

This above chapter was updated from the original article Kimple AJ *et al.*, "A High-Throughput Fluorescence Polarization Assay for Inhibitors of the GoLoco Motif/G-alpha Interaction" *Comb Chem High Throughput Screen*. 2008 June; 11(5): 396-409.

REFERENCES

[1]   Gilman, A.G., G proteins: transducers of receptor-generated signals. *Annu Rev Biochem,* **1987**, *56*, 615-649.

[2]   Overington, J.P.; Al-Lazikani, B.; Hopkins, A.L., How many drug targets are there? *Nat Rev Drug Discov,* **2006**, *5*, (12), 993-996.

[3]   Siderovski, D.P.; Diverse-Pierluissi, M.; De Vries, L., The GoLoco motif: a Galphai/o binding motif and potential guanine-nucleotide exchange factor. *Trends Biochem Sci,* **1999**, *24*, (9), 340-341.

[4]   Willard, F.S.; Kimple, R.J.; Siderovski, D.P., Return of the GDI: the GoLoco motif in cell division. *Annu Rev Biochem,* **2004**, *73*, 925-951.

[5]   De Vries, L.; Fischer, T.; Tronchere, H.; Brothers, G.M.; Strockbine, B.; Siderovski, D.P.; Farquhar, M.G., Activator of G protein signaling 3 is a guanine dissociation inhibitor for Galpha i subunits. *Proc Natl Acad Sci U S A,* **2000**, *97*, (26), 14364-14369.

[6]   Kimple, R.J.; De Vries, L.; Tronchere, H.; Behe, C.I.; Morris, R.A.; Gist Farquhar, M.; Siderovski, D.P., RGS12 and RGS14 GoLoco motifs are G alpha(i) interaction sites with guanine nucleotide dissociation inhibitor Activity. *J Biol Chem,* **2001**, *276*, (31), 29275-29281.

[7]   Kimple, R.J.; Kimple, M.E.; Betts, L.; Sondek, J.; Siderovski, D.P., Structural determinants for GoLoco-induced inhibition of nucleotide release by Galpha subunits. *Nature,* **2002**, *416*, (6883), 878-881.

[8]   Kimple, R.J.; Willard, F.S.; Hains, M.D.; Jones, M.B.; Nweke, G.K.; Siderovski, D.P., Guanine nucleotide dissociation inhibitor activity of the triple GoLoco motif protein G18: alanine-to-aspartate mutation restores function to an inactive second GoLoco motif. *Biochem J,* **2004**, *378*, (Pt 3), 801-808.

[9]   McCudden, C.R.; Willard, F.S.; Kimple, R.J.; Johnston, C.A.; Hains, M.D.; Jones, M.B.; Siderovski, D.P., G alpha selectivity and inhibitor function of the multiple GoLoco motif protein GPSM2/LGN. *Biochim Biophys Acta,* **2005**, *1745*, (2), 254-264.

[10]   Natochin, M.; Lester, B.; Peterson, Y.K.; Bernard, M.L.; Lanier, S.M.; Artemyev, N.O., AGS3 inhibits GDP dissociation from galpha subunits of the Gi family and rhodopsin-dependent activation of transducin. *J Biol Chem,* **2000**, *275*, (52), 40981-40985.

[11]   Peterson, Y.K.; Bernard, M.L.; Ma, H.; Hazard, S., 3rd; Graber, S.G.; Lanier, S.M., Stabilization of the GDP-bound conformation of Gialpha by a peptide derived from the G-protein regulatory motif of AGS3. *J Biol Chem,* **2000**, *275*, (43), 33193-33196.

[12]   Takesono, A.; Cismowski, M.J.; Ribas, C.; Bernard, M.; Chung, P.; Hazard, S., 3rd; Duzic, E.; Lanier, S.M., Receptor-independent activators of heterotrimeric G-protein signaling pathways. *J Biol Chem,* **1999**, *274*, (47), 33202-33205.

[13]   Willard, F.S.; Low, A.B.; McCudden, C.R.; Siderovski, D.P., Differential G-alpha interaction capacities of the GoLoco motifs in Rap GTPase activating proteins. *Cell Signal,* **2007**, *19*, (2), 428-438.

[14]   Siderovski, D.P.; Willard, F.S., The GAPs, GEFs, and GDIs of heterotrimeric G-protein alpha subunits. *Int J Biol Sci,* **2005**, *1*, (2), 51-66.

[15]   Du, Q.; Stukenberg, P.T.; Macara, I.G., A mammalian Partner of inscuteable binds NuMA and regulates mitotic spindle organization. *Nat Cell Biol,* **2001**, *3*, (12), 1069-1075.

[16]   Izumi, Y.; Ohta, N.; Hisata, K.; Raabe, T.; Matsuzaki, F., Drosophila Pins-binding protein Mud regulates spindle-polarity coupling and centrosome organization. *Nat Cell Biol,* **2006**, *8*, (6), 586-593.

[17]   Nipper, R.W.; Siller, K.H.; Smith, N.R.; Doe, C.Q.; Prehoda, K.E., Galphai generates multiple Pins activation states to link cortical polarity and spindle orientation in Drosophila neuroblasts. *Proc Natl Acad Sci U S A,* **2007**, *104*, (36), 14306-14311.

[18]   Schaefer, M.; Shevchenko, A.; Shevchenko, A.; Knoblich, J.A., A protein complex containing Inscuteable and the Galpha-binding protein Pins orients asymmetric cell divisions in Drosophila. *Curr Biol,* **2000**, *10*, (7), 353-362.

[19]   Afshar, K.; Willard, F.S.; Colombo, K.; Johnston, C.A.; McCudden, C.R.; Siderovski, D.P.; Gonczy, P., RIC-8 is required for GPR-1/2-dependent Galpha function during asymmetric division of C. elegans embryos. *Cell,* **2004**, *119*, (2), 219-230.

[20]   Colombo, K.; Grill, S.W.; Kimple, R.J.; Willard, F.S.; Siderovski, D.P.; Gonczy, P., Translation of polarity cues into asymmetric spindle positioning in Caenorhabditis elegans embryos. *Science,* **2003**, *300*, (5627), 1957-1961.

[21]   Wiser, O.; Qian, X.; Ehlers, M.; Ja, W.W.; Roberts, R.W.; Reuveny, E.; Jan, Y.N.; Jan, L.Y., Modulation of basal and receptor-induced GIRK potassium channel activity and neuronal excitability by the mammalian PINS homolog LGN. *Neuron,* **2006**, *50*, (4), 561-573.

[22]   Webb, C.K.; McCudden, C.R.; Willard, F.S.; Kimple, R.J.; Siderovski, D.P.; Oxford, G.S., D2 dopamine receptor activation of potassium channels is selectively decoupled by Galpha-specific GoLoco motif peptides. *J Neurochem,* **2005**, *92*, (6), 1408-1418.

[23]   Willard, M.D.; Willard, F.S.; Li, X.; Cappell, S.D.; Snider, W.D.; Siderovski, D.P., Selective role for RGS12 as a Ras/Raf/MEK scaffold in nerve growth factor-mediated differentiation. *Embo J,* **2007**, *26*, (8), 2029-2040.

[24]   Sambi, B.S.; Hains, M.D.; Waters, C.M.; Connell, M.C.; Willard, F.S.; Kimple, A.J.; Pyne, S.; Siderovski, D.P.; Pyne, N.J., The effect of RGS12 on PDGFbeta receptor signaling to p42/p44 mitogen activated protein kinase in mammalian cells. *Cell Signal,* **2006**, *18*, (7), 971-981.

[25]   Cho, H.; Kehrl, J.H., Localization of Gi alpha proteins in the centrosomes and at the midbody: implication for their role in cell division. *J Cell Biol,* **2007**, *178*, (2), 245-255.

[26]   Kimple, R.J.; Willard, F.S.; Siderovski, D.P., The GoLoco motif: heralding a new tango between G protein signaling and cell division. *Mol Interv,* **2002**, *2*, (2), 88-100.

[27]   Burke, T.J.; Loniello, K.R.; Beebe, J.A.; Ervin, K.M., Development and application of fluorescence polarization assays in drug discovery. *Comb Chem High Throughput Screen,* **2003**, *6*, (3), 183-194.

[28]   Drees, B.E.; Weipert, A.; Hudson, H.; Ferguson, C.G.; Chakravarty, L.; Prestwich, G.D., Competitive fluorescence polarization assays for the detection of phosphoinositide kinase and phosphatase activity. *Comb Chem High Throughput Screen,* **2003**, *6*, (4), 321-330.

[29]   Nikiforov, T.T.; Simeonov, A.M., Application of fluorescence polarization to enzyme assays and single nucleotide polymorphism genotyping: some recent developments. *Comb Chem High Throughput Screen,* **2003**, *6*, (3), 201-212.

[30]   Owicki, J.C., Fluorescence polarization and anisotropy in high throughput screening: perspectives and primer. *J Biomol Screen,* **2000**, *5*, (5), 297-306.

[31]   Parker, G.J.; Law, T.L.; Lenoch, F.J.; Bolger, R.E., Development of high throughput screening assays using fluorescence polarization: nuclear receptor-ligand-binding and kinase/phosphatase assays. *J Biomol Screen,* **2000**, *5*, (2), 77-88.

[32]   Prystay, L.; Gagne, A.; Kasila, P.; Yeh, L.A.; Banks, P., Homogeneous cell-based fluorescence polarization assay for the direct detection of cAMP. *J Biomol Screen,* **2001**, *6*, (2), 75-82.

[33]   Lakowicz, J.R. *Principles of fluorescence spectroscopy.* 2nd ed. Kluwer Academic/Plenum: New York, **1999**.

[34]   Szmacinski, H.; Terpetschnig, E.; Lakowicz, J.R., Synthesis and evaluation of Ru-complexes as anisotropy probes for protein hydrodynamics and immunoassays of high-molecular-weight antigens. *Biophys Chem,* **1996**, *62*, (1-3), 109-120.

[35]   Terpetschnig, E.; Szmacinski, H.; Malak, H.; Lakowicz, J.R., Metal-ligand complexes as a new class of long-lived fluorophores for protein hydrodynamics. *Biophys J,* **1995**, *68*, (1), 342-350.

[36]   Youn, H.J.; Terpetschnig, E.; Szmacinski, H.; Lakowicz, J.R., Fluorescence energy transfer immunoassay based on a long-lifetime luminescent metal-ligand complex. *Anal Biochem,* **1995**, *232*, (1), 24-30.

[37]   Akula, N.; Chen, Y.S.; Hennessy, K.; Schulze, T.G.; Singh, G.; McMahon, F.J., Utility and accuracy of template-directed dye-terminator incorporation with fluorescence-polarization detection for genotyping single nucleotide polymorphisms. *Biotechniques,* **2002**, *32*, (5), 1072-1076, 1078.

[38]   Bonin, P.D.; Erickson, L.A., Development of a fluorescence polarization assay for peptidyl-tRNA hydrolase. *Anal Biochem,* **2002**, *306*, (1), 8-16.

[39]   Duan, W.; Sun, L.; Liu, J.; Wu, X.; Zhang, L.; Yan, M., Establishment and application of a high throughput model for rho kinase inhibitors screening based on fluorescence polarization. *Biol Pharm Bull,* **2006**, *29*, (6), 1138-1142.

[40]   Hsu, T.M.; Chen, X.; Duan, S.; Miller, R.D.; Kwok, P.Y., Universal SNP genotyping assay with fluorescence polarization detection. *Biotechniques,* **2001**, *31*, (3), 560, 562, 564-568, passim.

[41]   Zhang, T.T.; Huang, Z.T.; Dai, Y.; Chen, X.P.; Zhu, P.; Du, G.H., High-throughput fluorescence polarization method for identifying ligands of LOX-1. *Acta Pharmacol Sin,* **2006**, *27*, (4), 447-452.

[42]   Inglese, J.; Auld, D.S.; Jadhav, A.; Johnson, R.L.; Simeonov, A.; Yasgar, A.; Zheng, W.; Austin, C.P., Quantitative high-throughput screening: a titration-based approach that efficiently identifies biological activities in large chemical libraries. *Proc. Nat. Acad. Sci.USA,* **2006**, *103*, (31), 11473-11478.

[43]   Yasgar, A.; Shinn, P.; Michael, S.; Zheng, W.; Jadhav, A.; Auld, D.; Austin, C.; Inglese, J.; Simeonov, A., Compound Management for Quantitative High-Throughput Screening. *JALA,* **2008**, *13*, 79-89.

[44]   Gasteiger, E.; Hoogland, C.; Gattiker, A.; Duvaud, S.; Wilkins, M.R.; Appel, R.D.; Bairoch, A. In *The Proteomics Protocols Handbook.* Walker, J.M., Ed.; Humana Press, **2005**, pp 571-607.

[45]   Willard, F.S.; Kimple, A.J.; Johnston, C.A.; Siderovski, D.P., A direct fluorescence-based assay for RGS domain GTPase accelerating activity. *Anal Biochem,* **2005**, *340*, (2), 341-351.

[46]   Willard, F.S.; Siderovski, D.P., Purification and *in vitro* functional analysis of the Arabidopsis thaliana regulator of G-protein signaling-1. *Methods Enzymol,* **2004**, *389*, 320-338.

[47]   Cleveland, P.H.; Koutz, P.J., Nanoliter dispensing for uHTS using pin tools. *Assay Drug Dev Technol,* **2005**, *3*, (2), 213-225.

[48]   Niles, W.D.; Coassin, P.J., Piezo- and Solenoid Valve-Based Liquid Dispensing for Miniaturized Assays. . *Assay Drug. Devel. Technol. ,* **2005**, *3*, 189-202.

[49]   Hill, A.V., The Possible Effects of the Aggregation of the Molecule of Haemoglobin on its Dissociation Curves. *J. Physiol. (London),* **1910**, *40*, 4-7.

[50]   Huang, X., Fluorescence polarization competition assay: the range of resolvable inhibitor potency is limited by the affinity of the fluorescent ligand. *J Biomol Screen,* **2003**, *8*, (1), 34-38.

[51]   Seethala, R.; Fernandes, P.B. *Handbook of drug screening*. Marcel Dekker: New York, **2001**.

[52]   Zhang, J.H.; Chung, T.D.; Oldenburg, K.R., A Simple Statistical Parameter for Use in Evaluation and Validation of High Throughput Screening Assays. *J Biomol Screen,* **1999**, *4*, (2), 67-73.

[53]   Simeonov, A.; Jadhav, A.; Thomas, C.; Wang, Y.; Huang, R.; Southall, N.; Shinn, P.; Smith, J.; Austin, C.; Auld, D.; Inglese, J., Fluorescent Spectroscopic Profiling of Compound Libraries. *J. Med. Chem.,* *51*, 2363-2371.

CHAPTER 8

G-Protein Activation State-Selective Binding Peptides as New Tools for Probing Heterotrimeric G-protein Subunit Signaling Dynamics

Christopher A. Johnston[1], Francis S. Willard[1,¶], Kevin Ramer[2,§], Rainer Blaesius[2,†], Natalia Roques[2,‡] and David P. Siderovski[*1]

[1]*Department of Pharmacology, Lineberger Comprehensive Cancer Center, and UNC Neuroscience Center, School of Medicine, The University of North Carolina at Chapel Hill, Chapel Hill, NC, 27599-7365 USA;* [2]*Karo Bio USA, Durham, NC, 27703, USA*

Abstract: Heterotrimeric G-proteins, comprising Gα, Gβ, and Gγ subunits, are molecular switches that regulate numerous signaling pathways involved in cellular physiology. This characteristic is achieved by the adoption of two principal states: an inactive state in which GDP-bound Gα is complexed with the Gβγ dimer, and an active state in which GTP-bound Gα is freed of its Gβγ binding partner. Structural studies have illustrated the basis for the distinct conformations of these states which are regulated by alterations in three precise 'switch regions' of the Gα subunit. Discrete differences in conformation between GDP- and GTP-bound Gα underlie its nucleotide-dependent protein-protein interactions (*e.g.*, with Gβγ/receptor and effectors, respectively) that are critical for maintaining their proper nucleotide cycling and signaling properties. Recently, several screening approaches have been used to identify peptide sequences capable of interacting with Gα (and free Gβγ) in nucleotide-dependent fashions. These peptides have demonstrated applications in direct modulation of the nucleotide cycle, assessing the structural basis for aspects of Gα and Gβγ signaling, and serving as biosensor tools in assays for Gα activation including high-throughput drug screening. In this review, we highlight some of the methods used for such discoveries and discuss the insights that can be gleaned from application of these identified peptides.

Keywords: Biosensors, G protein-coupled receptors, heterotrimeric G proteins, nucleotide binding, peptides.

---

*Address correspondence to David P. Siderovski: WVU Department of Physiology and Pharmacology, Robert C. Byrd Health Sciences Center, West Virginia University, One Medical Center Drive, Morgantown, WV 26506-9229 USA; Tel: 1-304-293-4991; Fax: 1-304-293-2850; E-mail: dpsiderovski@hsc.wvu.edu

Current addresses: ¶Lilly Research Laboratories, Eli Lilly and Company, Indianapolis, IN, USA; §Biogen Idec Inc., 5000 Davis Drive, Research Triangle Park, NC 27709-4627 USA; †Becton Dickenson, 21 Davis Drive, Research Triangle Park, NC 27709 USA; ‡Kenan-Flagler Business School, UNC-Chapel Hill, Chapel Hill, NC 27599-2400 USA

INTRODUCTION

Diverse extracellular signals, including hormones, neurotransmitters, growth factors, and sensory stimuli, transmit information intracellularly by activation of plasma membrane-bound receptors. The largest class of such receptors is the superfamily of heptahelical G protein-coupled receptors (GPCRs). GPCRs transmit signals by activating heterotrimeric G-proteins that normally exist in an inactive state of Gα·GDP bound to Gβγ subunits. In the traditional model (Fig. **1**), agonist activation of GPCRs induces incompletely defined conformational changes within the receptor, which subsequently catalyze the exchange of GDP for GTP on the Gα subunit by inducing conformational changes within Gαβγ that lower the affinity for GDP allowing for nucleotide release and subsequent GTP binding [1-3]. By this means, GPCRs serve as guanine nucleotide exchange factors (GEFs) for Gα·GDP/Gβγ complexes. Although the exact mechanism by which GPCRs exert their GEF activity remains to be fully elucidated [3], this action is critical to the commencement of G protein signaling, as GDP release is the rate-limiting step of the Gα guanine nucleotide cycle [4]. Subsequent to GDP release, GTP, a nucleotide present in a relative excess, binds Gα and induces a conformational change in three flexible 'switch regions' of the Gα subunit, which deforms the Gβγ binding interface leading to both the dissociation of the Gβγ dimer as well as the adoption of the conformation capable of interacting with effectors [1,5]. Activated Gα·GTP and liberated Gβγ both signal to a diverse family of downstream effectors including ion channels, adenylyl cyclases, phosphodiesterases, and phospholipases, producing second messenger molecules that regulate cellular responses underlying physiological processes [2]. Based on their sequence homology and differential regulation of effectors, G-proteins are grouped in four classes: $G\alpha_s$, $G\alpha_{i/o}$, $G\alpha_q$, and $G\alpha_{12/13}$ [6]. GPCRs have the ability to couple selectively to members of one or more of these G-protein subfamilies, thus allowing selective modulation of signaling cascades by particular GPCR ligands. G-protein signaling is terminated by the intrinsic GTPase activity of the Gα subunit, which occurs at a rate that varies among the G-protein subfamilies. GTP hydrolysis rates can be dramatically enhanced by members of a superfamily of "regulators of G-protein signaling" (RGS) proteins [7-9] that serve as GTPase-accelerating proteins (or "GAPs"). This deactivation reaction results in conversion back to the inactivated, GDP-bound Gα that subsequently reassociates with Gβγ to complete the cycle. Because this represents a true *cycle* of activation (by nucleotide exchange and subunit dissocation) and deactivation (by GTP hydrolysis and subunit reassociation), heterotrimeric G-proteins serve as molecular switches and are critical to defining the spatial and temporal aspects of cellular responses to external stimuli.

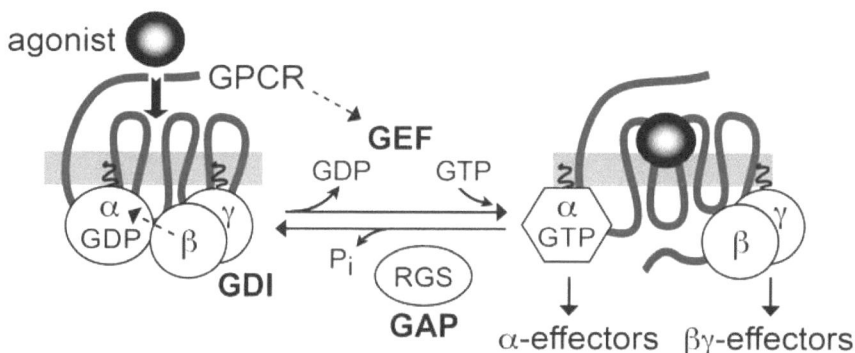

Fig. (1). The traditional model of the guanine nucleotide exchange and hydrolysis cycle governing the receptor-mediated activation of heterotrimeric G protein-coupled signal transduction. GPCRs bind, *via* their intracellular loops, to the heterotrimeric G-protein consisting of Gα (with bound GDP) associated with the Gβγ dimer. The isoprenylated Gβγ dimer aids in association of the heterotrimer with the plasma membrane, participates in receptor coupling, and serves as a guanine nucleotide dissociation inhibitor (GDI) preventing spontaneous activation of the Gα subunit. Agonist-bound receptors act as guanine nucleotide exchange factors (GEFs) by provoking conformational changes in Gαβγ resulting in the release of GDP and binding of GTP by the Gα subunit. Binding of GTP induces changes in three flexible switch regions within the Gα subunit, leading to Gβγ dimer dissociation. Both Gα·GTP and freed Gβγ dimer subsequently regulate downstream effectors, either alone or in a coordinated fashion. The GPCR/heterotrimer complex returns to the inactive state by the intrinsic GTP hydrolysis activity of Gα, cleaving the terminal γ-phosphate from GTP and rendering Gα again bound to GDP and reassociated with the Gβγ dimer, thus mutually terminating the signaling capacity of both subunits of the heterotrimer. GTP hydrolysis is greatly enhanced by the "regulator of G-protein signaling" (RGS) family of proteins, which serve as GTPase-accelerating proteins (GAPs) for the Gα subunit.

Biochemical and structural analyses over the past two decades have advanced our understanding of the mechanics underlying G-protein regulation and the guanine nucleotide cycle [3,10]. As previously alluded to, the structural basis for this activation cycle is governed principally by three Gα switch regions -- flexible segments that change conformation upon GTP binding and hydrolysis. Indeed, the switch regions contain high sequence conservation among G-proteins, especially in those residue positions that are directly involved in nucleotide binding, GTP hydrolysis, and protein-protein interactions (*e.g.* Gβγ, RGS proteins, effectors) [3,10]. Given that conformational changes to these switch regions are crucial to the G-protein signaling cycle, further defining the structural and biochemical nature of these distinct signaling states of Gα should provide further insight into G-protein signaling dynamics and greatly facilitate assay development and target validation for drug discovery.

Screening methods have been developed to identify randomized peptide sequences capable of interacting with target proteins as a means of creating novel binding partners for signaling proteins of interest [11-13]. Recently, such screening methods have been utilized to develop peptides capable of interacting with Gα and Gβγ subunits. In the case of Gα, nucleotide-state specific peptides have been identified by selectively screening against both inactive and active conformations. These peptides have been used to gain additional biochemical and structural insights into the signaling dynamics of heterotrimeric G-proteins. Additionally, several diverse applications have been made of these peptides towards the identification of small molecule drugs that modulate G-protein activity both directly and indirectly, with the ultimate goal of facilitating novel GPCR screening methods. This review will highlight the current successes in the screening for, and early uses of, G-protein binding peptides.

SCREENING METHODS

Several *in vitro* methods have been developed to identify and evolve peptide sequences that interact specifically with target proteins of interest (*e.g.*, refs. [11-13]). Although technically different in several design aspects, these various methods all entail certain fundamental criteria. Each process begins with the assembly of a diverse, randomized cDNA library suitable for translation into diverse peptide sequences of desired length (although certain *a priori* constraints on sequence complexity can be employed if desired). A method for expressing this library must then be developed to physically link the cDNA sequence to its associated, encoded peptide. Popular methods currently employed for this process include peptides-on-plasmids [14], phage display [13,15,16], and mRNA display

[17]. Once such a library of nucleic acid-tethered peptides is generated, the next step involves affinity selection against a desired protein target that is typically immobilized for retaining bound peptides (and their tethered genetic information) and removing non-specific peptides through extensive washing procedures. Sequences found to interact selectively with target proteins are then amplified, thereby producing an enriched library. This library is then typically subjected to further rounds of the selection process to enrich further the target library. Moreover, sequences identified can also be used for further targeted evolution of second-generation libraries. This entire process can be repeated until particular polypeptide sequences of desired quality (*e.g.*, affinity and selectivity) are identified. Cloning and DNA sequencing ultimately defines the amino acid sequence of binding peptides. These *in vitro* techniques can effectively screen libraries as diverse as 10^8 to 10^{13} unique peptide sequences [18,19]. Numerous alternative approaches to identify affinity reagents, such as mass spectrometry/proteomic approaches to peptide library screening, have been discussed in depth elsewhere [20].

The discoveries highlighted herein focus primarily on the techniques of phage display and mRNA display, both recently successful in identifying both Gα- and Gβγ-binding peptides. These peptides represent useful new tools for investigating the biochemical and structural dynamics of G-protein signal transduction.

BIOCHEMICAL INSIGHTS

KB-752 – a Gα·GDP-Selective Phage-Display Peptide

Recent studies using phage display analysis have identified novel Gα binding peptides [16,21-23]. Some of these peptides are capable of interacting selectively with either the inactive (GDP-bound) or active (GTP-bound) conformation of Gα [21-23]. A GDP-selective peptide, KB-752 (Table **1**), was initially demonstrated to interact most avidly with $G\alpha_{i1-3}$ family members [23]. KB-752 interaction with $G\alpha_{i1}$·GDP results in an enhancement of spontaneous nucleotide exchange; thus, KB-752 serves as a $G\alpha_{i1}$ GEF [23]. This result illustrates the potential usefulness of identified peptides in biochemical studies of G-protein signaling, as the native GEFs for Gα (*i.e.*, GPCRs) are notoriously difficult to produce given that they are integral membrane proteins. Subsequent studies revealed that KB-752 was also capable of binding the inactive conformation of $G\alpha_s$ [22]. Interestingly, interaction with $G\alpha_s$ results in guanine nucleotide dissociation inhibitor (GDI) activity, the opposite action to that elicited towards $G\alpha_{i1}$. By virtue of these synergistic activities on both $G\alpha_i$ and $G\alpha_s$, KB-752 inhibits GTPγS-stimulated adenylyl cyclase activation in cell membranes [22]. Together, these results demonstrate several unique biochemical properties of KB-752 potentially useful to the investigation of G-protein mediated signaling pathways. For example, sequence similarity between KB-752 and a C-terminal motif within the protein GIV/Girdin was instrumental in the discovery of GIV/Girdin as a non-receptor GEF for $G\alpha_i$ subunits involved in Akt signaling and actin cytoskeleton remodeling [64].

Table 1.   Specific G-Protein Subunit Binding Peptides Discussed in this Review

Name	Peptide Sequence (N-Terminal to C-Terminal)	Binding Specificity and Relative Affinities	Reference(s)
KB-752	SRVTWYDFLMEDTKSR	$G\alpha_{i1/i2/i3}$·GDP > $G\alpha_s$·GDP > $G\alpha_o$·GDP	[22,23,64,65]
KB-1753	SSRGYYHGIWVGEEGRLSR	$G\alpha_{i1/i2/i3/t}$·GDP·AlF$_4^-$ (Fig. **4A, D**)	[21,65]
KB-1746	SSSYSEHCQRWGCYARLSR	$G\alpha_{i1/i2/i3/t/o}$·GDP·AlF$_4^-$ (Fig. **4B, E**)	This study
KB-1755	SSRLCPEWICPWEWPASSR	$G\alpha_t$·GDP·AlF$_4^-$ > $G\alpha_{i1/i2/i3}$·GDP·AlF$_4^-$(Fig. **4C, F**)	This study
R6A	MSQTKRLDDQLYWWEYL	$G\alpha_{i1/i2/i3}$·GDP	[35]
R6A-1	DQLYWWEYL	$G\alpha$·GDP	[32,39]
SIRK	SIRKALNILGYPDYD	$G\beta_1\gamma_2$	[40]
SIGK	SIGKAFKILGYPDYD	$G\beta_1\gamma_2$	[45]

KB-1753 and Two Other Peptide Families with Affinity for Activated Gα Subunits

Our phage display screening technique has also yielded peptides that specifically bind the activated conformation of Gα. We recently published on the identification of KB-1753 (Table **1**), which interacts with selective Gα subunits in both the GTP-bound conformation and (especially) the transition-state mimetic, GDP·AlF$_4^-$-bound conformation [21]. This ability of KB-1753 to discriminate between active and inactive states of Gα was invaluable to our recent discovery that the P-loop G42R mutation within Gα subunits, previously assumed to force adoption of an activated state, instead prevents achievement of an active conformation [65]; this discovery has led to a re-evaluation of a sizeable body of literature that previously employed the G42R mutation in genetic studies of fungal heterotrimeric G-protein signal transduction pathways.

In vitro experiments demonstrated that, by binding to GTP-bound Gα, KB-1753 blocks the interaction of Gα$_t$ with its effector, PDEγ, and prevents stimulation of cyclic GMP degradation. Binding of the transition state Gα results in an analogous inhibition of RGS protein interaction and resultant GAP activity [21]. These results highlight the usefulness of KB-1753 as a potential tool for blocking both Gα signaling to effector molecules as well as RGS protein activity for Gα. Direct inhibition of Gα signaling has been suggested as an alternative therapeutic target to traditional modulation of receptor activity [24,25]. Inhibitors of RGS proteins, including both peptidomimetics and small molecules, represent another potentially attractive route of drug discovery targeting G-protein signaling pathways [7,26,27] (cf. [28]). KB-1753 may therefore serve as a useful template for future design of such strategies (see below). In addition to these biochemical properties, KB-1753 has been modified to act as a biosensor for activated Gα [21]. By tagging the peptide with YFP, KB-1753-YFP served as the acceptor to the Gα$_{i1}$-CFP donor in a FRET-based biosensor pair *in vitro*. Such probes, along with other dye-based approaches, provide a starting point for the design of *in vivo* applicable probes that may help unravel the spatiotemporal dynamics of G-protein signaling in living cells [29,30].

In our original phage display screen [23], we identified two additional groups of activated-state-dependent, Gα$_i$-binding peptides with distinct sequence homology compared to the KB-1753 group. These two groups have been termed the KB-1746 and KB-1755 groups (Fig. **2**) after the representative peptide sequence from each that was further characterized (Table **1**). In contrast to the G$^V/_I$WxG motif of the KB-1753 group [21], the KB-1746 and KB-1755 groups display consensus motifs of CxGWxCY and $^I/_V$CPW$^E/_D$, respectively (Fig. **2**). These results suggest that distinct sequence motifs within multiple peptide groups can each selectively recognize the activated conformation of Gα in common.

A   1753 group: $G{}^{V}_{I}WxG$

```
KB-796      SRMGDSVLPYGGVWLGPS-R
KB-G9       SSNLDGCFTSGGVWSGCS-R
KB-353            WDGGVWMGPA-S
KB-G11         SSWDGGVWWWGQ-YGSR
KB-G22         SRYGGVWLGPEG-NSR
KB-382        LGYDINGVWIG
KB-1753        SSRGYYHGIWVGEEGRLSR
```

B   1746 group: C x G W x C Y

```
KB-792      SVLSSSEMCFGWACY
KB-366            SEMCFGWACY
KB-1746     SSSYSEHCQRWGCYARLSR
KB-1744     SSKGPEICRGWGCYTRESR
KB-1745     SSVLDDVCIGWGCYNYSSR
KB-1751     SSTRLDVCKGWECYVPRSR
KB-1752     SSAGPDSCFGWSCYVGGSR
KB-1747     SRHDYEVCKGWGCYLGQSR
KB-1750     SRGEMEYCLGWQCYLGQSR
KB-1749     SREGPEFCKGWGCYATGSR
KB-G33       SNARPCQGWHCYLPSQ
```

C   1755 group: ${}^{I}_{V}CPW{}^{E}_{D}$

```
KB-1756     SSERCPRWVCPWDYS-SDSR
KB-1757     SSGRCPRTICPWDYM-GDSR
KB-1758     SRTYCPQWICPWEYQ-EFSR
KB-1759     SRWSCPPFICPWESD-GISR
KB-1755     SSRLCPEWICPWEWP-ASSR
KB-817       SSACGPAICPWDFMPQLSR
KB-806       SSICD--IIPWEES--CSR
```

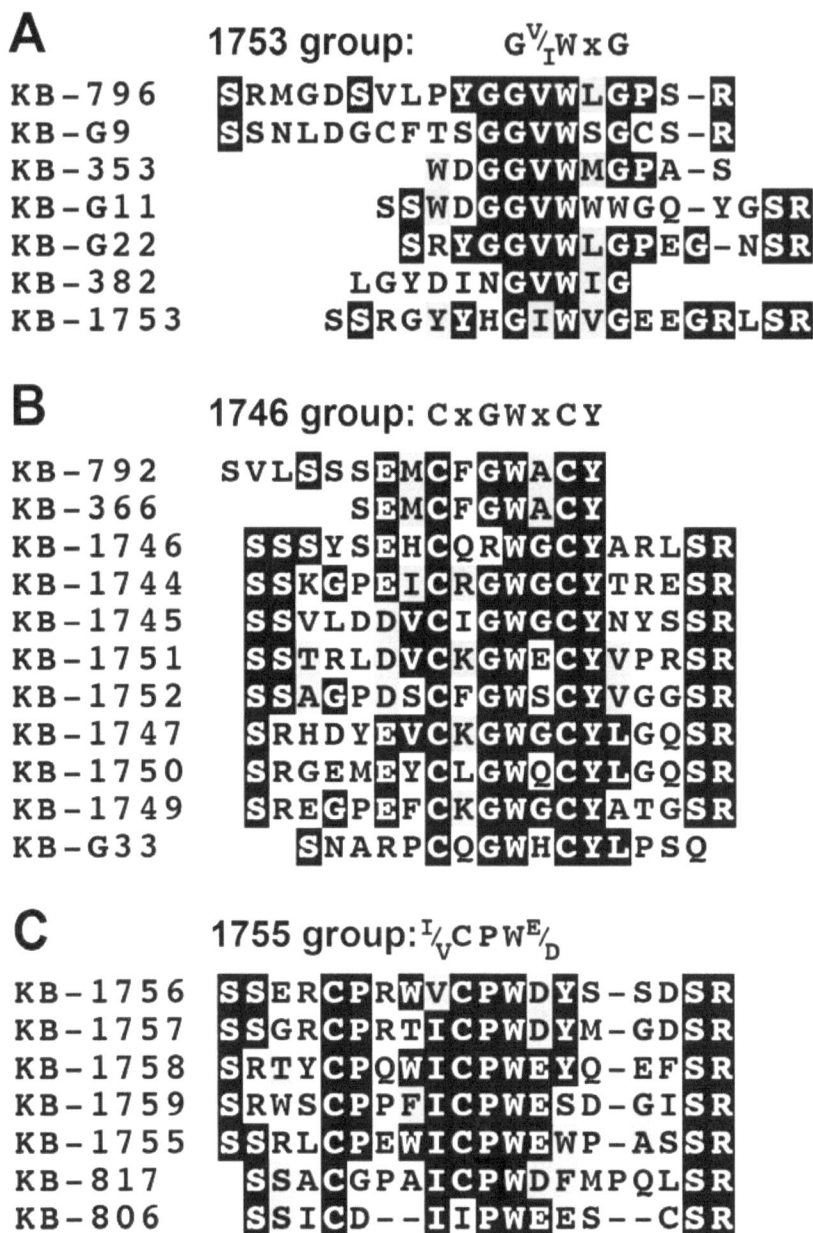

Fig. (2). Multiple sequence alignments of the (**A**) KB-1753, (**B**) KB-1746, and (**C**) KB-1755 groups of phage-display peptide sequences with affinity for activated $G\alpha_i$ subunits, with consensus sequence elements denoted above each grouping. Sequence alignments were created using ClustalW and BOXSHADE programs. (Note that the N-terminal and C-terminal SS and SR dipeptide sequences are generated during cloning and thus typically not involved in the peptide/$G\alpha$ interaction; *e.g.*, refs. [21,23]).

To quantitate the affinity of the interaction of KB-1746 and KB-1755 with $G\alpha$, we used both fluorescence anisotropy and surface plasmon resonance (SPR) assays. To determine binding affinities in a solution-based assay, we first synthesized N-terminal FITC-labeled KB-1746 and KB-1755 peptides and investigated their interaction with $G\alpha$ using fluorescence anisotropy. KB-1746 and KB-1755 interacted with $G\alpha_{i1} \cdot GDP \cdot AlF_4^-$ and $G\alpha_{t/i1} \cdot GDP \cdot AlF_4^-$ with binding affinities (K_D values) of approximately 1.0 and 1.8 µM, respectively (Fig. **3**) – both values comparable to the previously determined binding affinity of KB-1753

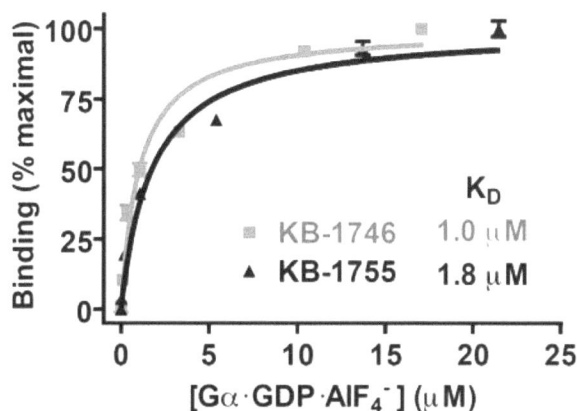

Fig. (3). Fluorescence anisotropy was used to measure the equilibrium affinity of interaction between $G\alpha$ subunits and 10 nM of FITC-KB-1746 or FITC-KB-1755 peptides. Equilibrium fluorescence anisotropy was measured in the presence of increasing concentrations of $G\alpha_{i1} \cdot GDP \cdot AlF_4^-$ (for KB-1746; *grey*) or $G\alpha_{t/i1} \cdot GDP \cdot AlF_4^-$ (for KB-1755; *black*). 95% confidence intervals for K_D determinations were: KB-1746, 0.86 - 1.2 µM; KB-1755, 1.4 - 2.1 µM.

for its $G\alpha_{i1}$ target (K_D of 1.2 µM; ref. [21]). Follow-up SPR experiments involved the injection of various $G\alpha$ isoforms (in specific nucleotide-bound forms) over N-terminally biotinylated KB-series peptides immobilized on a streptavidin-coated gold biosensor chip (Fig. **4**). Observed interactions were in general agreement with the nucleotide-state-selective nature of these binding events first observed in the initial phage screening and phage ELISA results [23], although the KB-1746 peptide displayed more modest nucleotide selectivity in showing some affinity also for the inactive, GDP-bound conformation of $G\alpha_{i1}$ (*e.g.*, Fig. **4B**). As

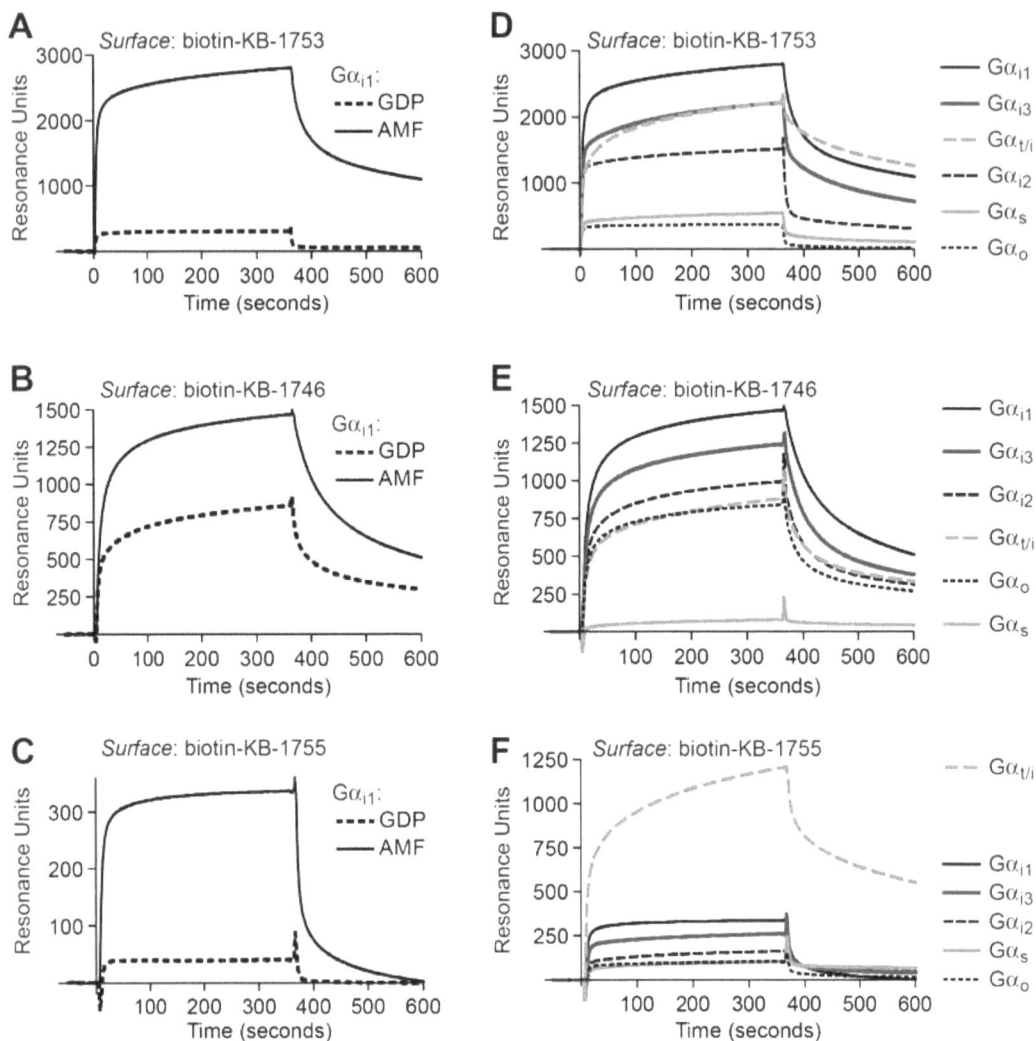

Fig. (4). Surface plasmon resonance (SPR) biosensor assays demonstrating nucleotide-selective (**A-C**) and $G\alpha$ subfamily-selective (**D-F**) binding to immobilized KB-series peptides. SPR assays were conducted as previously described (refs. [21-23]), with biotinylated KB-series peptides immobilized on separate streptavidin biosensor surfaces to relative surface densities of ~500 RU and indicated $G\alpha$ subunit analytes (each at 5 μM) injected for 360 seconds (in GDP or GDP·AlF$_4^-$ ["AMF"] nucleotide states as indicated in panels A-C; solely in GDP·AlF$_4^-$ nucleotide state in panels **D-F**).

originally reported [21], KB-1753 interacted solely with $G\alpha$ subunits in their active conformation (Fig. **4A**, **D**), including $G\alpha_{i1}$, $G\alpha_{i2}$, $G\alpha_{i3}$, and $G\alpha$-transducin (the latter actually an *E. coli*-purifiable $G\alpha_{t/i1}$ chimera that closely mimics the biology of Gα-transducin; refs. [21,31]). KB-1746 displayed a similar Gα subfamily selectivity profile to that of KB-1753; however, a robust interaction with Gα$_o$ was also detected (Fig. **4E**). This Gα$_o$/KB-1746 interaction is unique, as neither KB-1753 nor KB-1755 is capable of binding this particular Gα subunit despite its high degree of sequence similarity to Gα$_i$ subunits in the predicted peptide-binding site [21]. Although KB-1755 showed significant binding to Gα$_{i1}$, Gα$_{i2}$, and Gα$_{i3}$, its interaction with Gα$_t$ was even more robust (Fig. **4F**), suggesting that this peptide has a relative Gα$_t$ selectivity unseen in the other two peptide groups. Overall, these results highlight the similar preference all three peptide groups possess for the activated Gα conformation, as well as differing Gα subunit preferences of specific peptides.

Our previous biochemical and structural studies of KB-1753 demonstrated a mode of binding to activated Gα analogous to that of effectors, with the specific site of interaction comprising the highly conserved hydrophobic α2/α3 cleft of Gα [21]. To investigate whether KB-1746 and KB-1755 peptides bind to the same site on Gα$_{i1}$, we carried out competition binding analyses again using surface plasmon resonance. SPR surfaces of biotinylated KB-1753 (as a control), KB-1746, and KB-1755 were created and their interaction with activated Gα$_{i1}$ was examined in the absence and presence of increasing concentrations of non-biotinylated KB-1753 peptide (Fig. **5**). As expected, unlabelled wildtype KB-1753 peptide was capable of competing for Gα$_{i1}$ binding to the immobilized KB-1753 surface. A loss-of-function mutant KB-1753 peptide (Ile-9 and Trp-10 mutated to alanine; "I9A/W10A"), which does not bind Gα$_{i1}$ [21], was ineffective at competing for Gα$_{i1}$ binding (Fig. **5A**). Wildtype KB-1753 peptide was also capable of precluding binding of Gα$_{i1}$ to the immobilized KB-1755 surface, suggesting that the binding site for KB-1755 at least partially overlaps with that of KB-1753 (Fig. **5C**). However, preincubation with KB-1753 did not significantly alter the binding of Gα$_{i1}$ to the immobilized KB-1746 surface (Fig. **5B**), suggesting that KB-1746 interacts through a novel binding site distinct from that of KB-1753.

To more completely investigate the interaction of KB-1746 and KB-1755 with activated Gα and map the interaction site, we conducted fluorescence-based binding experiments in the presence of potential competitors for Gα binding.

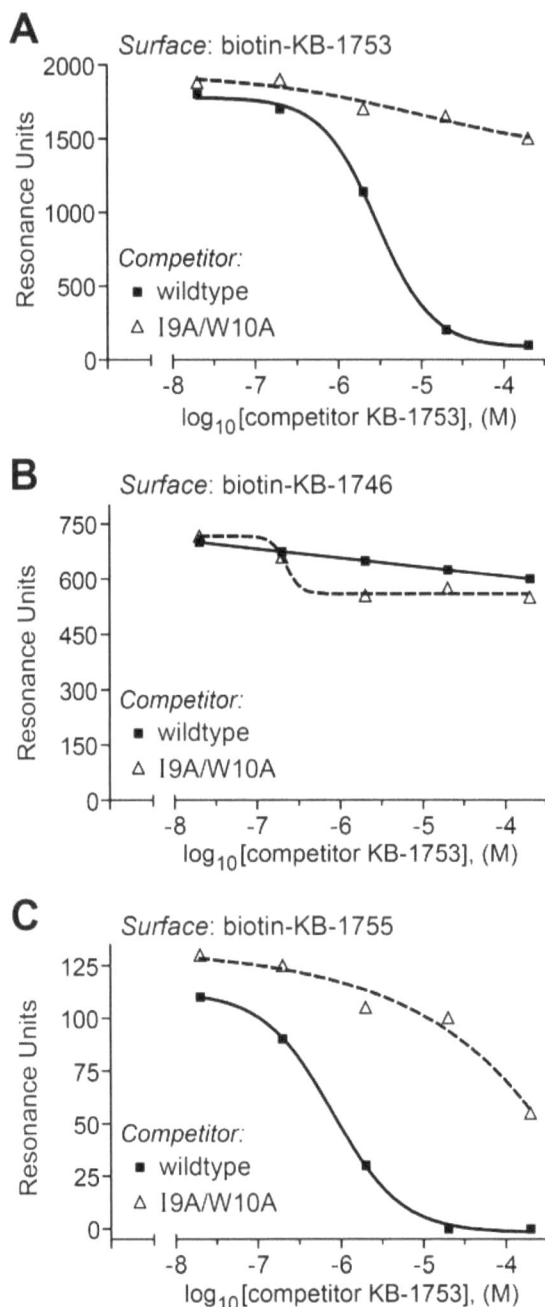

Fig. (5). SPR biosensor competition binding assays, conducted in a similar fashion to panels **D-F** of Fig. (**4**) and using 1 μM $G\alpha_{i1}\cdot GDP\cdot AlF_4^-$ preincubated with the indicated concentrations of wildtype or loss-of-function (I9A/W10A mutant) KB-1753 peptide prior to injection over indicated immobilized peptide surfaces.

Analogous to our prior studies with FITC-labelled KB-752 and R6A-1 peptides [22,32], we observed that FITC-labelled KB-1755 undergoes a significant (~30%) and reproducible enhancement of fluorescence quantum yield upon binding to activated $G\alpha$ (Fig. **6**). Using this approach, we conducted competition binding studies and found that all three classes of activation-state-selective peptides compete with FITC-KB-1755 for binding to $G\alpha_{i1} \cdot GDP \cdot AlF_4^-$ (Fig. **6A**). The KB-1753 result is in agreement with the SPR analysis described above (Fig. **5C**). The ability of KB-1746 to compete with FITC-KB-1755, despite its inability to be competed by KB-1753 (Fig. **5B**), suggests that KB-1755 shares contacts on $G\alpha$ with the two distinct binding sites created by KB-1746 and KB-1753.

To further explore the binding of KB-1755, we also assessed the ability of an RGS protein (RGS16) and an effector peptide (PDEγ) to compete for binding to $G\alpha_{t/i1}$. RGS proteins and effectors both bind the active conformation of $G\alpha$ subunits, but have distinct, non-overlapping binding sites [33,34]. RGS16 and PDEγ each competed with FITC-KB-1755 for $G\alpha_{t/i1}$ binding, reducing nearly equally the observed EC_{50} of the latter, fluorophore-labeled peptide (Fig. **6B**). Thus, the KB-1755 peptide likely interacts with $G\alpha$ in a mode that extends partially through both the effector- and RGS protein-binding sites. As KB-1753 binds predominantly to the effector binding site of $G\alpha$ [21], this suggests that KB-1746 potentially interacts predominantly with the RGS protein binding site of $G\alpha$.

Derivatives of the GoLoco Motif

As an alternative approach to phage display, mRNA display has also been used for identifying novel $G\alpha$ binding sequences. Ja and Roberts recently described a class of $G\alpha$ binding peptides with the prototype sequence termed R6A ([35]; Table **1**). In contrast to the completely random screening approach employed in our phage display studies, the mRNA display of Ja and Roberts was done using an *a priori* input sequence: namely, that of the consensus motif for the GoLoco motif (also known as the GPR motif) [36]. The naturally-occuring GoLoco motif, found in known GPCR signaling regulators such as RGS12 and RGS14, binds $G\alpha_{i/o}$ subunits preferentially in the GDP-bound form and inhibits spontaneous nucleotide release (*i.e.*, guanine nucleotide dissociation inhibitor or "GDI"

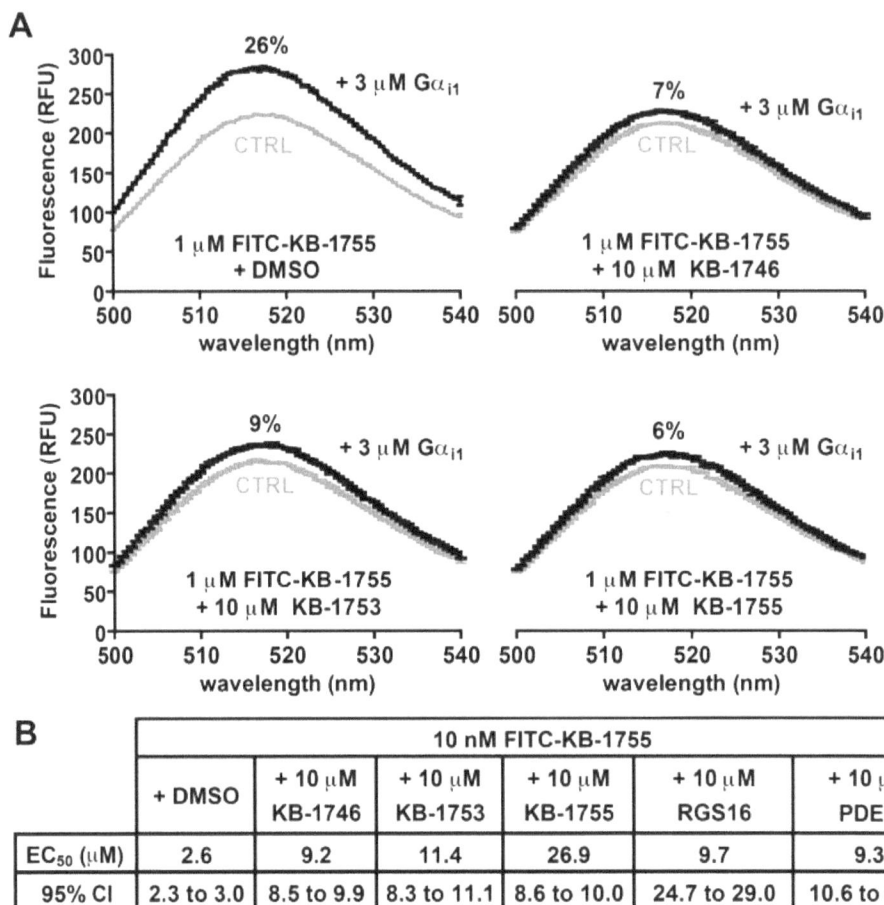

Fig. (6). Fluorescence spectroscopic assays were used to measure competitive binding interactions between peptides (and proteins) that bind to activated Gα subunits. (**A**) FITC-labeled KB-1755 peptide was used as a conformation-sensitive reporter for Gα binding. The fluorescence emission spectrum of 1 µM FITC-KB-1755 was measured in the presence of DMSO vehicle or 10 µM competitor peptides: biotin-KB-1746, biotin-KB-1753, or biotin-KB-1755 (control ["CTRL"] spectra in *grey*). 3 µM of $G\alpha_{i1} \cdot GDP \cdot AlF_4^-$ was then added and fluorescence emission was re-measured after 5 minutes (*black* spectra). Data are presented as the average of triplicate determinations. The net fluorescence enhancement was calculated in percent. Percent fluorescence enhancement (± SEM): DMSO (26.2 ± 1%), KB-1746 (6.9 ± 1%), KB-1753 (9.4 ± 1%), KB-1755 (6.0 ± 1%). RFU = relative fluorescence units. (**B**) Equilibrium fluorescence anisotropy assays were conducted using 10 nM FITC-KB-1755 and increasing concentrations of $G\alpha_{i/i1} \cdot GDP \cdot AlF_4^-$. Titrations were conducted in the presence of DMSO vehicle or indicated concentration of competitor proteins/peptides: biotin-KB-1746, biotin-KB-1753, biotin-KB-1755, RGS16 (aa 53-190, spanning the minimal RGS domain), or biotin-PDEγ(aa 63-87). Data are presented as EC_{50} values (with 95% confidence intervals) of the fluorescence increase derived from the FITC-KB-1755/$G\alpha_{i/i1} \cdot GDP \cdot AlF_4^-$ interaction.

activity; [37,38]. Not surprisingly, R6A and other peptides patterned after the GoLoco motif also interact with Gα·GDP and have apparent binding affinities of 60~200 nM [35]. Despite containing a non-conservative replacement of the critical arginine residue in the signature Asp-Gln-Arg motif of the GoLoco motif [36], the R6A family of peptides was initially reported to have GDI activity similar to that of the parent GoLoco motif peptide [35]. However, more careful exploration using multiple modes of measuring nucleotide binding and cycling suggested that, while the R6A-1 peptide does interact with Gα·GDP, it does not elicit any significant GDI activity [32]. The R6A-1 peptide, a minimal 9-mer representing the R6A family (Table **1**) and possessing sequence similarity to KB-752 [32], was further reported to bind to all Gα family members in their GDP-liganded forms and this binding was mutually exclusive with Gβγ binding [39]. One could imagine the R6A family of peptides, as well as their naturally-occuring GoLoco motif progenitors, may prove useful as Gα-interacting peptides that occlude Gβγ interaction for studying the regulation of heterotrimer association/dissociation related to the temporal aspects of G-protein signaling.

Gβγ-Interacting Peptides

In addition to screens for Gα-interacting peptides, the isolated Gβγ subunit has also been used as the bait for discovery of novel binding sequences. Smrcka and colleagues used randomized peptide libraries in a phage display screen to identify four distinct groups of Gβγ binding peptides [40]. Peptides in 'group I' had a consensus motif with homology to two native Gβγ binding proteins, PLCβ and phosducin, while 'groups II-IV' had little homology to known proteins or to that of group I peptides ([40] and see Table **1**). Interestingly, a peptide with the strongest correlation to the group I consensus, termed SIRK, was capable of selectively inhibiting Gβγ-mediated activation of PLCβ while not affecting Gβγ stimulation of voltage-gated calcium channels or type I adenylyl cyclases. Despite divergent sequences, competition experiments revealed peptides from all four groups bound the same site on Gβγ, which was termed the Gβγ 'hot spot' that is likely used for selectivity among effector interactions. [40]. This elegant study has thus produced inhibitory peptides capable of dissecting among several Gβγ-mediated signaling pathways, which will undoubtedly prove to be valuable tools for studying these pathways both *in vitro* and *in vivo*.

STRUCTURAL INSIGHTS

KB-752/Gα_{i1}·GDP Complex

We recently elucidated the structural determinants of Gα_{i1} binding by the GDP-selective phage peptide KB-752 to high resolution using x-ray diffraction crystallography [23]. In mimicking the biochemical activity of a receptor (*i.e.*, GEF activity), and representing a much more tractable target for structural studies than a membrane-bound GPCR, the KB-752 peptide was instrumental to understanding mechanistic details of Gα activation, which remains poorly understood [3]. The crystal structure of the Gα_{i1}/KB-752 dimer revealed a peptide binding site centered on the hydrophobic $\alpha2/\alpha3$ cleft of the Gα Ras-like domain [23] (Fig. **7A**). Binding of KB-752 levers the conformationally-flexible switch II ($\alpha2$) helix in a direction away from the GDP binding site. In turn, this alteration causes the preceding $\beta3/\alpha2$ loop to assume a novel conformation, stabilized by interactions with KB-752, that removes it from its normal disposition within the G$\alpha_{i1}\beta_1\gamma_2$ heterotrimer. In the heterotrimer structure [41], as well as the Gα_{i1}/RGS14-GoLoco structure [38], this $\beta3/\alpha2$ loop sits directly in front of the bound GDP molecule. It is thought that this conformation, when bound to either of these GDI proteins, allows the $\beta3/\alpha2$ loop to serve as an 'occlusive barrier' to the release of GDP and, furthermore, that receptors use G$\beta\gamma$ as a 'lever' to remove this barrier allowing for nucleotide exchange [3,42,43]. That KB-752 induces such a conformational change in the $\beta3/\alpha2$ loop likely explains its GEF activity for Gα_{i1}. Thus, KB-752, identified using phage display, has provided insight into the mechanism of G-protein activation by giving structural support to a previously proposed model of receptor-mediated activation [42,43]. The KB-752 peptide, in its ability to stabilize inactive conformations of Gα_i subunits, has also proven to be an invaluable tool for Gα subunit crystallography; for example, KB-752 was recently used to establish crystallization conditions for the P-loop G42R mutant of Gα_{i1} [65].

KB-1753/Gα_{i1}·GDP·AlF$_4^-$ Complex

We have also elucidated the high-resolution structural determinants of the interaction of the activation-state-dependent phage peptide KB-1753 bound to Gα_{i1} in its transition-state (GDP·AlF$_4^-$) conformation [21]. KB-1753 was found to bind the same $\alpha 2/\alpha 3$ cleft within Gα as was seen with KB-752 (Fig. **7B**). The stringent reversal in nucleotide selectivity between these two peptides (Table **1**) is due to the difference in peptide residues that insert in this same hydrophobic groove: the larger tryptophan in KB-752 requires a flexible $\alpha 2$ helix found only in the GDP-bound form, whereas the smaller isoleucine in KB-1753 is perfectly accommodated by the more closed and rigid conformation of the $\alpha 2$ helix found

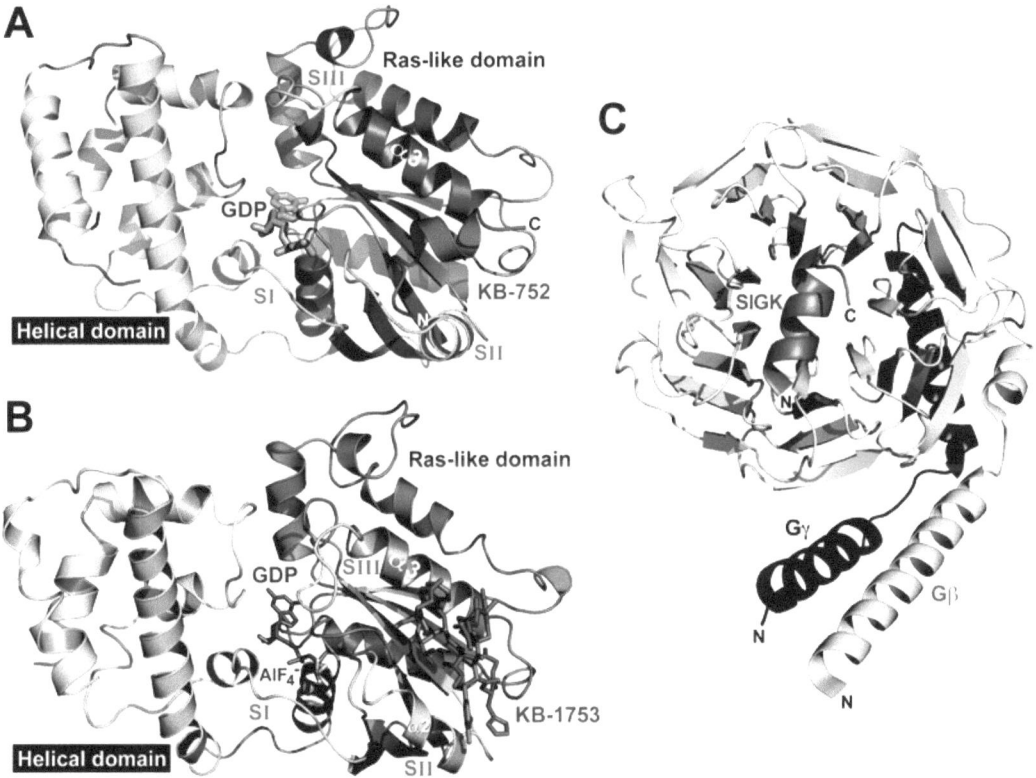

Fig. (7). Structures of G-protein subunit-selective peptides and their G-protein partners. (**A, B**) Overall structural features of the KB-752/Gα_{i1}·GDP (panel **A**) and KB-1753/ Gα_{i1}·GDP·AlF$_4^-$ (panel **B**) complexes as obtained by X-ray diffraction crystallography [21,23]. Both KB-752 and KB-1753 peptides (*red*) are found situated in a groove between the $\alpha 2$ ("switch II") and $\alpha 3$ helices of the Gα Ras-like domain (*blue*; switch regions in *green*); no contacts are made by either peptide to the Gα all-helical domain (*yellow*) nor GDP moiety (*magenta*). Bound magnesium and aluminum tetrafluoride ion are colored *orange* and *grey*, respectively, in panel **B**. (**C**) Overall structural features of the SIGK/G$\beta_1\gamma_2$ complex as obtained by X-ray diffraction crystallography [45]. The SIGK peptide (*red*) lies in an α-helical conformation over the top of the β-propeller (or "torus") structure of the Gβ subunit (*cyan*); no contacts are made to the Gγ subunit (*black*).

in the active conformation ([21,23] and see Fig. **7A, B**). The Gα/KB-1753 complex revealed a striking similarity to native Gα/effector complexes and, thus, represents the first structural glimpse of an 'effector-like molecule' bound to Gα$_{i1}$. Although the Gα residues contacted by KB-1753 are strictly conserved among all Gα families, KB-1753 displays a narrow selectivity profile for specific Gα$_i$ subfamily members [21], even discriminating within this subfamily against Gα$_o$ (Fig. **4D**). These results have implications for the nature of specificity determinants within Gα/effector interactions, supporting the notion that specificity for particular Gα subfamily members can be determined by differences, even subtle, in effector sequence and structure [34]. Structural information on the Gα/KB-1753 complex will undoubtedly aid in the rational design of future biosensors as previously described above. Another application made possible by this crystal structure is structure-based design and screening of small molecule inhibitors of Gα signaling. Analogous efforts using Gβγ have already proven fruitful ([44] and see discussion below).

SIGK/Gβ$_1$γ$_2$ Complex

A crystal structure of Gβ$_1$γ$_2$ bound to a SIRK peptide derivative (denoted "SIGK"; Table **1**) has been determined, depicting the molecular basis for this peptide's unique regulatory properties [45]. The SIGK peptide assumes a partial α-helical conformation reminiscent of the switch II helix of Gα within the Gαβγ heterotrimer (Fig. **7C**). Moreover, the SIGK helical peptide binds the same face of Gβ$_1$γ$_2$ as does switch II. This common binding site likely explains the apparent ability of this peptide to induce G-protein heterotrimer dissociation [46,47]. It is interesting to note that Gβ$_1$γ$_2$ uses both polar and nonpolar residues in the overall SIGK binding interface, suggesting that this combination of binding determinants dictates the ability of Gβ$_1$γ$_2$ to recognize a variety of downstream effector binding partners. In support of this notion, mutation of selective Gβ$_1$γ$_2$ residues within the Gβ$_1$γ$_2$/SIGK interface [45] resulted in loss of binding to different subset(s) of the four distinct groups of Gβγ-binding peptides originally isolated by Smrcka and colleagues [40]. Thus, effectors likely share the common overall SIGK-footprint for interaction with Gβ$_1$γ$_2$, yet distinct sets of interactions within this Gβγ 'hot spot' dictate the specificity of each given complex [45].

'Hot spots' have been proposed to represent unique areas within proteins for guiding targeted drug design efforts [48]. Data from the SIGK family of peptides has defined such an area within $G\beta_1\gamma_2$. The ability of these phage display peptides to selectively perturb $G\beta_1\gamma_2$ signaling nodes [40] further implies that drug design to this particular 'hot spot' may lead to the identification of small molecules capable of inhibiting specific $G\beta\gamma$-driven signal transduction pathways. Such a discovery has recently been reported using the $G\beta_1\gamma_2$/SIGK interface as the basis for computational docking screens of virtual compound libraries [44]. Several $G\beta_1\gamma_2$ binding compounds were identified in this *in silico* screen and subsequently shown to selectively modulate $G\beta_1\gamma_2$ signaling both *in vitro* and *in vivo*. M119, a fluorescein-like lead compound from this virtual screening, was capable of blocking $G\beta_1\gamma_2$/SIGK interaction, inhibiting $G\beta_1\gamma_2$ activation of purified PLCβ2, preventing chemoattractant-induced calcium signaling in intact immune cells, and sensitizing mice to the antinociceptive effects of morphine *in vivo* [44]. These sensational findings demonstrate perhaps the ultimate application of peptide screening studies, given that the original finding of the SIRK phage-display peptide formed the basis for the discovery of the small molecule $G\beta\gamma$-inhibitor M119. Analogous screens using the $G\alpha_{i1}$/KB-1753 complex could be envisioned to identify useful small molecule inhibitors of $G\alpha$ signaling. Heterotrimeric G-protein alpha subunits may emerge as drug targets themselves, for example in the case of their mutational activation (as for *GNAS* in endocrine tumors, ref. [49]; or for *GNAQ* and *GNA11* in ocular melanoma, ref. [66]). There is a real need for chemical biology approaches to target heterotrimeric G-protein function and thus further delineate the biology of these signaling molecules as well as facilitate new screening paradigms. A germane example is the $G\alpha$q-selective inhibitor YM-254890, which has provided novel biological insights [50,51,67].)

APPLICATION IN HIGH-THROUGHPUT SCREENING FOR GPCR MODULATORS

The original intent of our efforts in identifying nucleotide-state-selective $G\alpha$-binding peptides was to employ them in developing rapid, non-radioactive means for assaying heterotrimeric G-protein activation by GPCRs. In particular, peptides that show specificity towards the active, GTP-bound state of $G\alpha$ subunits (such as KB-1753, KB-1746, and KB-1755) or freed $G\beta\gamma$ subunits (such as SIRK and

SIGK) should be valuable tools in detecting receptor-mediated heterotrimer activation when coupled to a real-time, sensitive readout of their specific G-protein subunit interaction (*e.g.*, the FITC-labelled KB-1755 shown in Fig. **6A**). In pilot studies previously described [52], we have shown that activation-state-selective Gα-binding peptides are able to faithfully report on several different GPCR/heterotrimer complexes when their activation status has been modulated by specific agonists and antagonists (Fig. **8**). In examining the G_i-heterotrimer coupled M2 muscarinic acetylcholine receptor (M2-mAChR), we have observed that binding of the $Gα_i$·GTP-specific peptide KB-1755 to immobilized membrane preparations is significantly higher in the presence of the receptor-specific agonist carbachol than in its absence; furthermore, this binding was inhibited in a dose-dependent manner by the receptor-specific antagonist atropine, a clear indication that the effect is receptor promoted (Fig. **8A**).

$Gα_i$·GTP-specific peptides, besides being useful in detecting activation or inactivation of a G_i-coupled GPCR, can also be employed in conjunction with chimeric Gα subunits (reviewed in [53]) to detect activation or inactivation of a GPCR which normally couples to a different G-protein such as $Gα_q$ or $Gα_s$. For example, to demonstrate utility in measuring β2-adrenergic receptor (β2-AR) status upon treatment with agonist isoproterenol and inverse agonist ICI-118,551 (Fig. **8B**), we employed a $Gα_{i6s}$ chimera [54] in which the last 6 residues of $Gα_{i1}$ (KDCGLF) were replaced with the last 6 residues of $Gα_s$ (RQYELL), thus allowing functional coupling of a $Gα_{i6s}/Gβγ$ heterotrimer to the normally G_s-coupled β2-AR [55]. We further confirmed the utility of this chimeric Gα strategy in detecting dose-dependent activation of the dopamine D1-receptor (another normally Gs-coupled receptor [56]; Fig. **8C**). We have also used this peptide-based assay to recapitulate the known relative efficacies and rank order of potencies of the β2-AR agonists isoproterenol, salbutamol, and dobutamine (refs. [52,57]; Fig. **8D**); all activation signals from these agonists were inhibited by the inverse agonist ICI-118,551 (data not shown). These latter results indicate that the known pharmacological properties of the β2-adrenergic receptor are faithfully preserved in this $Gα_i$·GTP-specific peptide-based activation assay. Additional screening for activation-state-selective phage peptides specifically directed to $Gα_s$ and other Gα subfamilies (*e.g.*, $Gα_q$, $Gα_{12/13}$) should obviate the need to reconstitute GPCR/heterotrimer complexes with chimeric $Gα_i$ subunits.

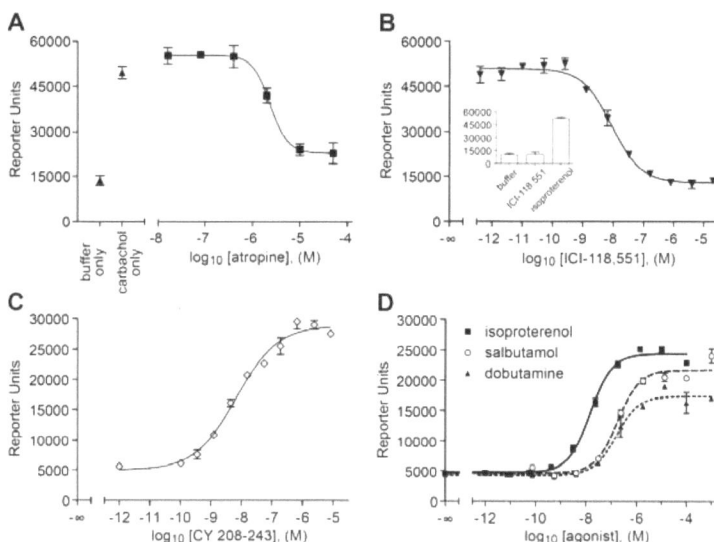

Fig. (8). Examples of using activation-state-selective, Gα-binding peptides to measure GPCR/heterotrimer activation state upon agonist and antagonist treatment. (**A**) Peptide-based assay of M2-muscarinic acetylcholine receptor (M2-mAChR) stimulation and inhibition. Membranes of Sf9 insect cells co-infected with baculoviruses expressing M2-mAChR, $G\alpha_{i1}$, and $G\beta_1\gamma_2$ subunits were prepared by nitrogen cavitation; 10 µg of membrane protein per well (in 100 µL of immobilization buffer containing 100 mM NaCl, 20 mM HEPES pH 7.4, 5 mM $MgCl_2$, 3 mM EDTA) was covalently immobilized *via* primary amines using N-oxysuccinimide surface chemistry ("DNA-Bind" 96-well plates from Corning Costar; Kennebunk, ME). After five 200 µL rinses with a wash-buffer containing 150 mM NaCl, 20 mM Tris pH 8, 0.05% Tween-20 and 0.1% BSA, immobilized membranes were incubated with 100 µL of buffer only (immobilization buffer containing an optimized concentration of GTPγS) or buffer containing the agonist carbachol (100 µM) in the absence or presence of indicated concentrations of the antagonist atropine. Guanine nucleotide exchange was allowed to proceed for 20 minutes at room temperature, followed by quenching with the addition of 100 µL detection buffer (150 mM NaCl, 100 µM GDP, 0.05% Tween-20, 0.1% BSA, and 1 nM of N-terminally biotinylated KB-1755 peptide pre-adsorbed to Neutravidin-Alkaline Phosphatase (NA-AP; Pierce). After 1 hour incubation, wells were rinsed five times with wash-buffer and bound peptide/NA-AP complex detected with the addition of 100 µL CDP-Star (Perkin-Elmer) and luminescence detection using a BMG Lumistar plate-reader. (**B**) Peptide-based assay of β2-adrenergic receptor (β2-AR) stimulation and inhibition. Membranes of Sf9 insect cells co-infected with baculoviruses expressing β2-AR, chimeric $G\alpha_{i6s}$ (see text for details), and $G\beta_1\gamma_2$ subunits were stimulated with the specific adrenergic agonist isoproterenol (0.2 µM) in the absence or presence of indicated concentrations of the inverse agonist ICI-118,551. (**C**) Peptide-based assay of D1-dopamine receptor (D1-R) stimulation. Membranes of Sf9 insect cells co-infected with baculoviruses expressing D1-R, chimeric $G\alpha_{i6s}$, and $G\beta_1\gamma_2$ subunits were stimulated with indicated concentrations of the specific agonist CY 208-243. (**D**) Peptide-based assay of β2-AR stimulation by full and partial agonists at the indicated concentrations, conducted as in panel (**B**).

We conducted a pilot screening assay in 96-well plates using baculovirus-infected Sf9 cells to investigate the applicability of the M2-mAChR assay for high-throughput screening. The screen was performed on the Library of Pharmacologically Active Compounds (LOPAC; Sigma-Aldrich), with compounds applied at a final concentration of 1 μM. Twelve documented cholinergic agonists were detected with responses greater than 50% of the signal observed with a saturating dose (10 nM) of oxotremorine. Representative examples including the agonists arecoline, dioxolane, carbachol and the antagonist atropine are annotated in Fig. (**9**). The Z-factor of the assay was calculated as 0.62, indicative of a robust and sensitive screening assay [58]. The compound TMB-8 (3,4,5-trimethoxybenzoic acid 8-[diethylamino]octyl ester) was also identified as an M2-mAChR agonist in our pilot screen. The peptide-detection assay was then used to directly compare TMB-8 and carbachol for M2-mAChR activation (Fig. **10A**). In terms of G-protein agonism, the two compounds

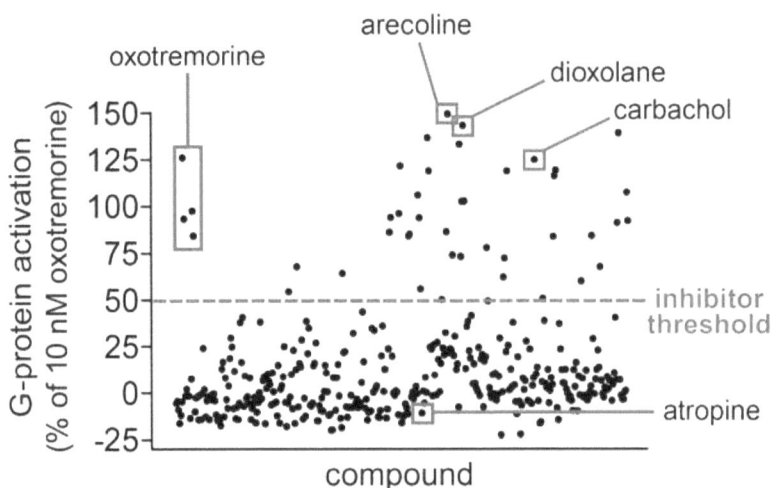

Fig. (9). A pilot, peptide-based screening assay for activators of the M2-muscarinic acetylcholine receptor (M2-mAChR). Intact Sf9 cells were used as the source membrane material. Sf9 cells infected with baculoviruses encoding the M2-mAChR receptor, $G\alpha_{i1}$, and $G\beta_1\gamma_2$ subunits were washed with phosphate-buffered saline and applied to DNA-Bind 96-well plates at a density of 500,000 cells per well. Compounds from the LOPAC library (Sigma-Aldrich) were applied at a final concentration of 1 μM. Compound additions and washing steps were performed with an automated liquid handling device (Tecan, Durham, NC). Results are depicted graphically, with a 'hit' being defined as a response greater than 50% of the signal observed with a saturating dose of oxotremorine (10 nM).

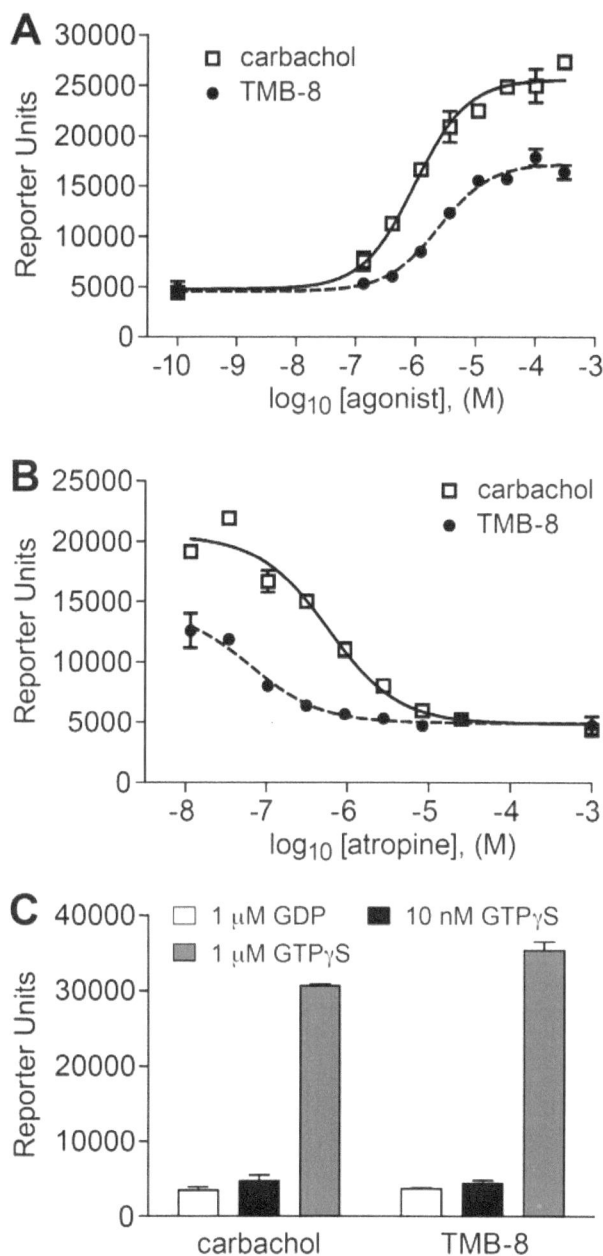

Fig. (10). Peptide-based measurements of M2-mAChR activation confirming agonist activity of TMB-8. (**A**) TMB-8 is an equipotent partial agonist of the M2-mAChR as compared to carbachol. (**B**) Dose-dependent atropine antagonism of carbachol- and TMB-8-agonist activity (10 μM of either agonist). Assays were performed essentially as described in Fig. (**8A**). (**C**) Nucleotide (and concentration) dependence of the peptide-based measurement of agonist activity on the M2-mAChR by carbachol and TMB-8.

were similar in potency, while the traditional agonist carabachol had a two-fold greater efficacy. Antagonist competition experiments were consistent with this data (Fig. **10B**). As a positive control, we examined the guanine nucleotide selectivity of the assay and demonstrated that G-protein activation could only be detected in the presence of micromolar GTPγS but not GDP (Fig. **10C**). TMB-8 has previously been characterized as a muscarinic antagonist [59] and an allosteric modulator of the M2 muscarinic acetylcholine receptor [60]. TMB-8 slows the rate of binding and the dissociation of pre-bound acetylcholine from the M2 muscarinic receptor [61]. Our data are interesting in that they demonstrate that TMB-8 has canonical agonist-like properties at the M2-mAChR with regard to G-protein activation (Fig. **10A**). However, we have not conducted the mechanistic studies necessary to confirm or refute TMB-8 as an allosteric modulator of the M2-mAChR. In this regard, it must be noted that TMB-8 has multiple pharmacologic effects and targets, having been originally characterized as a calcium anatagonist [62,63]. Therefore our discovery *via* this pilot, peptide-based GPCR screen that TMB-8 is an agonist at the M2 muscarinic acetylcholine receptor is a novel pharmacological finding.

CONCLUDING REMARKS

G-protein subunit binding peptides, identified by various screening methods, have presented themselves as powerful research tools for studying target proteins within vital signal transduction pathways. As described herein, these peptides have numerous applications including:

(1)   Serving as direct modulators of target protein activity.

(2)   Providing structural insights into previously undefined mechanisms of target protein function.

(3)   Supplying a template for designing biosensors capable of reporting the spatiotemporal dynamics of target protein signaling.

(4)   Presenting a model for screening small molecule libraries to identify inhibitors of target G-protein subunit activity and,

(5)   Serving as the starting material for novel, non-radioactive assays for measuring agonist and antagonist activity on GPCR/heterotrimer complexes.

Recently, screening methods including phage-display and mRNA-display techniques have successfully identified such peptides targeting both Gα and Gβγ subunits [12,16,21,23,35,39,40]. These peptides have already given valuable insights into various aspects of G-protein signaling and have been elegantly applied in various scenarios just described. Continued efforts to identify novel peptide sequences to additional G-protein subfamilies should afford further discoveries into heterotrimer nucleotide cycling as well as future assay platforms for receptor-mediated G-protein signaling.

ABBREVIATIONS

GDI   =   Guanine nucleotide dissociation inhibitor

GDP   =   Guanosine diphosphate

GEF   =   Guanine nucleotide exchange factor

GPCR   =   G protein-coupled receptor

GTP   =   Guanosine triphosphate

KB   =   Karo Bio

ACKNOWLEDGEMENTS

Work described in this chapter was funded in part by grants R01 GM074268 (to D.P.S.) and F32 GM076944 (to C.A.J.) from the U.S. National Institute of General Medical Sciences.

CONFLICT OF INTEREST

The authors state that there is no conflict of interest.

DISCLOSURE

Primary data, its analysis, and literature surveys described in this chapter were updated from a previously published article in *Combinatorial Chemistry & High Throughput Screening,* Volume 11, Issue 5, 2008, pp. 370 to 381.

REFERENCES

[1]   Hamm, H. E. The many faces of G protein signaling. *J Biol Chem* **1998**, *273*, 669-72.

[2]   McCudden, C. R.; Hains, M. D.; Kimple, R. J.; Siderovski, D. P.; Willard, F. S. G-protein signaling: back to the future. *Cell Mol Life Sci* **2005**, *62*, 551-77.

[3]   Johnston, C. A.; Siderovski, D. P. Receptor-mediated activation of heterotrimeric G-proteins: current structural insights. *Mol Pharmacol* **2007**, *72*, 219-30.

[4]   Ferguson, K. M.; Higashijima, T.; Smigel, M. D.; Gilman, A. G. The influence of bound GDP on the kinetics of guanine nucleotide binding to G proteins. *J Biol Chem* **1986**, *261*, 7393-9.

[5]   Wall, M. A.; Posner, B. A.; Sprang, S. R. Structural basis of activity and subunit recognition in G protein heterotrimers. *Structure* **1998**, *6*, 1169-83.

[6]   Offermanns, S. G-proteins as transducers in transmembrane signaling. *Prog Biophys Mol Biol* **2003**, *83*, 101-30.

[7]   Neubig, R. R.; Siderovski, D. P. Regulators of G-protein signaling as new central nervous system drug targets. *Nat Rev Drug Discov* **2002**, *1*, 187-97.

[8]   Siderovski, D. P.; Hessel, A.; Chung, S.; Mak, T. W.; Tyers, M. A new family of regulators of G-protein-coupled receptors. *Curr Biol* **1996**, *6*, 211-2.

[9]   Siderovski, D. P.; Strockbine, B.; Behe, C. I. Whither goest the RGS proteins? *Crit Rev Biochem Mol Biol* **1999**, *34*, 215-51.

[10]   Sprang, S. R. G protein mechanisms: insights from structural analysis. *Annu Rev Biochem* **1997**, *66*, 639-78.

[11]   Alluri, P. G.; Reddy, M. M.; Bachhawat-Sikder, K.; Olivos, H. J.; Kodadek, T. Isolation of protein ligands from large peptoid libraries. *J Am Chem Soc* **2003**, *125*, 13995-4004.

[12]   Ja, W. W.; Roberts, R. W. G-protein-directed ligand discovery with peptide combinatorial libraries. *Trends Biochem Sci* **2005**, *30*, 318-24.

[13]   Scholle, M. D.; Kehoe, J. W.; Kay, B. K. Efficient construction of a large collection of phage-displayed combinatorial peptide libraries. *Comb Chem High Throughput Screen* **2005**, *8*, 545-51.

[14]   Cull, M. G.; Miller, J. F.; Schatz, P. J. Screening for receptor ligands using large libraries of peptides linked to the C terminus of the lac repressor. *Proc Natl Acad Sci U S A* **1992**, *89*, 1865-9.

[15]   Scott, J. K.; Smith, G. P. Searching for peptide ligands with an epitope library. *Science* **1990**, *249*, 386-90.

[16]   Hessling, J.; Lohse, M. J.; Klotz, K. N. Peptide G protein agonists from a phage display library. *Biochem Pharmacol* **2003**, *65*, 961-7.

[17]   Roberts, R. W.; Szostak, J. W. RNA-peptide fusions for the *in vitro* selection of peptides and proteins. *Proc Natl Acad Sci U S A* **1997**, *94*, 12297-302.

[18]   Dower, W. J.; Mattheakis, L. C. *In vitro* selection as a powerful tool for the applied evolution of proteins and peptides. *Curr Opin Chem Biol* **2002**, *6*, 390-8.

[19]   Lipovsek, D.; Pluckthun, A. *In vitro* protein evolution by ribosome display and mRNA display. *J Immunol Methods* **2004**, *290*, 51-67.

[20]   Turk, B. E.; Cantley, L. C. Peptide libraries: at the crossroads of proteomics and bioinformatics. *Curr Opin Chem Biol* **2003**, *7*, 84-90.

[21]   Johnston, C. A., Lobanova, E.S., Shavkunov, A.S., Low, J., Ramer, J.K., Blaesius, R., Fredericks, Z., Willard, F.S., Kuhlman, B., Arshavsky, V.Y., Siderovski, D.P. Minimal determinants for binding activated G alpha from the structure of a G alpha(i1)-peptide dimer. *Biochemistry* **2006**, *45*, 11390-400.

[22]   Johnston, C. A.; Ramer, J. K.; Blaesius, R.; Fredericks, Z.; Watts, V. J.; Siderovski, D. P. A bifunctional Galphai/Galphas modulatory peptide that attenuates adenylyl cyclase activity. *FEBS Lett* **2005**, *579*, 5746-50.

[23]   Johnston, C. A.; Willard, F. S.; Jezyk, M. R.; Fredericks, Z.; Bodor, E. T.; Jones, M. B.; Blaesius, R.; Watts, V. J.; Harden, T. K.; Sondek, J.; Ramer, J. K.; Siderovski, D. P. Structure of Galpha(i1) bound to a GDP-selective peptide provides insight into guanine nucleotide exchange. *Structure* **2005**, *13*, 1069-80.

[24]   Freissmuth, M.; Waldhoer, M.; Bofill-Cardona, E.; Nanoff, C. G protein antagonists. *Trends Pharmacol Sci* **1999**, *20*, 237-45.

[25]   Holler, C.; Freissmuth, M.; Nanoff, C. *Cell Mol Life Sci* **1999**, *55*, 257-70.

[26]   Jin, Y.; Zhong, H.; Omnaas, J. R.; Neubig, R. R.; Mosberg, H. I. Structure-based design, synthesis, and pharmacologic evaluation of peptide RGS4 inhibitors. *J Pept Res* **2004**, *63*, 141-6.

[27]   Roman, D. L.; Talbot, J. N.; Roof, R. A.; Sunahara, R. K.; Traynor, J. R.; Neubig, R. R. Identification of small-molecule inhibitors of RGS4 using a high-throughput flow cytometry protein interaction assay. *Mol Pharmacol* **2007**, *71*, 169-75.

[28]   Kimple, A. J.; Willard, F. S.; Giguere, P. M.; Johnston, C. A.; Mocanu, V.; Siderovski, D. P. The RGS protein inhibitor CCG-4986 is a covalent modifier of the RGS4 Galpha-interaction face. *Biochim Biophys Acta* **2007**, *1774*, 1213-20.

[29]   Nalbant, P.; Hodgson, L.; Kraynov, V.; Toutchkine, A.; Hahn, K. M. Activation of endogenous Cdc42 visualized in living cells. *Science* **2004**, *305*, 1615-9.

[30]   Pertz, O.; Hodgson, L.; Klemke, R. L.; Hahn, K. M. Spatiotemporal dynamics of RhoA activity in migrating cells. *Nature* **2006**, *440*, 1069-72.

[31]   Pereira, R.; Cerione, R. A. A switch 3 point mutation in the alpha subunit of transducin yields a unique dominant-negative inhibitor. *J Biol Chem* **2005**, *280*, 35696-703.

[32]   Willard, F. S.; Siderovski, D. P. The R6A-1 peptide binds to switch II of Galphai1 but is not a GDP-dissociation inhibitor. *Biochem Biophys Res Commun* **2006**, *339*, 1107-12.

[33]   Slep, K. C.; Kercher, M. A.; He, W.; Cowan, C. W.; Wensel, T. G.; Sigler, P. B. Structural determinants for regulation of phosphodiesterase by a G protein at 2.0 Å. *Nature* **2001**, *409*, 1071-7.

[34]   Tesmer, V. M.; Kawano, T.; Shankaranarayanan, A.; Kozasa, T.; Tesmer, J. J. Snapshot of activated G proteins at the membrane: the Galphaq-GRK2-Gbetagamma complex. *Science* **2005**, *310*, 1686-90.

[35]   Ja, W. W.; Roberts, R. W. *In vitro* selection of state-specific peptide modulators of G protein signaling using mRNA display. *Biochemistry* **2004**, *43*, 9265-75.

[36]   Willard, F. S.; Kimple, R. J.; Siderovski, D. P. Return of the GDI: the GoLoco motif in cell division. *Annu Rev Biochem* **2004**, *73*, 925-951.

[37]   Kimple, R. J.; De Vries, L.; Tronchere, H.; Behe, C. I.; Morris, R. A.; Gist Farquhar, M.; Siderovski, D. P. RGS12 and RGS14 GoLoco motifs are G alpha(i) interaction sites with guanine nucleotide dissociation inhibitor activity. *J Biol Chem* **2001**, *276*, 29275-81.

[38]   Kimple, R. J.; Kimple, M. E.; Betts, L.; Sondek, J.; Siderovski, D. P. Structural determinants for GoLoco-induced inhibition of nucleotide release by Galpha subunits. *Nature* **2002**, *416*, 878-81.

[39]   Ja, W. W.; Adhikari, A.; Austin, R. J.; Sprang, S. R.; Roberts, R. W. A peptide core motif for binding to heterotrimeric G protein alpha subunits. *J Biol Chem* **2005**, *280*, 32057-60.

[40]   Scott, J. K.; Huang, S. F.; Gangadhar, B. P.; Samoriski, G. M.; Clapp, P.; Gross, R. A.; Taussig, R.; Smrcka, A. V. Evidence that a protein-protein interaction 'hot spot' on heterotrimeric G protein betagamma subunits is used for recognition of a subclass of effectors. *Embo J* **2001**, *20*, 767-76.

[41]   Wall, M. A.; Coleman, D. E.; Lee, E.; Iniguez-Lluhi, J. A.; Posner, B. A.; Gilman, A. G.; Sprang, S. R. The structure of the G protein heterotrimer Gi alpha 1 beta 1 gamma 2. *Cell* **1995**, *83*, 1047-58.

[42]   Iiri, T.; Farfel, Z.; Bourne, H. R. Mutant G protein alpha subunit activated by Gbeta gamma: a model for receptor activation? *Nature* **1998**, *394*, 35-8.

[43]   Rondard, P.; Iiri, T.; Srinivasan, S.; Meng, E.; Fujita, T.; Bourne, H. R. *Proc Natl Acad Sci U S A* **2001**, *98*, 6150-5.

[44]   Bonacci, T. M.; Mathews, J. L.; Yuan, C.; Lehmann, D. M.; Malik, S.; Wu, D.; Font, J. L.; Bidlack, J. M.; Smrcka, A. V. Differential targeting of Gbetagamma-subunit signaling with small molecules. *Science* **2006**, *312*, 443-6.

[45]   Davis, T. L.; Bonacci, T. M.; Sprang, S. R.; Smrcka, A. V. Structural and molecular characterization of a preferred protein interaction surface on G protein beta gamma subunits. *Biochemistry* **2005**, *44*, 10593-604.

[46]   Goubaeva, F.; Ghosh, M.; Malik, S.; Yang, J.; Hinkle, P. M.; Griendling, K. K.; Neubig, R. R.; Smrcka, A. V. Stimulation of cellular signaling and G protein subunit dissociation by G protein betagamma subunit-binding peptides. *J Biol Chem* **2003**, *278*, 19634-41.

[47]   Malik, S.; Ghosh, M.; Bonacci, T. M.; Tall, G. G.; Smrcka, A. V. Ric-8 enhances G protein betagamma-dependent signaling in response to betagamma-binding peptides in intact cells. *Mol Pharmacol* **2005**, *68*, 129-36.

[48]   Tesmer, J. J. Pharmacology. Hitting the hot spots of cell signaling cascades. *Science* **2006**, *312*, 377-8.

[49]   Weinstein, L. S.; Liu, J.; Sakamoto, A.; Xie, T.; Chen, M. Minireview: GNAS: normal and abnormal functions. *Endocrinology* **2004**, *145*, 5459-64.

[50]   Takasaki, J.; Saito, T.; Taniguchi, M.; Kawasaki, T.; Moritani, Y.; Hayashi, K.; Kobori, M. A novel Galphaq/11-selective inhibitor. *J Biol Chem* **2004**, *279*, 47438-45.

[51]   Uemura, T.; Kawasaki, T.; Taniguchi, M.; Moritani, Y.; Hayashi, K.; Saito, T.; Takasaki, J.; Uchida, W.; Miyata, K. Biological properties of a specific Galpha q/11 inhibitor, YM-254890, on platelet functions and thrombus formation under high-shear stress. *Br J Pharmacol* **2006**, *148*, 61-9.

[52]   Fowlkes, D. M.; Christensen, D. J.; Hamilton, P. T.; Blaesius, R.; Ramer, J. K.; Hyde-DeRuyscher, R.; Duffin, D.; Fredericks, Z. Synthetic or partially purified peptides which

can bind to specific subunits of G proteins and uses. *Organization, W. I. P.* **2004**, PCT/EP2003/000352.

[53] Broach, J. R.; Thorner, J. High-throughput screening for drug discovery. *Nature* **1996**, *384*, 14-6.

[54] Fong, C. W.; Bahia, D. S.; Rees, S.; Milligan, G. Selective activation of a chimeric Gi1/Gs G protein alpha subunit by the human IP prostanoid receptor: analysis using agonist stimulation of high affinity GTPase activity and [35S]guanosine-5'-O-(3-thio)triphosphate binding. *Mol Pharmacol* **1998**, *54*, 249-57.

[55] Liggett, S. B. Update on current concepts of the molecular basis of beta2-adrenergic receptor signaling. *J Allergy Clin Immunol* **2002**, *110*, S223-7.

[56] Senogles, S. E.; Amlaiky, N.; Berger, J. G.; Caron, M. G. Biochemical properties of D1 and D2 dopamine receptors. *Adv Exp Med Biol* **1988**, *235*, 33-41.

[57] Lee, T. W.; Seifert, R.; Guan, X.; Kobilka, B. K. Restricting the mobility of Gs alpha: impact on receptor and effector coupling. *Biochemistry* **1999**, *38*, 13801-9.

[58] Zhang, J. H.; Chung, T. D.; Oldenburg, K. R. A Simple Statistical Parameter for Use in Evaluation and Validation of High Throughput Screening Assays. *J Biomol Screen* **1999**, *4*, 67-73.

[59] Leipziger, J.; Thomas, J.; Rubini-Illes, P.; Nitschke, R.; Greger, R. 8-(N,N-diethylamino)octyl 3,4,5-trimethoxybenzoate (TMB-8) acts as a muscarinic receptor antagonist in the epithelial cell line HT29. *Naunyn Schmiedebergs Arch Pharmacol* **1996**, *353*, 295-301.

[60] Ellis, J.; Seidenberg, M. Two allosteric modulators interact at a common site on cardiac muscarinic receptors. *Mol Pharmacol* **1992**, *42*, 638-41.

[61] Gnagey, A.; Ellis, J. Allosteric regulation of the binding of [3H]acetylcholine to m2 muscarinic receptors. *Biochem Pharmacol* **1996**, *52*, 1767-75.

[62] Chiou, C. Y.; Malagodi, M. H. Studies on the mechanism of action of a new Ca-2+ antagonist, 8-(N,N-diethylamino)octyl 3,4,5-trimethoxybenzoate hydrochloride in smooth and skeletal muscles. *Br J Pharmacol* **1975**, *53*, 279-85.

[63] Palmer, F. B.; Byers, D. M.; Spence, M. W.; Cook, H. W. Calcium-independent effects of TMB-8. Modification of phospholipid metabolism in neuroblastoma cells by inhibition of choline uptake. *Biochem J* **1992**, *286 (Pt 2)*, 505-12.

[64] Garcia-Marcos, M.; Ghosh, P.; Farquhar, M. G. GIV is a nonreceptor GEF for G alpha i with a unique motif that regulates Akt signaling. *Proc Natl Acad Sci U S A* **2009**, *106*, 3178-83.

[65] Bosch, D. E.; Willard, F. S.; Ramanujam, R.; Kimple, A. J.; Willard, M. D.; Naqvi, N. I.; Siderovski, D. P. A P-loop mutation in Gα subunits prevents transition to the active state: implications for G-protein signaling in fungal pathogenesis. *PLoS Pathog* **2012**, *8*, e1002553.

[66] Kimple, A. J.; Bosch, D. E.; Giguere, P. M.; Siderovski, D. P. Regulators of G-protein signaling and their Gα substrates: promises and challenges in their use as drug discovery targets. *Pharmacol Rev* **2011**, *63*, 728-49.

[67] Nishimura, A.; Kitano, K.; Takasaki, J.; Taniguchi, M.; Mizuno, N.; Tago, K.; Hakoshima, T.; Itoh, H. Structural basis for the specific inhibition of heterotrimeric Gq protein by a small molecule. *Proc Natl Acad Sci U S A* **2010**, *107*, 13666-71.

INDEX

F

G

H

K

L

T

U

V

W

www.ingramcontent.com/pod-product-compliance
Lightning Source LLC
Chambersburg PA
CBHW050818220326
41598CB00006B/249

9781608057467